VOLUME EIGHTY NINE

ADVANCES IN BOTANICAL RESEARCH

Molecular Physiology and Biotechnology of Trees

ADVANCES IN BOTANICAL RESEARCH

Series Editor

Jean-Pierre Jacquot
Professor, Membre de l'Institut Universitaire de France, Unité Mixte de Recherche INRA, UHP 1136 "Interaction Arbres Microorganismes", Université de Lorraine, Faculté des Sciences, Vandoeuvre, France

VOLUME EIGHTY NINE

ADVANCES IN BOTANICAL RESEARCH

Molecular Physiology and Biotechnology of Trees

Volume Editor

FRANCISCO M. CÁNOVAS
Department of Molecular Biology and Biochemistry, Faculty of Science, University of Málaga, Spain

ELSEVIER

ACADEMIC PRESS
An imprint of Elsevier

Academic Press is an imprint of Elsevier
125 London Wall, London, EC2Y 5AS, United Kingdom
The Boulevard, Langford Lane, Kidlington, Oxford OX5 1GB, United Kingdom
50 Hampshire Street, 5th Floor, Cambridge, MA 02139, United States
525 B Street, Suite 1650, San Diego, CA 92101, United States

First edition 2019

ISBN: 978-0-12-815465-6
ISSN: 0065-2296

Publisher: Zoe Kruze
Acquisition Editor: Sam Mahfoudh
Editorial Project Manager: Teresa Pons-Ferrer
Production Project Manager: James Selvam
Cover Designer: Miles Hitchen

Typeset by SPi Global, India

Dedication

This book is dedicated to the memory of Professor Carl Douglas.

Contents

Contributors

Malin Abrahamsson
SweTree Technologies AB, Umeå, Sweden

Concepción Ávila
Departamento de Biología Molecular y Bioquímica, Universidad de Málaga, Málaga, Spain

Adeline Becquer
CSIRO Agriculture, Canberra, ACT, Australia; Eco&Sols, Univ. Montpellier, CIRAD, INRA, IRD, Montpellier SupAgro, Montpellier, France

David Bolonio
Centro de Biotecnología y Genómica de Plantas (CBGP UPM-INIA), Universidad Politécnica de Madrid, Madrid, Spain

Amy M. Brunner
Department of Forest Resources and Environmental Conservation, Virginia Tech, Blacksburg, VA, United States

Heike Bücking
Biology and Microbiology Department, South Dakota State University, McFadden Biostress, Brookings, SD, United States

Eduardo L.O. Camargo
Laboratoire de Recherche en Sciences Végétales, Université de Toulouse, CNRS, UPS, Castanet-Tolosan, France; Laboratório de Genômica e Expressão, Departamento de Genética, Evolução e Bioagentes, Instituto de Biologia, Universidade Estadual de Campinas, Campinas, Brazil

Rafael A. Cañas
Departamento de Biología Molecular y Bioquímica, Universidad de Málaga, Málaga, Spain

Francisco M. Cánovas
Departamento de Biología Molecular y Bioquímica, Universidad de Málaga, Málaga, Spain

Hua Cassan-Wang
Laboratoire de Recherche en Sciences Végétales, Université de Toulouse, CNRS, UPS, Castanet-Tolosan, France

David Clapham
Department of Plant Biology, Uppsala BioCenter, Swedish University of Agricultural Sciences, Uppsala, Sweden

Gary D. Coleman
Department of Plant Science and Landscape Architecture, University of Maryland, College Park, MD, United States

Angela Contreras
Centro de Biotecnología y Genómica de Plantas (CBGP UPM-INIA), Universidad Politécnica de Madrid, Madrid, Spain

Fernando N. de la Torre
Departamento de Biología Molecular y Bioquímica, Universidad de Málaga, Málaga, Spain

Janice L. Eibensteiner
Biology and Microbiology Department, South Dakota State University, McFadden Biostress, Brookings, SD, United States

Kevin Garcia
Department of Crop and Soil Sciences, North Carolina State University, Raleigh, NC, United States

Luis Gómez
Centro de Biotecnología y Genómica de Plantas (CBGP UPM-INIA); Forestry School, Universidad Politécnica de Madrid, Madrid, Spain

Jacqueline Grima-Pettenati
Laboratoire de Recherche en Sciences Végétales, Université de Toulouse, CNRS, UPS, Castanet-Tolosan, France

Carmen Guerrero-Galán
BPMP, Univ Montpellier, CNRS, INRA, SupAgro, Montpellier, France

Scott A. Harding
Warnell School of Forestry and Natural Resources; Department of Genetics; Department of Plant Biology, University of Georgia, Athens, GA, United States

Gabriella Houdinet
BPMP, Univ Montpellier, CNRS, INRA, SupAgro, Montpellier, France

Edward G. Kirby
Department of Biological Sciences, Rutgers University, Newark, NJ, United States

Gen Li
Department of Plant Science and Landscape Architecture, University of Maryland, College Park, MD, United States

Irene Merino
Centro de Biotecnología y Genómica de Plantas (CBGP UPM-INIA), Universidad Politécnica de Madrid, Madrid, Spain

Juan Jesús Molina-Rueda
Department of Biological Sciences, Rutgers University, Newark, NJ, United States

Fabien Mounet
Laboratoire de Recherche en Sciences Végétales, Université de Toulouse, CNRS, UPS, Castanet-Tolosan, France

Luis Oñate-Sánchez
Centro de Biotecnología y Genómica de Plantas (CBGP UPM-INIA), Universidad Politécnica de Madrid, Madrid, Spain

Mª Belén Pascual
Departamento de Biología Molecular y Bioquímica, Universidad de Málaga, Málaga, Spain

Raphaël Ployet
Laboratoire de Recherche en Sciences Végétales, Université de Toulouse, CNRS, UPS, Castanet-Tolosan, France

Julia Quintana
Centro de Biotecnología y Genómica de Plantas (CBGP UPM-INIA), Universidad Politécnica de Madrid, Madrid, Spain

Xiaoyan Sheng
Department of Forest Resources and Environmental Conservation, Virginia Tech, Blacksburg, VA, United States

Nathaniel Robert Street
Umeå Plant Science Centre, Department of Plant Physiology, Umeå University, Umeå, Sweden

Rita Teresa Teixeira
Department of Forest Resources and Environmental Conservation, Virginia Tech, Blacksburg, VA, United States

Chung-Jui Tsai
Warnell School of Forestry and Natural Resources; Department of Genetics; Department of Plant Biology, University of Georgia, Athens, GA, United States

Sara von Arnold
Department of Plant Biology, Uppsala BioCenter, Swedish University of Agricultural Sciences, Uppsala, Sweden

Sabine D. Zimmermann
BPMP, Univ Montpellier, CNRS, INRA, SupAgro, Montpellier, France

Preface

Twenty-five years ago I attended an IUFRO Meeting in Segovia (Spain) entitled "Biotechnology of Trees". Only a few papers in that conference included molecular approaches to study tree biology. From that time, rapid developments in plant molecular biology and sequencing capacity have revolutionized the way we currently study physiological processes in plants, including trees (Tsai, Harding, & Cooke, 2018). A crucial advance was the sequencing of the poplar genome (*Populus trichocarpa*) over 10 years ago (Tuskan et al., 2006). Afterwards, with the development of second-generation high-throughput sequencing, the genomes of an increasing number of angiosperm trees have been sequenced and it has even been possible to complete drafts for the megagenomes of several gymnosperm species (De la Torre et al., 2014; Street, Chapter 1; Cañas et al., Chapter 2). Overall, the reviews included in this book largely reflect the extraordinary impact that the above-mentioned technologies have had to accelerate the exponential progress in our current understanding of essential processes in tree biology.

The aim of this book is to present an overview of recent advances in the molecular physiology and biotechnology of forest trees. The 10 chapters of the book cover important aspects of fundamental tree biology which are addressed by leading scientists in the field. The authors present the current status and future directions in research related to structural and functional genomics, nutrition, growth and development, wood and cell wall regulation, stress tolerance, and the use of trees for phytoremediation purposes. A brief introduction to the contents is presented below.

In Chapter 1, Street presents the current sequencing technologies and how they are used for different applications in the genomic analysis of forest trees. The decreasing cost of high-throughput sequencing has facilitated the routine characterization of transcriptomes by RNA sequencing, studies of gene expression variation, and the resequencing of genomes (Ingvarsson, Hvidsten, & Street, 2016). Useful guidelines are provided for genome assembly, validation, annotation, transcriptome assembly, and profiling as well as genomic approaches for epigenetic and population studies. New approaches are addressed to the understanding of adaptive genetic variation in natural populations.

To elucidate the function of key genes and networks controlling essential processes a variety of functional genomics approaches are now available.

Cañas et al. consider present-day technologies and online resources for functional genomic studies in conifers (Chapter 2). Recent methods have been developed to investigate how metabolic processes are organized in specialized tissues and cell types (Cañas et al., 2017). Homologous and heterologous systems are used for gene validation and characterization of regulatory regions. Recently, platforms for functional analysis in transgenic conifers have also been established via somatic embryogenesis, a technology that is described in Chapter 5.

Trees acquire nutrients to incorporate essential elements that are needed to complete their life cycles. Because of their low abundance in the soil, trees have established symbiotic associations to enhance nutrient use efficiency. Membrane proteins involved in nutrient transport between fungi and colonized roots constitute the so-called "transportome" (García, Doidy, Zimmermann, Wipf, & Courty, 2016). In Chapter 3, Becquer et al. reviewed the recent progress in understanding how ectomycorrhizas contribute to tree nutrition. The molecular mechanisms involved in the transport of macronutrients, micronutrients, and water are examined in the mutualistic interaction with their fungal partners. The evolution and ecological relevance of the transportome are also considered.

Nitrogen is an essential macronutrient and frequently a major limiting factor for tree growth. In addition to uptake and assimilation from the soil, internal nitrogen cycling is largely influencing the efficient use of nitrogen in woody plant species exhibiting long life cycles with annual periods of seasonal growth and development (Babst & Coleman, 2018). In Chapter 4, Li and Coleman present an updated description of the molecular and physiological basis of nitrogen cycling in forest and fruit trees but with a special focus on *Populus*, a model tree in which most of the recent knowledge has been generated. Major players involved in nitrogen cycling are described as well as potential mechanisms controlling this process and its environmental regulation.

The development of single cells to mature embryos is a complex process that is strictly regulated. Tree embryogenesis is a developmental programme of paramount importance from a scientific and applied point of view. In Chapter 5, von Arnold et al. provided an updated overview of the developmental patterning and its regulation during zygotic and somatic embryogenesis in conifers. Recent advances include the identification and functional analyses of key genes involved in the process (Alvarez et al., 2015). This knowledge is of great interest in fundamental tree biology but also may have an immediate biotechnological application because somatic embryogenesis in combination with cryopreservation is one of the most relevant technologies for mass propagation of conifer trees.

The relationships and coordinated regulation of shoot apical, axillary, and cambial meristems determine tree growth and shape. Meristematic activities are largely influenced by environmental signals and seasonal changes adapting tree species to changes in their living habitats (Brunner, Evans, Hsu, & Sheng, 2014). Using *Populus* as a model, Teixeira et al. review the coordination of meristematic activity during the transition to secondary growth and wood formation (Chapter 6). The roles of signalling proteins, sugars, and hormones in long-distance communication between apical and vascular meristems are examined and discussed. Genes involved in the molecular regulation of meristematic activity by photoperiod and temperature have been identified. However, further work is necessary to unravel the complex pathways involved in long-distance signalling as well as the molecular components of the responses in the target tissues.

Wood, the most abundant lignocellulosic biomass, plays essential structural and functional roles in plants and represents an important raw material for many different purposes. Genome-wide analysis has been made to identify key genes involved in the biosynthesis of wood components in forest trees (Carocha et al., 2015). In Chapter 7, Camargo et al. reviewed the recent advances to elucidate the mechanisms of wood formation in trees, highlighting the similarities and differences found in the secondary xylem differentiation described in herbaceous model plants. The functional analysis of the regulatory gene networks governing wood formation has revealed a high level of complexity involving sophisticated mechanisms of control and cross-regulation.

Plant cell wall composition and organization change during development and in response to environmental cues (Harding et al., 2018). Although of low abundance in lignocellulosic biomass, the pectin fraction of the cell wall usually has negative effects in the postprocessing of woody plant biomass for paper production or saccharification. Harding and Tsai discussed how pectin interacts with other components of the cell wall during cell expansion and the lignification process (Chapter 8). A better knowledge of pectin biology is needed to improve our current understanding of cell wall organization and function. Recent findings in this research area are providing new insights into the biotechnological manipulation of pectin biosynthesis and degradation.

Poplar has been extensively used to study gene function by its amenability to transformation and easy generation of transgenic lines for molecular and physiological analyses. Trees with increased tolerance to a wide range of environmental stresses are needed to cope with a changing climate and at the same time maintaining the productivity of forest and fruit plantations (Merino et al., 2014). Transgenic approaches to develop and understand drought tolerance in poplar are reviewed in Chapter 9. Transgenic poplars with enhanced nitrogen assimilation also exhibit increased tolerance to drought stress (Molina-Rueda & Kirby, 2015). It has been found that genes involved in the protection against reactive oxygen species appear to be critical in the observed tolerance to drought. However, further studies will be necessary to investigate the connection between nitrogen metabolism and enhanced stress tolerance.

Human activities are consubstantial with the release of a vast variety of pollutants into the environment. The use of plants to restore clean environmental conditions is based on their favourable and unique characteristics to extract, store, and even metabolize a wide range of pollutants. Gómez et al. examined the potential of trees in phytoremediation of contaminated areas (Chapter 10). Some fast-growing trees used as phytoremediators include additional advantages such as longevity, the production of high levels of biomass, complex pathways in secondary metabolism, and stress tolerance. Nowadays, the phytoremediation capacity of selected genotypes can be accelerated and improved using advances in genomics and biotechnology. On the other hand, tree plantations for cleaning up polluted areas represent a valuable source of feedstocks for biorefineries and bioenergy applications.

I am extremely grateful to all authors for contributing with their excellent reviews to this volume. We are confident that the book will be a useful reference for researchers and professionals interested in an updated monography on the functional genomics of forest trees. We also hope that the book will be useful for advanced students and young researchers interested in the molecular aspects of tree biology. Finally, I want to thank the kind invitation of Professors Jacquot and Gadal to coordinate this project.

References

Alvarez, J. M., Sohlberg, J., Engström, P., Zhu, T., Englund, M., Moschou, P. N., et al. (2015). The WUSCHEL-RELATED HOMEOBOX 3 gene PaWOX3 regulates lateral organ formation in Norway spruce. *The New Phytologist*, *208*, 1078–1088.

Babst, B. A., & Coleman, G. D. (2018). Seasonal nitrogen cycling in temperate trees: Transport and regulatory mechanisms are key missing links. *Plant Science*, *270*, 268–277.

Brunner, A. M., Evans, L. M., Hsu, C. Y., & Sheng, X. (2014). Vernalization and the chilling requirement to exit bud dormancy: Shared or separate regulation? *Frontiers in Plant Science*, *5*, 732.

Cañas, R. A., Li, Z., Pascual, M. B., Castro-Rodríguez, V., Ávila, C., Sterck, L., et al. (2017). The gene expression landscape of pine seedling tissues. *The Plant Journal*, *91*, 1064–1087.

Carocha, V., Soler, M., Hefer, C., Cassan-Wang, H., Fevereiro, P., Myburg, A. A., et al. (2015). Genome-wide analysis of the lignin toolbox of *Eucalyptus grandis*. *The New Phytologist*, *206*, 1297–1313.

De la Torre, A., Birol, I., Bousquet, J., Ingvarsson, P., Jansson, S., Jones, S. J. M., et al. (2014). Insights into conifer giga-genomes. *Plant Physiology*, *166*, 1724–1732.

García, K., Doidy, J., Zimmermann, S. D., Wipf, D., & Courty, P. E. (2016). Take a trip through the plant and fungal transportome of mycorrhiza. *Trends in Plant Science*, *21*, 937–950.

Harding, S. A., Hu, H., Nyamdari, B., Xue, L. J., Naran, R., & Tsai, C. J. (2018). Tubulins, rythms and cell walls in poplar leaves: It's all in timing. *Tree Physiology*, *38*, 397–408.

Ingvarsson, P. K., Hvidsten, T. R., & Street, N. R. (2016). Towards integration of population and comparative genomics in forest trees. *The New Phytologist*, *212*, 338–344.

Merino, I., Contreras, A., Jing, Z. P., Gallardo, F., Cánovas, F. M., & Gómez, L. (2014). Plantation forestry under global warming: Hybrid poplars with improved thermotolerance provide new insights on the in vivo function of small heat shock protein chaperones. *Plant Physiology*, *164*, 978–991.

Molina-Rueda, J. J., & Kirby, E. G. (2015). Transgenic poplar expressing the pine GS1a show alterations in nitrogen homeostasis during drought. *Plant Physiology and Biochemistry*, *94*, 181–190.

Tsai, C. J., Harding, S. A., & Cooke, J. E. K. (2018). Branching out: A new era of investigating physiological processes in forest trees using genomic tools. *Tree Physiology*, *38*, 303–310.

Tuskan, G. A., Difazio, S., Jansson, S., Bohlmann, J., Grigoriev, I., Hellsten, U., et al. (2006). The genome of black cottonwood, *Populus trichocarpa* (Torr. & Gray). *Science*, *313*, 1596–1604.

CHAPTER ONE

Genomics of forest trees

Nathaniel Robert Street*
Umeå Plant Science Centre, Department of Plant Physiology, Umeå University, Umeå, Sweden
*Corresponding author: e-mail address: nathaniel.street@umu.se

Contents

Abstract

Over the past decade increasing availability and decreasing costs of "next" generation sequencing has revolutionised our ability to assay numerous aspects of genome function. Initially, massively parallel high throughput sequencing technologies resulted in a rapid increase in the number of published tree genomes and, to an even greater extent, transcriptome studies. Whereas previous genomics efforts were concentrated on a small number of model species, these new sequencing technologies have liberated the choice of species, particularly so assaying gene expression using RNA Sequencing and, more recently, for assays of epigenomics including genome structure and accessibility. These advances in sequencing throughput and cost enabled the first draft assemblies of the large (~20 Gbp) genomes of a number of conifer species in addition to an ever-increasing number of angiosperm tree species, with more than 40 genomes now publicly available. The falling cost of sequencing has also enabled genome resequencing, genome wide association studies and population gene expression studies,

Advances in Botanical Research, Volume 89
ISSN 0065-2296
https://doi.org/10.1016/bs.abr.2018.11.001

the results of which are providing new insight into the developmental programs associated with wood formation, stress and disease tolerance, adaptive potential, population genetics and evolutionary history of tree species.

1. Overview

Over the past decade increasing availability and decreasing costs of "next" generation sequencing (NGS) has revolutionised our ability to assay numerous aspects of genome function. Initially, second generation massively parallel high throughput sequencing (HTS) technologies resulted in a rapid increase in the number of published tree genomes and, to an even greater extent, transcriptome studies. Whereas previous genomics efforts were concentrated on a small number of model species, HTS has liberated the choice of species, particularly so for the use of HTS to assay gene expression using RNA Sequencing (RNA-Seq) and, more recently, for assays of genome structure and accessibility. These advances in sequencing throughput and cost enabled the first draft assemblies of the large (~20 Gbp) genomes of a number of conifer species (Birol et al., 2013; De La Torre et al., 2014; Nystedt et al., 2013; Wegrzyn et al., 2014) in addition to an ever-increasing number of angiosperm tree species, with more than 40 genomes now publicly available (Ingvarsson, Hvidsten, & Street, 2016; Neale, Martínez-García, De La Torre, Montanari, & Wei, 2017). The falling cost of sequencing has also enabled genome resequencing, genome wide association studies (GWAS) and population RNA-Seq studies (Evans et al., 2014; Mähler et al., 2017; Porth et al., 2013; Wang et al., 2018).

Tree species represent a range of unique biology including extensive secondary growth, longevity, delayed flowering, seasonal senescence and dormancy, extensive fungal and ecological interactions in additional to being of commercial importance and essential components of the global carbon cycle. As such, tree species are of extensive importance to humans despite the substantially lower research investment than in agricultural species. Genomics approaches are providing important advancements in our understanding of the evolution of tree form, advancing our ability to address the question "What makes a tree a tree?" (Groover, 2005). Genomics is also providing new insight into the adaptive evolution of tree species, including responses to biotic and abiotic stresses, which will be key to understanding the future adaptive potential of species in the face of continuing climate change. Finally, genomics methods hold vast potential for application in breeding programs through approaches such as genomic selection or targeted breeding informed by Genome Wide Association Studies (GWAS), and in assisted migration to mitigate climate change impact.

Tree species are physically large, affording greater spatial resolution, making them useful systems for multi-omics approaches, for which sample volumes can often be limiting. As such, tree species are ideal systems for elucidating the molecular and genetic basis of complex traits such as wood formation and quality, yield productivity and interactions with the biotic and abiotic environment. There is also considerable interest in exploring the genetic architecture of adaptive traits in natural populations (Lind, Menon, Bolte, Faske, & Eckert, 2018) and the decreasing costs of sequencing are rendering such population scale studies increasingly feasible.

In this chapter we provide a brief overview of current technologies for performing genomics analyses, including technical details and considerations for genome assembly and RNA-Seq profiling of gene expression as these represent the most common uses of current sequencing technologies for research on tree species. Although there are a large number of tools available to perform the analyses we discuss, we point readers to current common options. However, this is a rapidly advancing field, so an up-to-date literature review is always advised. We consider applications at the individual genotype and population scales, including assays of genome sequence, structure and gene expression and focus on forest tree species. We additionally provide an overview of current resources for exploring genomics data from tree species. There have been a number of comprehensive, recent reviews of forest tree genomics and we direct readers to these (Holliday et al., 2017; Neale & Ingvarsson, 2008; Neale & Kremer, 2011; Parent, Raherison, Sena, & MacKay, 2015; Plomion et al., 2016) while providing here discussion of example experiments or topics. Although there was much excellent previous work performed using Sanger EST sequencing and microarrays, we have not included that work here as it has been covered in the above, and other, reviews and we focus here specifically on applications of high throughput sequencing for genomics analyses.

2. Genome assembly and annotation

2.1 Assembling genomes from short reads

During the past 5 years the Illumina family of sequencing platforms has come to dominate the market for high throughput, short read sequencing with other earlier technologies, such as Roche 454, having been superseded by technology improvements. Today the range of Illumina sequencing platforms ranges from machines generating ~eight million paired end reads per machine run (iSeq) to >20 billion paired-end reads (NovaSeq 6000), with maximum read lengths differing between platforms, but with a current maximum length of 2×300 bp (MiSeq). The larger platforms provide cost-efficient generation of vast quantities of data, effectively rendering all genomes permissible for genome assembly. Illumina sequencing is based on sequencing by synthesis and involves global amplification of genomic DNA fragments in a bridge PCR reaction.

The huge number of short sequence reads generated presents a computational challenge for genome assembly. Whereas the original approaches to assembly of Sanger sequencing data performed pairwise alignment of all possible sequence read pairs, this is intractable for the number of sequences generated using HTS for even relatively small tree genomes. A number of alternative approaches to genome assembly of short read data have therefore been developed, most commonly utilising a de Bruijn graph-based approach, which deconstruct sequencing reads to a non-redundant set of *k*mers that are subsequently used to identify overlaps and to construct contiguous sequences. The short read lengths necessitate greater sequencing depth than was common for Sanger sequencing (where ~8× was sufficient) to ensure adequate coverage of all genomic regions—typically a good rule of thumb is to generate a minimum 40× coverage per haplotype for a heterozygous species. As some raw sequence data will be removed by quality control filtering steps, this often translates into generating 100× coverage of raw sequence data. Performing assemblies of these quantities of data mandates the use of high performance compute facilities with large memory and storage resources. The majority of genome assemblies generated using short read sequencing data are based on high coverage paired-end data for contig formation that is often complemented by lower coverage mate-pair (jumping) reads that provide long-range (typically up to 40 kbp) links that are used to connect and order contigs through the process of scaffolding. A large number of reviews of genome assembly from short read sequencing data are available (Baker, 2012; Basantani et al., 2017; Ekblom & Wolf, 2014; Lin et al., 2011; Miller, Koren, & Sutton, 2010).

The large quantities of sequence data that can be produced by these platforms means that generating adequate coverage of effectively all plant genomes is both technically and economically feasible, but a number of key characteristics render assembly of that data into highly contiguous sequences problematic. Short read length is a primary factor limiting contiguity, with the severity of this being determined by the extent of repetitive DNA, heterozygosity, genome duplication (whole genome or tandem and segmental duplications) and polyploidy. Plant genomes also contain many highly expanded gene families containing numerous genes with highly similar sequence, which renders them intractable to short read assembly in some cases. Higher levels of any one of these components will result in less contiguous assemblies. Many tree species have high levels of exactly these characteristics, for example, high heterozygosity as a result of the obligate outbreeding nature of many species. As a result, genome assemblies generated using only short read data are often highly fragmented (De La Torre et al., 2014). A number of sequencing projects attempted to overcome, or alleviate, these issues by producing sequencing libraries from haploid tissue (*1n*) or through use of fosmid pools that were independently assembled and subsequently merged (Alexeyenko et al., 2014; Neale, Martínez-García, et al., 2017; Nystedt et al., 2013). In the case of the currently available conifer genomes, the degree of fragmentation extends to numerous examples of single genes being split across two or more assembled contigs (Nystedt et al., 2013), which present substantial challenges to downstream utilisation of the corresponding genome annotation.

2.2 Single molecule sequencing

In recent years new "third generation" sequencing platforms have been commercialised, primarily from Pacific Biosciences (PacBio) and Oxford Nanopore (ON). Both technologies perform sequencing of single DNA molecules and can produce multi-kilobase sequence reads (Basantani et al., 2017), with the majority of reads being >10 Kbp and with ON reads currently reaching maximum lengths approaching 1 Mbp. As such, individual sequence reads can often unambiguously span regions such as individual Long Terminal Repeat (LTR) retrotransposons that remained elusive to assembly using short read technologies. In contrast to short read NGS platforms the technologies do not involve DNA amplification and sequencing read length distribution is highly dependent on the length of extracted DNA fragments and also requiring greater input DNA quantities than for short read library construction. This places substantially greater importance on molecular biology (wet lab) protocol optimisation to ensure that adequate quantities of high molecular weight (HMW) DNA that is free from common contaminants, such as polyphenolics and carbohydrates, are used as input to generate the sequencing libraries. This commonly involves protocols that isolate intact nuclei as an initial step before lysis and DNA isolation. Size selection of HMW DNA is a common final step, which can further assist in removal of contaminants that can interfere with the sequencing reaction. Unfortunately, tree species typically contain high levels of contaminants that can interfere with sequencing, particularly polyphenolics, and per-species protocol optimisation is often required before high quality HMW DNA can be obtained. While the sequence reads produced by both platforms are orders of magnitude longer than those of short read technologies, they suffer from substantially higher error rates (currently ~15%). Fortunately, those errors are random and, as a result, generating adequately high coverage can overcome the high error rate as most sequence reads will have a correct base called at a position where one read contains a sequencing error. It is too early for a consensus rule-of-thumb for adequate coverage to have been established, but a minimum 30 × coverage is required to obtain reasonable quality assemblies using current assembly algorithms. Despite these limitations, both sequencing platforms have been used either alone or in combination to generate high quality and highly contiguous assemblies of large, complex genomes including Axolotl (Nowoshilow et al., 2018) and human (Jain et al., 2018). At the time of writing there are no available tree genomes assembled using these platforms, although a number of projects are in progress. There is active and rapid development of assembly algorithms for such long read, noisy data but commonly used tools designed specifically for such data currently include HGAP (https://github.com/PacificBiosciences/Bioinformatics-Training/wiki/HGAP), FALCON (Chin et al., 2016) and MARVEL (https://github.com/schloi/MARVEL), while a number of existing tools including Abyss (Jackman et al., 2017), MaSuRCA (Zimin et al., 2013) and Canu (Koren et al., 2017) have been adapted to make use of such data, often in combination with short read sequence data in a hybrid assembly approach.

2.3 Guidelines for genome assembly

Given the current state of sequencing technologies a suggested guideline for initiating a new genome assembly project would be to generate 30–50 × coverage of PacBio sequencing data. ON could alternatively be used but, at the time of writing, the technology is less mature and there has been considerably less protocol development and optimisation to date, particularly for plant species. However, sequencing technologies are rapidly advancing and therefore these recommendations should be considered in light of subsequent developments. The long-read coverage can be complemented with 50–80× coverage of short paired-end read data that can be used to identify single base errors in the assembled sequence or to perform assembly quality validation (see below) to identify and correct or break assembly errors. The optimal choice of assembly algorithm is highly dependent on the specific characteristics of the data generated, the target genome and available compute resources, but for pure PacBio or ON assemblies the MARVEL assembler performs well for large, complex genomes containing high levels of heterozygosity or high repeat content, although this is extremely compute-resource demanding. For simpler genomes the FALCON assembler performs well. For hybrid assemblies we suggest either Canu or Abyss. However, given the rapid development of new tools and approaches to assembly we suggest always performing a literature search for recent genome assemblies to ascertain which tools are newly available. We direct readers to Ekblom and Wolf (2014) and Dominguez Del Angel et al. (2018) for a primer on how to approach a genome assembly project.

2.4 Assembly scaffolding and pseudochromosome scale assembly

It was common practice for genome assembly projects utilising short read data to incorporate the use of long-range mate-pair libraries, where the two sequencing reads of a read-pair were originally separated by a relatively long distance in the genome—commonly up to ~40 Kbp. This long-range linkage was particularly useful for identifying connections between short, assembled contigs that were then arranged into scaffolds, therefore increasing the contiguity of the assembly, albeit with gaps of unknown sequence between the connected contigs but for which gap lengths could be inferred (Sahlin, Street, Lundeberg, & Arvestad, 2012). The shift to using third generation sequencing platforms has negated the usefulness of such libraries as individual sequence reads are often longer than the maximum distance separating the two reads of a mate-pair. High quality mate-pair libraries were also notoriously difficult to produce. There has therefore been a shift towards the use of optical mapping technologies and chromosome conformation capture (Hi-C) to perform scaffolding. Optical maps can additionally be used to perform assembly verification and correction by detecting false joins in assembled contigs.

The most commonly used optical mapping technology is from Bionano, with the latest iteration being the Saphyr system, which can generate linearised fragments of DNA up to 2 Mbp. Optical map construction involves extraction of HMW DNA, the use of enzymes to introduce single strand nicks that are fluorescently labelled, linearisation of the DNA in microfluidic channels and imaging. These steps create a unique molecular "barcode" of sequence-specific nick patterns along the DNA fragments that are utilised to perform *de novo* map assembly and to link the map to an *in silico* nicked version of a draft assembly (Levy-Sakin & Ebenstein, 2013). Optical maps have been used to improve assembly quality and contiguity for a number of plant genomes including maize (Jiao, Peluso, et al., 2017), *Medicago truncatula* (Moll et al., 2017) and relatives of *Arabidopsis thaliana* (Jiao, Accinelli, et al., 2017).

An alternative approach to assembly scaffolding utilises proximity ligation-based methods such as Hi-C (Belton et al., 2012; Dekker, Marti-Renom, & Mirny, 2013; Lieberman-Aiden et al., 2009) in combination with short-read sequencing to assay the three-dimensional structure of chromosomes. After fixation of DNA, segments in close proximity within the nucleus have a higher probability to be ligated together than distant regions. This pattern is then reflected in the frequency of those ligated segments being sequenced as a pair. There is therefore a frequency distribution formed for the number of read pairs between intra-chromosomal regions, with a decreasing frequency as genomic distances increase. This information can be used to perform scaffolding of contigs in genome assemblies as Hi-C read pairs provide evidence of linkage between sequence regions present on different assembled contigs. The frequency distribution of read pairs provides inference of the genomic distance between the contigs (Putnam et al., 2016).

A final approach to scaffolding and long-range contiguity improvement to attain pseudo-chromosome-scale assemblies is the use of sequence marker-based genetic maps. HTS has made sequence based marker approaches far more feasible with common approaches including sequence capture using capture probes (often exome capture), whole-genome resequencing and restriction site associated DNA sequencing (RAD-Seq; Davey & Blaxter, 2010). In all cases SNP markers are most commonly identified and genotyped within the target mapping population and this genotype information is used to construct a genetic map. While it is possible to identify and utilise INDEL markers, these are substantially harder to genotype, and their use is therefore much less common at present.

As INDEL and SNP markers will often be in linkage there is also little to no additional benefit to using combinations of the two marker types. For the purposes of assisting genome assembly, the location of SNP markers used to construct a genetic map is determined within the set of assembled contigs. The ordering of SNPs within the genetic map is then used to order and orient contigs along linkage groups, representing pseudo-chromosomes.

2.5 Assembly validation

Regardless of the assembly strategy employed, it is essential that the accuracy and completeness of the resultant assembly is validated. A number of tools have been developed for assessing the correctness of *de novo* assemblies by use of read alignment of paired-end and mate-pair reads to the assembled set of contigs. These tools (for example, Feature Response Curves; Vezzi, Narzisi, & Mishra, 2012) consider whether reads from a pair map back to the assembly with the expected distance, identifying regions displaying conflation (reads align too far apart), contraction (read align too close together) and whether reads map in the expected orientation. Optical maps are also highly suited to assembly correctness assessment (Udall & Dawe, 2018) as given the rapidly decreasing costs of generating optical maps, their inclusion should become standard practice for new genome assembly projects. Genetic maps can also be used to similar effect (Fierst, 2015); however, the substantially higher investment required to generate a genetic map often negates their use unless a map is called for to enable additional genetic analyses.

While the above methods are useful for determining assembly accuracy, they do not provide insight into the completeness of the assembly. In particular, the most common downstream uses of an assembled genome will be primarily influenced to the completeness and contiguity of the gene space. A number of generic and plant-focused tools are available to assess the completeness of a gene space (for a review see Veeckman, Ruttink, & Vandepoele, 2016). Generic tools, of which BUSCO (Simão, Waterhouse, Ioannidis, Kriventseva, & Zdobnov, 2015) is the most common, comprise sets of core eukaryotic genes that are expected to be largely present in any eukaryotic genome. While BUSCO provides a useful first indicator, it can be relatively easy for even a highly fragmented assembly to achieve good completeness scores for such a restricted set of core genes. The PLAZA set of core plant gene families represents a more comprehensive set of highly conserved and largely single-copy genes (Van Bel et al., 2011).

Other approaches include comparative genomics assessments, for example, using a gold standard genome resource such as that for *A. thaliana* to determine how many reference genes have identifiable homologues within an assembly and to assess whether those homologues are present within single assembled contigs. However, as the gene set considered grows larger, comprehensive assessment criteria become harder due to the highly dynamic nature of plant genomes. An additional, and highly informative, approach is to generate RNA-Seq data for the target species and to align this to the assembly, assessing the extent to which RNA-Seq data aligns and using read-pair information to assess evidence for gene fragmentation across contigs.

One common problem faced by short-read assemblies in species with even moderate heterozygosity is assembly inflation due to the assembly of both haplotypes for heterozygous regions. In many cases these regions can include genes, leading to an over-estimation of the number of genes in the genome (for example, see the case of *Quercus lobate,* Née; Cartolano, Huettel, Hartwig, Reinhardt, & Schneeberger, 2016). Such haplotype splitting is not only problematic for gene space inflation, it can also hamper downstream uses such as genomic read alignment to detect SNPs or RNA-Seq gene expression analysis. In the case of SNP detection, genomic reads will align to the haploid from which the read was derived and for those regions of the genome, no SNPs will be detected on the basis of read alignments as a result. Regions representing split haplotypes can be detected on the basis of coverage when the original input sequencing data used for genome assembly is aligned to the assembled contigs in that such split regions will display half of the expected read coverage, while merged regions will have the expected read coverage. Regions can also be detected using self-self alignments of all assembled contigs; however, in either case the signature is often not clear-cut and in many cases it is not possible to unambiguously detect the two assembled haplotypes. When these regions contain genes, this will also influence RNA-Seq read alignment as RNA-Seq sequence reads will preferentially align to one haplotype or the other for positions where there are sequence variants between the two haplotypes or reads will align non-uniquely (i.e., equally well to both haplotypes) and many downstream methods will discard such non-uniquely mapped reads. Such issues also complicate gene family analyses, for example, comparative analyses to identify inflated or contracted gene families. As such, it is always important to be highly cautious when interpreting results based on draft assemblies.

2.6 Assembly annotation

Annotating protein coding genes, non-coding genes, repeat elements and other features within a genome is essential to enable downstream utilisation. Gene annotation typically involves two approaches employing *ab initio* predictions algorithms or RNA-Seq guided gene space annotation, with the most widely employed tools now combining the two (Yandell & Ence, 2012). One such commonly used tool for structural annotation of the gene space of plant genomes is the BRAKER pipeline (Hoff, Lange, Lomsadze, Borodovsky, & Stanke, 2016; http://bioinf.uni-greifswald.de/bioinf/braker/). There are also new tools under development that optimise use of long-read data (see below), such as Long Read Annotation (LoReAn; Cook et al., 2017). The use of RNA-Seq data substantially improves annotation of UTRs as well as the detection on long non-coding gene loci, which are currently not predicted by *ab initio* tools. While BRAKER can incorporate RNA-Seq evidence to improve gene space annotation, there are tools designed only for use with RNA-Seq data. One common choice is Cufflinks (Roberts, Pimentel, Trapnell, & Pachter, 2011), which annotates both genes and alternative splice variants (i.e., transcripts). Importantly for plants, the chloroplast and mitochondria require separate annotation and there are mature tools for both organelle genomes (Alverson et al., 2010; Schattner, Brooks, & Lowe, 2005; Tillich et al., 2017; Wyman, Jansen, & Boore, 2004). The chloroplast is typically easy to assemble and annotate (for a recent example, see Price & Arkin, 2017) while there is far greater variation in mitochondria size and genome structure (for example, Jackman et al., 2015).

There are additional tools optimised for the annotation of repetitive elements, of which the RepeatMasker suite of tools is the most common (Smit, Hubley, & Green, 1996), although others are under active development (Zeng, Kortschak, Raison, Bertozzi, & Adelson, 2018).

An important component of genome annotation is assigning functional information to the structural annotation of gene loci. As a genome matures, this information is developed by the community through empirical works, such as forward genetic screens. However, initial functional annotation is commonly assigned through sequence homology methods whereby knowledge from other species is transferred to the new genome annotation. There are inherent risks in this process, primarily that an assumption is made that sequence homology infers functional orthology, which is often not the case (Netotea, Sundell, Street, & Hvidsten, 2014). Beyond *A. thaliana* very few species have empirically derived functional gene information for anything more than a handful of genes. As such, the majority of functional annotation relies on sequence-homology based inference of gene function as determined in *A. thaliana*. In the case of tree species this may be particularly problematic as *A. thaliana* lacks many of the processes present in trees including seasonal growth, secondary wood formation, delayed flowering or prolonged growth. It is also important to be aware that even in *A. thaliana* the majority of genes with some degree of functional annotation have that information assigned either from gene expression evidence only (e.g., "expressed in response to water deficit") or from electronic evidence.

The two most commonly used tools for assigning functional annotation are currently BLAST2GO (Conesa & Götz, 2008) and InterproScan (Zdobnov & Apweiler, 2001). Both tools use sequence homology to transfer information from a set of target species to a new genome annotation with annotations including Gene Ontology (GO) and Protein family (PFAM). There are also tools that can assign descriptive annotations to genes, such as PaperBLAST (Price & Arkin, 2017).

2.7 New insights from genome sequencing of tree species

As for all plant species, trees span a huge range of genome sizes with currently available genomes spanning from the modest genomes of *Populus* species at ~480 Mbp through to the giant genomes of gymnosperms at >20 Gbp. The availability of a range of both tree and other plant genomes has enabled comparative genomics analyses, revealing extensive synteny among closely related species such as between the genomes of willow and poplar species (Dai et al., 2014; Hanley, Mallott, & Karp, 2006; Ma et al., 2013), including pan-genome analyses among *Populus* species (Pinosio et al., 2016). There are also examples of synteny being detected across wider phylogenetic distances among angiosperm trees (Luo et al., 2015). Analysis of available tree genomes has also revealed contrasting mechanisms underlying genome expansion, contraction and size variation with examples including the prominent, shared *Salicaceae* whole genome duplication (Tuskan et al., 2006), extensive tandem duplication blocks in Eucalyptus (Myburg et al., 2014) and the unique retention and massive expansion of transposable elements in conifer genomes (Kovach et al., 2010; Nystedt et al., 2013; Zuccolo, Scofield, De Paoli, & Morgante, 2015) as well as the presence of long introns (De La Torre et al., 2014; Gonzalez-Ibeas et al., 2016; Nystedt et al., 2013; Zimin et al., 2014). Remarkably, despite the huge variation in genome size, the number of protein coding genes is highly consistent (after accounting for artefacts from fragmented assembly), with the majority of genomes containing ~30,000–40,000 genes (Neale, Martínez-García, et al., 2017). Annotation of non-protein coding genes has received less attention and is also more variable in methodology, but in many cases extensive sets of non-coding genes have been identified (Nystedt et al., 2013; Sundell et al., 2017). In common with other species, these genes tend to be shorter, often single-exon, have highly tissue-specific expression and are more likely to be species or lineage unique. The ability to perform such comparative genomics analyses brings new perspective to the question "What makes a tree a tree?" (Groover, 2005), the results of which to date have not yielded a tree-specific set of genes that can explain the various biological features characteristic of tree species. The likely implication of this is that the repeated re-emergence of the tree lifeform and extensive secondary growth results from altered regulatory and expression control utilising a largely-common set of genes shared among all plant species.

3. Transcriptome assembly and profiling

The application of HTS to transcriptomics has been transformative in terms in both data quality and ease of applicability. Sequence-based assays of the transcriptome typically comprise RNA extraction, cDNA synthesis and subsequent sequencing of the cDNA. Current quantitative approaches involve amplification of the cDNA prior to sequencing and it is important to be aware that current methods provide relative, not absolute, measures of expression. In terms of data quality RNA-Sequencing (RNA-Seq) provides highly specific quantitation across a vast dynamic range and with extremely high technical reproducibility (Wang, Gerstein, & Snyder, 2009). Comparisons of microarray gene expression data to RNA-Seq established that ~10 million sequence reads (5 M paired-end reads) are adequate to provide gene expression data that has higher specificity and a wider dynamic range than microarray technologies (Wang et al., 2009). In terms of applicability, there are no pre-requisite requirements to be able to perform RNA-Seq analyses in a species of interest, which is in stark contrast to microarrays for which the sequence of target genes needed to be determined *a priori*. This liberation from a focus on model systems has enabled a broad range of questions in a diverse array of species to be addressed to date.

When designing an RNA-Seq study a primary factor in the study design is the availability of a reference genome. If a genome assembly for your study species is available, then analyses can be performed based on alignment of the RNA-Seq data generated to the reference genome and counting the number of reads that align within each annotated gene. Addition, novel, genes can optionally also be identified if desired, including non-coding genes that may have been missed by initial genome annotation. Alternatively, when a reference genome is not available, the transcriptome can be assembled *de novo* from the RNA-Seq data to form an assembly of transcripts. The original RNA-Seq data is then aligned to this transcript assembly to infer transcript expression levels. Such transcript assemblies are, however, notoriously noisy and highly fragmented due to regions of high similarity among members of gene families, unequal coverage due to differential transcript expression, the expression of alternative splice variants in additional to issues shared with whole-genome assembly including high heterozygosity and polyploidy. Transcriptome assembly is also computationally demanding, which can restrict applicability depending on available compute resources. Details of the initial analysis steps and tools used vary based on this fundamental divide between alignment-based or *de novo* assembly-based approaches, while later steps to determine gene (or transcript) differential abundance are common. For a tree-focused review see López de Heredia and Vázquez-Poletti (2016). Below we assume the use of Illumina short-read sequencing unless otherwise stated as this is the current *de facto* choice for performing RNA-Seq assays.

As for any experiment, good experimental design informed by sound statistical methodology and the question of interest is essential to conducting a successful RNA-Seq experiment. These considerations should guide decisions regarding sample size and replication in addition to sequencing depth (the number of sequence reads to be generated per sample), read type and length (primarily single or paired-end reads) and sampling approach. As many tree species lack mature reference genomes (this remains true even for the first available tree genome of *Populus trichocarpa*) we advise always generating paired-end data as this affords greater ability to differentiate between highly sequence similar regions of genes in cases where one read of a pair lies within the sequence-similar region and the second read is located in a more sequence-unique region. For the same reason we also advise generating the longest sequence reads possible within budget constraints. Longer sequence reads can be particularly beneficial for *de novo*-based studies. Common experimental design issues that can often be overlooked include recording metadata relating to sample collection and RNA extraction (for example, if more than one person performed the work or if samples were collected/extracted on different dates or times then record which person did what step for each sample and on which date/time). Typically, numerous samples are sequenced on a single machine run and it is important to record which samples were on which machine run to account for potential technical issues. For this same reason, samples from different experimental groups (for example, control and treated) should also be randomised at all stages, including sequencing. In a final analysis, all such metadata factors should initially be included in statistical models and their importance determined statistically. Non-significant factors can then be removed from the final model.

An essential part of any transcriptome study is decisions about what, where and when to sample. As gene expression is highly dynamic, sampling at the correct time is essential to observe expression associated with the trait or response of interest. This is particularly important in studies where a single sample point will be obtained, providing a snapshot measure of gene expression, as sampling at the wrong time will lead to false biological conclusions. Equally, sampling the wrong tissue or combination of cell types can result in mixed, ambiguous or uninformative expression measures. Imagine, for example, a gene that is highly expressed in cambium but that has low expression in phloem and xylem: a sample comprising all three cell types would give a measure of gene expression that was uninformative for any of the three cell types included; similarly, unequal representation of the three cell types among samples within an experiment would lead to incorrect biological conclusions. A number of sampling approaches can be used to overcome or alleviate these issues and to provide as specific an expression signal as possible. These include laser capture microdissection (LCM; for example, Blokhina et al., 2017; Canas, Canales, Gomez-Maldonado, Avila, & Canovas, 2014), cryogenic sectioning (for example, Immanen et al., 2016; Jokipii-Lukkari et al., 2017; Sundell et al., 2017) and spatial transcriptomics (Giacomello et al., 2017), among others. Although increasingly common in other biological systems, single cell transcriptome analysis remains challenging in plants primarily due to the presence of the cell wall.

It is also important to consider what type of transcripts are of interest. For example, typical RNA extraction protocols will eliminate short RNAs and many rely on poly-A enrichment to deplete ribosomal RNA (rRNA). If non-polyadenylated non-coding transcripts are of interest then RNA extraction and sequencing library preparation must be adjusted accordingly, for example, through use of rRNA depletion methods.

In common to either a reference-guided or *de novo* approach, there are a number of beneficial quality control steps that aim to identify technical problems, remove low quality data and remove partial sequences, such as adapters, remaining from the sequencing process. These include visual checks of the data to examine properties such as GC content, error distribution, high abundance sequences and read lengths, trimming of low quality bases and removal of any adapter sequence remnants, removal of rRNA and other contaminant sequences (this is particularly important for field material that will co-sample fungal and bacterial transcripts) and a final visual check of the cleaned data. These steps are overviewed in Delhomme, Mähler, et al. (2015), which includes recommendations for tools for these different stages.

Reference-based analysis of RNA-Seq data entails alignment of the quality-controlled sequence data to the reference genome. This is optimally performed using alignment algorithms that are splice-aware and that allow alignment of a single read to span across introns. While TopHat (Kim et al., 2013) is one of the most commonly used tools, other read aligners perform as well or better (Baruzzo et al., 2017; Costa-Silva, Domingues, & Lopes, 2017; Zhang, Zhang, Lin, & Zhao, 2017). STAR (Dobin et al., 2013) is a common choice that provides high alignment rates while being tolerant to the presence of heterozygosity and that consistently performs well in recent evaluations. When using such alignment tools, it is important to consider default parameters defined by the software for alignment, most notably expected intron length range (for example, there is a vast difference in the range of intron sizes between *Populus* species and all conifers). Setting the intron length too small will result in alignments being fragmented across exons of a single gene while setting the length too long will suggest alignment across genes. Most recently, there have also been new pseudo- and quasi-alignment based approaches that provide comparable results to traditional alignment algorithms but that are considerably faster and more computationally efficient (Bray, Pimentel, Melsted, & Pachter, 2016; Patro, Duggal, Love, Irizarry, & Kingsford, 2017; Patro, Mount, & Kingsford, 2014). For alignment-based tools the next step is to then count the number of reads aligned to each annotated gene locus. A common option for performing this step is HTSeq-Count (Anders, Pyl, & Huber, 2015). Importantly, this workflow provides expression quantification at the gene level rather than the transcript level. As such either a single representative transcript must be used per gene locus or, alternatively, a set of non-redundant pseudo-transcripts can be constructed representing all possible exons (Delhomme, Padioleau, Furlong, & Steinmetz, 2012). Expression quantification at the transcript level remains significantly challenging.

To perform *de novo* transcriptome analyses the first step is to use the set of input reads from the RNA-Seq experiment to perform a transcript assembly. We advise constructing a single assembly using all RNA-Seq data from all samples as this substantially simplifies downstream analysis steps by generating a single, common assembly. There are a large number of transcript assembly tools available (Geniza & Jaiswal, 2017; Moreton, Izquierdo, & Emes, 2016; Ungaro et al., 2017), of which Trinity (Haas et al., 2013) has become the most widely used. One highly recommended step is to use the provided *in silico* normalisation tool to reduce the input set of RNA-Seq reads to a non-redundant set. This can dramatically reduce computation resource demand and assembly time without compromising assembly quality. As for genome assembly, quality evaluation of the derived assembly is important and there are specific tools and guidelines available for this (Bushmanova, Antipov, Lapidus, Suvorov, & Prjibelski, 2016; Ghaffari et al., 2018; Ungaro et al., 2017). Transcriptome assembly from short-read data is extremely challenging and the resultant assemblies will often contain many times more (even by orders of magnitude) assembled transcripts than there are genes (or transcripts) in the target organism. While quality control filtering steps can remove many of the incorrectly assembled, artefactual transcripts, much of this assembly over-inflation results from the high degree of similarity among gene family members, alternative transcripts or maintained paralogs and these cannot be removed or resolved. An additional complication in many gymnosperm species is the extensive transcription of pseudogenes (Kovach et al., 2010; Nystedt et al., 2013; Warren et al., 2015). After an assembly has been produced, the input set of RNA-Seq reads are then aligned (or pseudo-aligned, see above) per sample to the assembled transcripts and these alignments are used for quantification counting. A key difference compared to reference-based analyses is that alignment does not need to account for the presence of introns and therefore standard genome alignment algorithms can be used.

While use of the set of RNA-Seq data generated from an experiment represents an efficient means of obtaining a transcriptome assembly for use in expression quantification, it is certainly not optimal. One increasingly popular and effectual alternative is to take advantage of the long-read lengths of PacBio or ON sequencing to generate full-length transcript sequences that can then be clustered and made non-redundant to obtain a reference transcriptome. As these sequencing technologies can generate read length in excess of all transcripts they effectively remove the need for any assembly and, importantly, provide unambiguous transcript (splice variant) sequences. Both technologies have specific protocols available that maximise the generation of full-length cDNA, with the PacBio IsoSeq approach (Cartolano et al., 2016) being by far the most common current solution. While this approach will provide a superior reference transcriptome for use in expression quantification, the data is not suitable for expression quantification due to the lower sequence read output of current long-read platforms, which restricts their dynamic range for expression quantification. As such, these methods are used to obtain the reference transcriptome and short-read, HTS is then used to assay each sample to obtain relative expression quantification.

Having obtained counts of the number of read aligning to each gene (or transcript) there are then a number of data scaling and normalisation steps that are required before performing downstream DE analysis or other approaches such as co-expression network calculation or data visualisation. One important factor that must be corrected for is the number of quality-filtered reads obtained per sample—and here we highly recommend removing rRNA contamination as this can sometimes have a large impact on the effective library size (the number of sequencing reads) available for gene expression quantification. For analyses where comparing the expression level of different genes to each other (rather than per gene among samples), the length of the gene sequence must also be controlled for. One common approach to controlling these effects is to calculate Transcripts Per kilobase Million values (for a review see Cartolano et al., 2016). TPM values are useful in that they enable comparison of expression values both among genes within a sample or for genes among samples, for example, making them ideal for plotting expression profiles or heat map visualisation. For statistical analysis methods additional normalisation of the data is needed to make the data heteroscedastic (i.e., to make variance independent of expression). There are a number of methods for achieving this including the common options of Variance Stabilising Transformation (VST) or regularised log (rlog) transformation as implemented in the DESeq2 package (Love, Huber, & Anders, 2014) and voom+limma (Law, Chen, Shi, & Smyth, 2014), both of which are packages for the R statistical language (R Core Team, 2013).

An important aspect of performing an RNA-Seq analysis is that all transcripts present in the sampled material are sequenced and represented. As such, novel genes or splice variants not currently annotated in a reference genome can be detected, annotated and their expression profiled. Common tools for identifying novel genes from RNA-Seq data include cufflinks (Roberts et al., 2011), Trinity in reference-guided mode

(Haas et al., 2013) and PASA (Haas, 2003). This also means that sequences from other organisms, particularly fungi and bacteria, will be present. While these should not align to a reference genome, they will be included in a *do novo* assembly and they should be identified and removed from the host-focused analysis. The transcripts deriving from such "contaminants" can, however, provide informative biological insights within a metatranscriptomics context (Delhomme, Sundström, et al., 2015) or can be used in a controlled and targeted manner to co-assay host and pathogen in an approach termed dual RNA-Seq (Hayden et al., 2014; Wolf, Kämmer, Brunke, & Linde, 2018).

It is also possible to sequence the short RNA component of cells, which is typically defined as RNA molecules <35 nt, the majority of which represent the various populations of micro-RNAs (miRNA), short interfering RNAs (siRNA) and *trans*-acting siRNAs (tasiRNAs) and other sRNAs. The vast majority of these are between 18 and 24 nt. For a review of sRNA biogenesis and function see Chen et al. (2010)). While miRNAs functional in transcriptional regulation, either by causing transcript cleavage or translational inhibition, various other sRNAs act as guides for epigenetic control, for example, the location of DNA cytosine methylation (Mosher & Melnyk, 2010). As a result of the short length of sRNAs, these are often sequenced using single end reads. While miRNAs tend to form the most abundant population of sRNAs and can be assayed with reasonable dynamic range using ~5 M reads, the far more diverse and lower abundance population of other sRNA classes requires considerably deeper coverage to provide accurate quantitative data. RNA-Seq can also be applied to specifically target transcripts that have been cleaved by miRNAs using an assay termed degradome sequencing (German, Luo, Schroth, Meyers, & Green, 2009), which provides important confirmatory evidence to support miRNA-target gene predictions, which are often noisy with a high error rate (Srivastava, Moturu, Pandey, Baldwin, & Pandey, 2014).

3.1 Applications of RNA-Seq in forest trees

There have been a large number of publications utilising RNA-Seq in tree species covering biological processes such as development (Torales et al., 2013; Torre et al., 2014), seasonal growth and dormancy (Hao et al., 2017; Khalil-Ur-Rehman et al., 2017; Paul et al., 2015; Shim et al., 2014; Tylewicz et al., 2018), wood formation (Jokipii-Lukkari et al., 2017; Sundell et al., 2017; Thavamanikumar, Southerton, & Thumma, 2014), biotic

(Foster, Pelletier, Tanguay, & Séguin, 2015; Hayden et al., 2014; Wolf et al., 2018) and abiotic (Chaires, Gupta, Joshee, Cooper, & Basu, 2017; Du, Ding, & Cai, 2018; George, Manoharan, Li, Britton, & Parida, 2017) stress response, the effects of genetic transformation (Felten et al., 2018; Ratke et al., 2018) and ecological factors such as sexual dimorphisms (Robinson et al., 2014; Song, Ma, Ci, Zhang, & Zhang, 2013), among others. Much of this work has been covered by previous reviews, although we have pointed to a few additional, recent examples. A notable feature of the example studies that we cite is that many are on non-model species, highlighting the wide applicability of RNA-Seq. The study by Thavamanikumar, Southerton, and Thumma (2014) demonstrated a further strength of having sequence-level information from RNA-Seq as they were able to detect and quantify DE among alleles of individual genes within an individual.

Wood formation is one of the defining characteristics of trees and has been the focus of a wide range of RNA-Seq studies, particularly in the more established and economically important systems of *Populus* and *Eucalyptus* but, increasingly, in other species including conifers. Two recent studies utilised the high spatial resolution afforded by sampling from mature tree to profile wood development, employing tangential cryogenic sectioning across the developing wood zone of European aspen (Sundell et al., 2017) and Norway spruce (Jokipii-Lukkari et al., 2017) and subsequent RNA-Seq analysis of the transcriptome. Both studies provided new insight into the developmental process, including in aspen showing that there is a major remodelling of the transcriptome between early xylem and late xylem, while also elucidating a continual transition of transcriptional modules throughout the developmental program. By employing comparative genomics methods based on orthologous relationships between the two widely separated lineages it was shown that there are examples of both conserved and diverged co-expression network modules (Sundell et al., 2017) in additional to specific examples of the NST and VND master regulators of wood formation that appear to have divergent regulatory functions in the two systems (Jokipii-Lukkari et al., 2017). Similarly, the transcriptomes of wood forming tissues of poplar and eucalyptus were compared to identified common pairs of genes with preferential expression in xylem in both model systems (Hefer, Mizrachi, Myburg, Douglas, & Mansfield, 2015). RNA-Seq has further been employed in Norway spruce to profile gene expression in phloem and xylem throughout a growing cycle, demonstrating expression patterns associated with dormancy and cambial arrest in addition to identifying examples of genes that are highly abundant during the winter period

(Jokipii-Lukkari et al., 2018). Both this study and one based on a Norway spruce tissue culture system (Laitinen et al., 2017) have employed RNA-Seq to provide new insight into the role and functional specialisation of laccase genes in lignin formation. A common component of these studies is the use of co-expression network and other systems biology methods to detect interactions among large sets of genes and to perform cross-species comparisons on those sets rather than on a gene-by-gene basis. As complex phenotypes and developmental programs involve the action and interaction of large numbers of genes, such systems-scale approaches are required in order to extract biological insight.

4. Epigenetics

Epigenomics is the genome-wide study of any epigenetic pattern including DNA cytosine (C) methylation and histone modifications (Plomion et al., 2016). There is a growing body of studies documenting the results of HTS assays of the methylome of tree species, with most work placed in the context of organ development and stress responses. Most commonly bisulfite sequencing (BS-Seq) is used to assay DNA C methylation, whereby genomic DNA is extracted and treated with bisulfite, which causes conversion of unmethylated C bases to uracil and these are then converted to thymine during subsequent PCR amplification. As such, CG base pairs in the genome where the C was not methylated are converted to TA while methylated CG positions remain unchanged. The treated DNA is amplified, sequenced using standard methods for genomic DNA (most often using paired-end libraries) and reads are aligned back to a reference genome to detect CG/TA miss-match positions (Chatterjee, Rodger, Morison, Eccles, & Stockwell, 2017; Kim, El Baidouri, & Jackson, 2014; Yong, Hsu, & Chen, 2016). Alternatively, C methylation can be assayed using of methylated DNA immunoprecipitation combined with HTS sequencing (MeDIP-Seq) to enrich for methylation genomic regions (Cortijo, Wardenaar, Colomé-Tatché, Johannes, & Colot, 2014; Wardenaar, Liu, Colot, Colomé-Tatché, & Johannes, 2013). Chromatin can also be assayed for epigenetic marks on histone residues, which can be either repressive or activating via their influence on whether DNA is packaged as euchromatin (open and accessible) or heterochromatin (closed, inaccessible). Histone residue modifications are assayed using chromatin immune precipitation sequencing (ChIP-Seq) in which antibodies specific to epigenetic modifications (methylation, phosphorylation, acetylation, ubiquitylation, and sumoylation) of specific histone protein amino acid residues are used to co-precipitate DNA attached to those histones containing the epigenetic modification (Lefrançois, Zheng, & Snyder, 2010; Massie & Mills, 2012; Nakato & Shirahige, 2017). Those DNA regions are then isolated and sequenced using standard genomic DNA sequencing library methods. Antibodies are available for a number of common repressing and activating epigenetic marks and as the sequence of histones is highly conserved, these typically work well across species. After sequencing, reads are aligned to the genome and the position of the assayed histone modification detected as a peak of aligned reads.

A number of studies have used HTS methods to investigate epigenetic signatures in broad-leaved forest tree species. Raj et al. (2011) examined DE of genes involved in the poplar drought response, showing significant epigenetic effects on drought response in specific genotypic backgrounds, such that adaptive somatic modifications affected drought-resilience in some genotypes. Lafon-Placette et al. (2018) have additionally considered the role of epigenetics in drought response. Other studies in poplar include work by Vining et al. (2012), who used 5-methyl-cytosine DNA immunoprecipitation and HTS to assay DNA methylation patterns in *P. trichocarpa* and reported methylation differences among tissues and Lafon-Placette et al. (2013) who described DNA methylation in developing shoot apical meristems of *P. trichocarpa*, reporting that methylation is omnipresent across the genome. ********More recently, Conde et al. (2017) used RNA-seq and BS-seq to investigate temperature-dependent regulation of bud break in poplar, finding that demethylation steps precede seasonal release from dormancy. Ding, Liang, Diao, Su, and Zhang (2018) documented genome-wide methylation patterns in *P. nigra* to identify methylation patterns corresponding to the diurnal sampling of their experiment, reporting DNA methylation across the genome and some temporal differences in the regulation of expression by methylated genes. This highlights that epigenetic changes can be highly dynamic. Also in broadleaves, Sollars and Buggs (2018), working on the genome of *Fraxinus excelsior*, reported similar levels of cytosine methylation to *P. trichocarpa*; they examined genotypes with high or low susceptibility to *Hymenoscyphus fraxineus*, the fungus causing ash dieback, and reported some differentially methylated regions between the sample groups, suggesting that epigenetics may influence biotic disease resistance. Su, Wang, He, Yang, and Wang (2014) investigated methylation together with gene expression in *Betula platyphylla* and compared methylation rates and gene expression patterns among genome contexts.

Aside from angiosperms, developmental studies of conifers have focused on the epigenome of somatic embryos using HTS technologies. Ausin et al. (2016) explored the DNA of *Picea abies* methylome using whole genome bisulfite sequencing in needles and somatic embryos, finding regions of differential methylation between somatic embryo culture tissue and a vegetative bud sample. These differences in methylation patterns are particularly interesting given the highly tissue-specific patterns of 24 nt siRNAs previously reported in gymnosperm species (Gonzalez-Ibeas et al., 2016; Nystedt et al., 2013) and the unique repeat content of these species compared to angiosperm trees with available genome, methylome and siRNA data. Yakovlev et al. (2014) assayed the transcriptome of *P. abies* somatic embryos to reveal differential expression suggestive of temperature-dependent memory in influencing seedling development, suggesting that this effect is most likely due to epigenetics.

More broadly, Bräutigam et al. (2013) reviewed ecological applications of epigenomics, highlighting that epigenetic effects extend beyond the individual, which is of particular importance in the case of trees as they are often keystone species.

Beyond direct assays of epigenetic modifications, insight to genome structure can also be obtained by examining patterns of DNA accessibility (Zhang, Zhang, Wu, & Jiang, 2014). This is achieved by use of enzymes that cleave DNA—typically DNase I (Song & Crawford, 2010) or micrococcal nuclease (MNase; Pajoro, Muiño, Angenent, & Kaufmann, 2018)—which are applied for short reaction periods during which open and accessible DNA will be cut and closed, inaccessible DNA is protected from enzymatic activity. DNA is subsequently size selected and sequenced using standard genomic DNA approaches and reads aligned to a reference genome. In this way, a footprint of DNA accessibility is produced, and these footprints can be compared among samples. Alternatively, similar information can be obtained through use of a transposon system using an assay termed Assay for Transposase-Accessible Chromatin (ATAC; Bajic, Maher, & Deal, 2018; Buenrostro, Wu, Chang, & Greenleaf, 2015), whereby a hyperactive, modified transposon is used to insert sequencing adapters into accessible chromatin regions and these are then utilised for subsequent genomic DNA sequencing library preparation. These methods can also be combined with cell or nuclei sorting methods to obtain cell-type specific chromatin profiles (Lu, Hofmeister, Vollmers, DuBois, & Schmitz, 2017; Wang & Deal, 2015), for example, by use of isolation of nuclei tagged in specific cell types (INTACT; Deal & Henikoff, 2011). INTACT is a powerful approach that can be utilised to obtain various genomics cell-type specific measures including transcriptomics (Deal & Henikoff, 2011; Reynoso et al., 2018) and chromatin state (Lu et al., 2017; Wang & Deal, 2015). These methods have been applied to a range of plant species, providing insights into chromatin dynamics in *A. thaliana* (Bajic et al., 2018; Deal & Henikoff, 2011; Pass et al., 2017; Zhang et al., 2014) and *Zea mays* (Rodgers-Melnick, Vera, Bass, & Buckler, 2016) as well as in a cross-species comparison (Maher et al., 2018). There are, however, few example applications in trees, with the exception of work in eucalyptus using ChIP-Seq to assay H3K4me3 and H3K27me3 and integration of these results within the context of gene expression patterns during wood formation (Hussey, Loots, van der Merwe, Mizrachi, & Myburg, 2017; Hussey, Mizrachi, Groover, Berger, & Myburg, 2015).

ChIP-Seq can also be utilised to profile transcription factor binding to DNA. While the approach has been widely employed for model plant systems, it has been used for only a limited number of studies in trees. One notable example is work in poplar where ChIP-Seq has been used to assay binding of Class I KNOX, Class III HD ZIPs and BELL-Like homeodomain family proteins (Liu, Missirian, Zinkgraf, Groover, & Filkov, 2014; Liu, Ramsay, et al., 2015; Liu, Zinkgraf, et al., 2015) and to explore their role in the control of wood formation and secondary growth.

5. Population scale genomics

Approaches to understanding the genomics of forest trees at the population level include expression quantitative trait locus (eQTL) studies and the application of systems genetics, which integrates phenotype genome wide association studies (GWAS) with eQTL or association mapping of other 'omics scale data (Civelek & Lusis, 2013). eQTL studies are those in which gene expression values in a population are mapped to underlying polymorphisms via GWAS in natural populations or linkage mapping using hybrid populations (i.e., QTL mapping). This enables inference of the regulatory mechanisms contributing to a phenotype of interest. A key difference in the use of natural or hybrid populations is the mapping resolution available, with large linkage blocks, containing hundreds of genes or more, in controlled cross populations. In natural populations the mapping resolution is typically higher, although this is highly influenced by rates of heterozygosity and breeding system type (Ingvarsson & Street, 2011). In both cases, resolution is limited by population size and all current studies in trees are underpowered, although the limitations to larger populations using common garden experiments are obvious. However, even small population sizes can yield highly informative results depending on the complexity (genetic architecture) of the target trait (Wang et al., 2018). Regulatory inference, in particular deriving whether associations are local or distant (or act in *cis* or *trans*), is also influenced by the contiguity of the genome assembly (Mähler et al., 2017).

5.1 eQTL studies in pedigrees and natural populations of trees

In the earliest studies of genetical genomics (Jansen & Nap, 2001) in forest trees, mRNA microarray data were used as an intermediate phenotype to map eQTL. Studies were performed in controlled cross pedigrees of *Eucalytpus grandis* (Kirst, Basten, Myburg, Zeng, & Sederoff, 2005) and an interspecific *Populus* hybrid pedigree (Drost et al., 2010) to examine the genetic architecture of gene expression control. Both studies identified genomic hotspots, representing single locations in the genome contributing to the control of expression variation for a large number of genes. Drost et al. (2010) further investigated the tissue-specific patterns of eQTLs within the context of the derived co-expression network and examined gene ontology and transcription factor over-enrichments analysis based on those networks. Interestingly, they found that the underlying genetic control of gene expression is dynamic among tissues.

In a natural population of *Populus tremula*, Mähler et al. (2017) performed eQTL mapping using RNA-Seq. In contrast to previous studies in controlled cross populations, they did not identify any genomic hotspots. Whether this is a feature of the population type or a characteristic of the tissue type and developmental stage assayed remains unclear. This study showed that SNPs within UTR and regulatory regions were particularly associated with genetically controlled variation in gene expression. Furthermore, the relationship between co-expression network characteristics and signatures of selection was explored, with the results suggesting that eQTLs are less likely to be present for genes in the core of the co-expression network or for gene that have experienced positive selection. However, the authors point out that these findings may be influenced by the genetic architecture of the trait for which tissue was sampled and RNA extracted. In this study, RNA was collected from flushing buds and there was no evidence of adaptive selection associated with this trait, as suggested by low population differentiation (Q_{ST}) and the lack of a longitudinal cline. This was reflected in patterns of gene expression within the sampled material (Mähler et al., 2017).

5.2 Systems genetics

Systems genetics integrate molecular level data, such as eQTLs, that are themselves quantitative phenotypes, with traditional phenotype GWAS, often also including co-expression or other network analyses. Each of the of the numerous intermediate process between a causal polymorphism and a phenotype can be mapped individually to dissect the complex nature of the trait in question (Civelek & Lusis, 2013) and networks constructed to infer functional associations among structural variants, transcripts, proteins, metabolites and phenotype. The key feature differentiating systems genetics from systems biology is that the latter dissects the large-scale responses of individuals from a common genetic background to experimental or genetic perturbations, whereas systems genetics utilises a population of genetically diverse individuals to interrogate molecular pathways and processes affecting the diversity in that population (Mizrachi & Myburg, 2016). Mizrachi et al. (2017) fostered a systems genetic approach in a hybrid population to investigate complex lignocellulosic traits in eucalyptus, with this representing the only true systems genetics study in trees to date, although other studies are underway and should be reported in the near future.

HTS has been pivotal in enabling GWAS studies in tree species. To date, the majority of studies have been conducted in *Populus* species, although studies in an ever-widening range of species and underway. In *Populus*, GWAS studies have been reported in natural populations of *P. trichocarpa* (Evans et al., 2014; Geraldes et al., 2014; McKown et al., 2014; Porth et al., 2015, 2013) and *P. tremula* (Wang et al., 2018). In both species, loci and genes associated with environmental and climatic adaptation have been identified, while in *P. trichocarpa*, genes associated with wood development have also been examined. In the case of *P. tremula* the work made use of a draft *de novo* genome assembly produced using HTS and with the results showing the importance of having a species-specific genome assembly, as revealed by genomic differences in the region surrounding the *FT2* locus between *P. trichocarpa* and *P. tremula*. The genome re-sequencing data from both species has also been used to perform population demography analyses, to examine patterns of recombination and linkage and to identify loci associated with environmental and climatic gradients (Wang, Street, Scofield, & Ingvarsson, 2016a, 2016b) and to perform the eQTL analysis detailed above (Mähler et al., 2017).

NGS data are also pivotal to a thorough understanding the evolutionary basis of neutral or adaptive genetic variation in natural tree populations

(Holliday et al., 2017) and its applications in ecological genomics, for example, population response to climate change. For example, in a collection of *Pinus taeda* spanning its natural distribution, Eckert et al. (2010) identified SNPs related to climatic variables with available gene annotations suggesting that climatic factors correlated to genes involved in responses to abiotic stress. Such approaches are also informative to applications in forest management, conservation, assisted migration and tree breeding.

6. Genomic resources

While there is now a large number of online resources for exploring and analysing genomics data from plant species, we concentrate here on those that have integrated data from tree species. Among these, one of the longest established resources is the TreeGene database (Wegrzyn, Lee, Tearse, & Neale, 2008), which hosts a diverse collection of genomics, genetics and population data. The database now hosts a number of North American tree species genomes and associated transcriptome and other genetic data, including the genomes produced by the PineRefSeq project (Gonzalez-Ibeas et al., 2016; Neale, McGuire, et al., 2017; Zimin et al., 2014). The PLAZA database contains orthology and gene family information for an extensive collection of plant species, including both angiosperm and gymnosperm trees (Proost et al., 2015). Similarly, the Phytozome resource hosts a wide range of plant genomes, including a number of angiosperm trees (Goodstein et al., 2012). The hardwood genomics genome project website (https://www.hardwoodgenomics.org/) provides links to genome and transcriptome data for a number of hardwood tree species. The PlantGenIE web resource (Sundell et al., 2015) includes sites specific to *Populus* and *P. abies* and has a focus on providing tools to explore gene expression data, including the results of the above-described aspen and Norway spruce wood development RNA-Seq studies (AspWood and NorWood, respectively). An associated tool, ComPlEx (Netotea et al., 2014), additionally enables cross-species co-expression network visualisation to identified expression-conserved orthologs. In general, tree species are not well represented in other genome resource sites, but there are data from *Populus* and *Vitis* included in ATTED-II (Obayashi, Aoki, Tadaka, Kagaya, & Kinoshita, 2017) and *Populus* in PlaNet (Mutwil et al., 2011).

References

Alexeyenko, A., Nystedt, B., Vezzi, F., Sherwood, E., Ye, R., Knudsen, B., et al. (2014). Efficient de novo assembly of large and complex genomes by massively parallel sequencing of Fosmid pools. *BMC Genomics*, *15*, 439.

Alverson, A. J., Wei, X., Rice, D. W., Stern, D. B., Barry, K., & Palmer, J. D. (2010). Insights into the evolution of mitochondrial genome size from complete sequences of Citrullus lanatus and Cucurbita pepo (Cucurbitaceae). *Molecular Biology and Evolution*, *27*, 1436–1448.

Anders, S., Pyl, P. T., & Huber, W. (2015). HTSeq—A Python framework to work with high-throughput sequencing data. *Cold Spring Harbor Laboratory Journals*, *31*, 166–169.

Ausin, I., Feng, S., Yu, C., Liu, W., Kuo, H. Y., Jacobsen, E. L., et al. (2016). DNA methylome of the 20-gigabase Norway spruce genome. *Proceedings of the National Academy of Sciences of the United States of America*, *113*, E8106–E8113.

Bajic, M., Maher, K. A., & Deal, R. B. (2018). Identification of open chromatin regions in plant genomes using ATAC-Seq. *Methods in Molecular Biology (Clifton, N.J.)*, *1675*, 183–201.

Baker, M. (2012). De novo genome assembly: What every biologist should know. *Nature Methods*, *9*, 333–337.

Baruzzo, G., Hayer, K. E., Kim, E. J., Di Camillo, B., FitzGerald, G. A., & Grant, G. R. (2017). Simulation-based comprehensive benchmarking of RNA-seq aligners. *Nature Methods*, *14*, 135–139.

Basantani, M. K., Gupta, D., Mehrotra, R., Mehrotra, S., Vaish, S., & Singh, A. (2017). An update on bioinformatics resources for plant genomics research. *Current Plant Biology*, *11–12*, 33–40.

Belton, J.-M., McCord, R. P., Gibcus, J. H., Naumova, N., Zhan, Y., & Dekker, J. (2012). Hi-C: A comprehensive technique to capture the conformation of genomes. *Methods (San Diego, Calif.)*, *58*, 268–276.

Birol, I., Raymond, A., Jackman, S. D., Pleasance, S., Coope, R., Taylor, G. A., et al. (2013). Assembling the 20 Gb white spruce (*Picea glauca*) genome from whole-genome shotgun sequencing data. *Bioinformatics*, *29*, 1492–1497.

Blokhina, O., Valerio, C., Sokołowska, K., Zhao, L., Kärkönen, A., Niittylä, T., et al. (2017). Laser capture microdissection protocol for xylem tissues of Woody plants. *Frontiers in Plant Science*, *07*, 1965.

Bräutigam, K., Vining, K. J., Lafon-Placette, C., Fossdal, C. G., Mirouze, M., Marcos, J. G., et al. (2013). Epigenetic regulation of adaptive responses of forest tree species to the environment. *Ecology and Evolution*, *3*, 399–415.

Bray, N., Pimentel, H., Melsted, P., & Pachter, L. (2016). Near-optimal RNA-Seq quantification. *Nature Biotechnology*, *34*, 525–527.

Buenrostro, J. D., Wu, B., Chang, H. Y., & Greenleaf, W. J. (2015). ATAC-seq: A method for assaying chromatin accessibility genome-wide. *Current Protocols in Molecular Biology*, *109*. 21.29.1-9.

Bushmanova, E., Antipov, D., Lapidus, A., Suvorov, V., & Prjibelski, A. D. (2016). rnaQUAST: A quality assessment tool for *de novo* transcriptome assemblies. *Bioinformatics*, *32*, 2210–2212.

Canas, R. A., Canales, J., Gomez-Maldonado, J., Avila, C., & Canovas, F. M. (2014). Transcriptome analysis in maritime pine using laser capture microdissection and 454 pyrosequencing. *Tree Physiology*, *34*, 1278–1288.

Cartolano, M., Huettel, B., Hartwig, B., Reinhardt, R., & Schneeberger, K. (2016). cDNA library enrichment of full length transcripts for SMRT long read sequencing. (Y Xing, Ed.). *PLoS One*, *11*, e0157779.

Chaires, M., Gupta, D., Joshee, N., Cooper, K. K., & Basu, C. (2017). RNA-seq analysis of the salt stress-induced transcripts in fast-growing bioenergy tree, *Paulownia elongata*. *Journal of Plant Interactions*, *12*, 128–136.

Chatterjee, A., Rodger, E. J., Morison, I. M., Eccles, M. R., & Stockwell, P. A. (2017). Tools and strategies for analysis of genome-wide and gene-specific DNA methylation patterns. *Methods in Molecular Biology (Clifton, N.J.), 1537*, 249–277.

Chen, D., Meng, Y., Ma, X., Mao, C., Bai, Y., Cao, J., et al. (2010). Small RNAs in angiosperms: Sequence characteristics, distribution and generation. *Bioinformatics (Oxford, England), 26*, 1391–1394.

Chin, C.-S., Peluso, P., Sedlazeck, F. J., Nattestad, M., Concepcion, G. T., Clum, A., et al. (2016). Phased diploid genome assembly with single-molecule real-time sequencing. *Nature Methods, 13*, 1050–1054.

Civelek, M., & Lusis, A. J. (2013). Systems genetics approaches to understand complex traits. *Nature Reviews. Genetics, 15*, 34–48.

Conde, D., Le Gac, A.-L., Perales, M., Dervinis, C., Kirst, M., Maury, S., et al. (2017). Chilling-responsive DEMETER-LIKE DNA demethylase mediates in poplar bud break. *Plant, Cell & Environment, 40*, 2236–2249.

Conesa, A., & Götz, S. (2008). Blast2GO: A comprehensive suite for functional analysis in plant genomics. *International Journal of Plant Genomics, 2008*, 1–13.

Cook, D., Valle-Inclan, J. E., Pajoro, A., Rovenich, H., Thomma, B., & Faino, L. (2017). Long read annotation (LoReAn): Automated eukaryotic genome annotation based on long-read cDNA sequencing. *Plant Physiology*. bioRxiv: 230359.

Cortijo, S., Wardenaar, R., Colomé-Tatché, M., Johannes, F., & Colot, V. (2014). Genome-wide analysis of DNA methylation in arabidopsis using MeDIP-Chip. *Methods in molecular biology (Clifton, NJ), 1112*, 125–149.

Costa-Silva, J., Domingues, D., & Lopes, F. M. (2017). RNA-Seq differential expression analysis: An extended review and a software tool. (Z Wei, Ed.). *PLoS One, 12*, e0190152.

Dai, X., Hu, Q., Cai, Q., Feng, K., Ye, N., Tuskan, G. A., et al. (2014). The willow genome and divergent evolution from poplar after the common genome duplication. *Cell Research, 24*, 1274–1277.

Davey, J. W., & Blaxter, M. L. (2010). RADSeq: Next-generation population genetics. *Briefings in Functional Genomics, 9*, 416–423.

De La Torre, A. R., Birol, I., Bousquet, J., Ingvarsson, P. K., Jansson, S., Jones, S. J. M., et al. (2014). Insights into conifer Giga-genomes. *Plant Physiology, 166*, 1724–1732.

Deal, R. B., & Henikoff, S. (2011). The INTACT method for cell type–specific gene expression and chromatin profiling in Arabidopsis thaliana. *Nature Protocols, 6*, 56–68.

Dekker, J., Marti-Renom, M. A., & Mirny, L. A. (2013). Exploring the three-dimensional organization of genomes: Interpreting chromatin interaction data. *Nature Reviews. Genetics, 14*, 390–403.

Delhomme, N., Mähler, N., Schiffthaler, B., Sundell, D., Hvidsten, C. M. T. R., & Street, N. R. (2015). Guidelines for RNA-Seq data analysis (prot 67). *Epigenesis*.

Delhomme, N., Padioleau, I., Furlong, E. E., & Steinmetz, L. M. (2012). easyRNASeq: A bioconductor package for processing RNA-Seq data. *Bioinformatics (Oxford, England), 28*, 2532–2533.

Delhomme, N., Sundström, G., Zamani, N., Lantz, H., Lin, Y.-C., Hvidsten, T. R., et al. (2015). Serendipitous meta-transcriptomics: The fungal Community of Norway Spruce (Picea abies) (W Arthofer, Ed.). *PLoS One, 10*, e0139080.

Ding, C.-J., Liang, L.-X., Diao, S., Su, X.-H., & Zhang, B.-Y. (2018). Genome-wide analysis of day/night DNA methylation differences in Populus nigra. (E Albertini, Ed.). *PLoS One, 13*, e0190299.

Dobin, A., Davis, C. A., Schlesinger, F., Drenkow, J., Zaleski, C., Jha, S., et al. (2013). STAR: Ultrafast universal RNA-seq aligner. *Bioinformatics (Oxford, England), 29*, 15–21.

Dominguez Del Angel, V., Hjerde, E., Sterck, L., Capella-Gutierrez, S., Notredame, C., Vinnere Pettersson, O., et al. (2018). Ten steps to get started in genome assembly and annotation. *F1000Research*, 7, 148.

Drost, D. R., Benedict, C. I., Berg, A., Novaes, E., Novaes, C. R. D. B., Yu, Q., et al. (2010). Diversification in the genetic architecture of gene expression and transcriptional networks in organ differentiation of Populus. *Proceedings of the National Academy of Sciences of the United States of America, 107*, 8492–8497.

Du, M., Ding, G., & Cai, Q. (2018). The transcriptomic responses of Pinus massoniana to drought stress. *Forests, 9*, 326.

Eckert, A. J., Bower, A. D., GonzÁlez-MartÍnez, S. C., Wegrzyn, J. L., Coop, G., & Neale, D. B. (2010). Back to nature: Ecological genomics of loblolly pine (Pinus taeda, Pinaceae). *Molecular Ecology, 19*, 3789–3805.

Ekblom, R., & Wolf, J. B. W. (2014). A field guide to whole-genome sequencing, assembly and annotation. *Evolutionary Applications*, 7, 1026–1042.

Evans, L. M., Slavov, G. T., Rodgers-Melnick, E., Martin, J., Ranjan, P., Muchero, W., et al. (2014). Population genomics of Populus trichocarpa identifies signatures of selection and adaptive trait associations. *Nature Genetics, 46*, 1089–1096.

Felten, J., Vahala, J., Love, J., Gorzsás, A., Rüggeberg, M., Delhomme, N., et al. (2018). Ethylene signaling induces gelatinous layers with typical features of tension wood in hybrid aspen. *The New Phytologist, 218*, 999–1014.

Fierst, J. L. (2015). Using linkage maps to correct and scaffold de novo genome assemblies: Methods, challenges, and computational tools. *Frontiers in Genetics, 6*, 220.

Foster, A. J., Pelletier, G., Tanguay, P., & Séguin, A. (2015). Transcriptome analysis of poplar during leaf spot infection with Sphaerulina spp. (M Stromvik, Ed.). *PLoS One, 10*, e0138162.

Geniza, M., & Jaiswal, P. (2017). Tools for building de novo transcriptome assembly. *Current Plant Biology, 11–12*, 41–45.

George, S., Manoharan, D., Li, J., Britton, M., & Parida, A. (2017). Transcriptomic responses to drought and salt stress in desert tree Prosopis juliflora. *Canadian Journal of Biotechnology, 1*, 291. 291.

Geraldes, A., Farzaneh, N., Grassa, C. J., M#ckown, A. D., Guy, R. D., Mansfield, S. D., et al. (2014). Landscape genomics of *populus trichocarpa*: The role of hybridization, limited gene flow, and natural selection in shaping patterns of population structure. *Evolution, 68*, 3260–3280.

German, M. A., Luo, S., Schroth, G., Meyers, B. C., & Green, P. J. (2009). Construction of parallel analysis of RNA ends (PARE) libraries for the study of cleaved miRNA targets and the RNA degradome. *Nature Protocols, 4*, 356–362.

Ghaffari, N., Arshad, O. A., Jeong, H., Thiltges, J., Criscitiello, M. F., Yoon, B.-J., et al. (2018). Examining De novo transcriptome assemblies via a quality assessment pipeline. *IEEE/ACM Transactions on Computational Biology and Bioinformatics, 15*, 494–505.

Giacomello, S., Salmén, F., Terebieniec, B. K., Vickovic, S., Navarro, J. F., Alexeyenko, A., et al. (2017). Spatially resolved transcriptome profiling in model plant species. *Nature Plants, 3*, 17061.

Gonzalez-Ibeas, D., Martinez-Garcia, P. J., Famula, R. A., Delfino-Mix, A., Stevens, K. A., Loopstra, C. A., et al. (2016). Assessing the gene content of the megagenome: Sugar pine (*Pinus lambertiana*). *G3 (Bethesda, Md.), 6*, 3787–3802.

Goodstein, D., Shu, S., Howson, R., Neupane, R., Hayes, R., Fazo, J., et al. (2012). Phytozome: A comparative platform for green plant genomics. *Nucleic Acids Research, 40*, D1178–D1186.

Groover, A. T. (2005). What genes make a tree a tree? *Trends in Plant Science, 10*, 210–214.

Haas, B. J. (2003). Improving the Arabidopsis genome annotation using maximal transcript alignment assemblies. *Nucleic Acids Research, 31*, 5654–5666.

Haas, B. J., Papanicolaou, A., Yassour, M., Grabherr, M., Blood, P. D., Bowden, J., et al. (2013). De novo transcript sequence reconstruction from RNA-seq using the trinity platform for reference generation and analysis. *Nature Protocols, 8*, 1494–1512.

Hanley, S. J., Mallott, M. D., & Karp, A. (2006). Alignment of a Salix linkage map to the Populus genomic sequence reveals macrosynteny between willow and poplar genomes. *Tree Genetics & Genomes*, *3*, 35–48.

Hao, X., Yang, Y., Yue, C., Wang, L., Horvath, D. P., & Wang, X. (2017). Comprehensive transcriptome analyses reveal differential gene expression profiles of Camellia sinensis axillary buds at Para-, Endo-, Ecodormancy, and bud flush stages. *Frontiers in Plant Science*, *8*, 553.

Hayden, K. J., Garbelotto, M., Knaus, B. J., Cronn, R. C., Rai, H., & Wright, J. W. (2014). Dual RNA-seq of the plant pathogen Phytophthora ramorum and its tanoak host. *Tree Genetics & Genomes*, *10*, 489–502.

Hefer, C. A., Mizrachi, E., Myburg, A. A., Douglas, C. J., & Mansfield, S. D. (2015). Comparative interrogation of the developing xylem transcriptomes of two wood-forming species: *Populus trichocarpa* and *Eucalyptus grandis*. *New Phytologist*, *206*, 1391–1405.

Hoff, K. J., Lange, S., Lomsadze, A., Borodovsky, M., & Stanke, M. (2016). BRAKER1: Unsupervised RNA-Seq-based genome annotation with GeneMark-ET and AUGUSTUS. *Bioinformatics*, *32*, 767–769.

Holliday, J. A., Aitken, S. N., Cooke, J. E. K., Fady, B., González-Martínez, S. C., Heuertz, M., et al. (2017). Advances in ecological genomics in forest trees and applications to genetic resources conservation and breeding. *Molecular Ecology*, *26*, 706–717.

Hussey, S. G., Loots, M. T., van der Merwe, K., Mizrachi, E., & Myburg, A. A. (2017). Integrated analysis and transcript abundance modelling of H3K4me3 and H3K27me3 in developing secondary xylem. *Scientific Reports*, 7, 3370.

Hussey, S. G., Mizrachi, E., Groover, A., Berger, D. K., & Myburg, A. A. (2015). Genome-wide mapping of histone H3 lysine 4 trimethylation in Eucalyptus grandis developing xylem. *BMC Plant Biology*, *15*, 117.

Immanen, J., Nieminen, K., Smolander, O.-P., Kojima, M., Alonso Serra, J., Koskinen, P., et al. (2016). Cytokinin and auxin display distinct but interconnected distribution and signaling profiles to stimulate cambial activity. *Current Biology*, *26*, 1990–1997.

Ingvarsson, P. K., Hvidsten, T. R., & Street, N. R. (2016). Towards integration of population and comparative genomics in forest trees. *The New Phytologist*, *212*, 338–344.

Ingvarsson, P. K., & Street, N. R. (2011). Association genetics of complex traits in plants. *The New Phytologist*, *189*, 909–922.

Jackman, S. D., Vandervalk, B. P., Mohamadi, H., Chu, J., Yeo, S., Hammond, S. A., et al. (2017). ABySS 2.0: Resource-efficient assembly of large genomes using a bloom filter. *Genome Research*, *27*, 768–777.

Jackman, S. D., Warren, R. L., Gibb, E. A., Vandervalk, B. P., Mohamadi, H., Chu, J., et al. (2015). Organellar genomes of white spruce (Picea glauca): Assembly and annotation. *Genome Biology and Evolution*, *8*, 29–41.

Jain, M., Koren, S., Miga, K. H., Quick, J., Rand, A. C., Sasani, T. A., et al. (2018). Nanopore sequencing and assembly of a human genome with ultra-long reads. *Nature Biotechnology*, *36*, 338–345.

Jansen, R. C., & Nap, J. P. J. (2001). Genetical genomics: The added value from segregation. *Trends in Genetics*, *17*, 388–391.

Jiao, W.-B., Accinelli, G. G., Hartwig, B., Kiefer, C., Baker, D., Severing, E., et al. (2017). Improving and correcting the contiguity of long-read genome assemblies of three plant species using optical mapping and chromosome conformation capture data. *Genome Research*, *27*, 778–786.

Jiao, Y., Peluso, P., Shi, J., Liang, T., Stitzer, M. C., Wang, B., et al. (2017). Improved maize reference genome with single-molecule technologies. *Nature*, *546*, 524.

Jokipii-Lukkari, S., Delhomme, N., Schiffthaler, B., Mannapperuma, C., Prestele, J., Nilsson, O., et al. (2018). Transcriptional roadmap to seasonal variation in wood formation of Norway spruce. *Plant Physiology*, *176*, 2851–2870. 01590.2017.

Jokipii-Lukkari, S., Sundell, D., Nilsson, O., Hvidsten, T. R., Street, N. R., & Tuominen, H. (2017). NorWood: A gene expression resource for evo-devo studies of conifer wood development. *New Phytologist, 216*, 482–494.

Khalil-Ur-Rehman, M., Sun, L., Li, C.-X., Faheem, M., Wang, W., & Tao, J.-M. (2017). Comparative RNA-seq based transcriptomic analysis of bud dormancy in grape. *BMC Plant Biology, 17*, 18.

Kim, K. D., El Baidouri, M., & Jackson, S. A. (2014). Accessing epigenetic variation in the plant methylome. *Briefings in Functional Genomics, 13*, 318–327.

Kim, D., Pertea, G., Trapnell, C., Pimentel, H., Kelley, R., & Salzberg, S. L. (2013). TopHat2: Accurate alignment of transcriptomes in the presence of insertions, deletions and gene fusions. *Genome Biology, 14*, R36.

Kirst, M., Basten, C. J., Myburg, A. A., Zeng, Z. B., & Sederoff, R. R. (2005). Genetic architecture of transcript-level variation in differentiating xylem of a Eucalyptus hybrid. *Genetics, 169*, 2295–2303.

Koren, S., Walenz, B. P., Berlin, K., Miller, J. R., Bergman, N. H., & Phillippy, A. M. (2017). Canu: Scalable and accurate long-read assembly via adaptive *k*-mer weighting and repeat separation. *Genome Research, 27*, 722–736.

Kovach, A., Wegrzyn, J. L., Parra, G., Holt, C., Bruening, G. E., Loopstra, C. A., et al. (2010). The Pinus taeda genome is characterized by diverse and highly diverged repetitive sequences. *BMC Genomics, 11*, 420.

Lafon-Placette, C., Faivre-Rampant, P., Delaunay, A., Street, N., Brignolas, F., & Maury, S. (2013). Methylome of DNase I sensitive chromatin in Populus trichocarpa shoot apical meristematic cells: A simplified approach revealing characteristics of gene-body DNA methylation in open chromatin state. *New Phytologist, 197*, 416–430.

Lafon-Placette, C., Le Gac, A.-L., Chauveau, D., Segura, V., Delaunay, A., Lesage-Descauses, M.-C., et al. (2018). Changes in the epigenome and transcriptome of the poplar shoot apical meristem in response to water availability affect preferentially hormone pathways. *Journal of Experimental Botany, 69*(3), 537–551. https://doi.org/10.1093/jxb/erx409.

Laitinen, T., Morreel, K., Delhomme, N., Gauthier, A., Schiffthaler, B., Nickolov, K., et al. (2017). A key role for apoplastic H 2 O 2 in Norway spruce phenolic metabolism. *Plant Physiology, 174*, 1449–1475.

Law, C. W., Chen, Y., Shi, W., & Smyth, G. K. (2014). Voom: Precision weights unlock linear model analysis tools for RNA-seq read counts. *Genome Biology, 15*, R29.

Lefrançois, P., Zheng, W., & Snyder, M. (2010). ChIP-Seq using high-throughput DNA sequencing for genome-wide identification of transcription factor binding sites. *Methods in enzymology, 470*, 77–104.

Levy-Sakin, M., & Ebenstein, Y. (2013). Beyond sequencing: Optical mapping of DNA in the age of nanotechnology and nanoscopy. *Current Opinion in Biotechnology, 24*, 690–698.

Lieberman-Aiden, E., van Berkum, N. L., Williams, L., Imakaev, M., Ragoczy, T., Telling, A., et al. (2009). Comprehensive mapping of long-range interactions reveals folding principles of the human genome. *Science (New York, N.Y.), 326*, 289–293.

Lin, Y., Li, J., Shen, H., Zhang, L., Papasian, C. J., & H-W, D. (2011). Comparative studies of de novo assembly tools for next-generation sequencing technologies. *Bioinformatics (Oxford, England), 27*, 2031–2037.

Lind, B. M., Menon, M., Bolte, C. E., Faske, T. M., & Eckert, A. J. (2018). The genomics of local adaptation in trees: Are we out of the woods yet? *Tree Genetics & Genomes, 14*, 29.

Liu, L., Missirian, V., Zinkgraf, M., Groover, A., & Filkov, V. (2014). Evaluation of experimental design and computational parameter choices affecting analyses of ChIP-seq and RNA-seq data in undomesticated poplar trees. *BMC Genomics, 15*(Suppl. 5), S3.

Liu, L., Ramsay, T., Zinkgraf, M., Sundell, D., Street, N. R., Filkov, V., et al. (2015). A resource for characterizing genome-wide binding and putative target genes of transcription factors expressed during secondary growth and wood formation in *Populus*. *The Plant Journal*, *82*, 887–898.

Liu, L., Zinkgraf, M., Petzold, H. E., Beers, E. P., Filkov, V., & Groover, A. (2015). The *Populus* ARBORKNOX1 homeodomain transcription factor regulates woody growth through binding to evolutionarily conserved target genes of diverse function. *New Phytologist*, *205*, 682–694.

López de Heredia, U., & Vázquez-Poletti, J. L. (2016). RNA-seq analysis in forest tree species: Bioinformatic problems and solutions. *Tree Genetics & Genomes*, *12*, 30.

Love, M. I., Huber, W., & Anders, S. (2014). Moderated estimation of fold change and dispersion for RNA-seq data with DESeq2. *Genome Biology*, *15*, 550.

Lu, Z., Hofmeister, B. T., Vollmers, C., DuBois, R. M., & Schmitz, R. J. (2017). Combining ATAC-seq with nuclei sorting for discovery of cis-regulatory regions in plant genomes. *Nucleic Acids Research*, *45*, e41.

Luo, M.-C., You, F. M., Li, P., Wang, J.-R., Zhu, T., Dandekar, A. M., et al. (2015). Synteny analysis in Rosids with a walnut physical map reveals slow genome evolution in long-lived woody perennials. *BMC Genomics*, *16*, 707.

Ma, T., Wang, J., Zhou, G., Yue, Z., Hu, Q., Chen, Y., et al. (2013). Genomic insights into salt adaptation in a desert poplar. *Nature Communications*, *4*, 2797.

Maher, K. A., Bajic, M., Kajala, K., Reynoso, M., Pauluzzi, G., West, D. A., et al. (2018). Profiling of accessible chromatin regions across multiple plant species and cell types reveals common gene regulatory principles and new control modules. *The Plant Cell*, *30*, 15–36.

Mähler, N., Wang, J., Terebieniec, B. K., Ingvarsson, P. K., Street, N. R., & Hvidsten, T. R. (2017). Gene co-expression network connectivity is an important determinant of selective constraint. (NM Springer, Ed.). *PLoS Genetics*, *13*, e1006402.

Massie, C. E., & Mills, I. G. (2012). Mapping protein–DNA interactions using ChIP-sequencing. *Methods in molecular biology (Clifton, N.J.)*, *809*, 157–173.

McKown, A. D., Guy, R. D., Klápště, J., Geraldes, A., Friedmann, M., Cronk, Q. C. B., et al. (2014). Geographical and environmental gradients shape phenotypic trait variation and genetic structure in Populus trichocarpa. *The New Phytologist*, *201*, 1263–1276.

Miller, J. R., Koren, S., & Sutton, G. (2010). Assembly algorithms for next-generation sequencing data. *Genomics*, *95*, 315–327.

Mizrachi, E., & Myburg, A. A. (2016). Systems genetics of wood formation. *Current Opinion in Plant Biology*, *30*, 94–100.

Mizrachi, E., Verbeke, L., Christie, N., Fierro, A. C., Mansfield, S. D., Davis, M. F., et al. (2017). Network-based integration of systems genetics data reveals pathways associated with lignocellulosic biomass accumulation and processing. *Proceedings of the National Academy of Sciences of the United States of America*, *114*, 1195–1200. 201620119.

Moll, K. M., Zhou, P., Ramaraj, T., Fajardo, D., Devitt, N. P., Sadowsky, M. J., et al. (2017). Strategies for optimizing BioNano and dovetail explored through a second reference quality assembly for the legume model, Medicago truncatula. *BMC Genomics*, *18*, 578.

Moreton, J., Izquierdo, A., & Emes, R. D. (2016). Assembly, assessment, and availability of de novo generated eukaryotic transcriptomes. *Frontiers in Genetics*, *6*, 361.

Mosher, R. A., & Melnyk, C. W. (2010). siRNAs and DNA methylation: Seedy epigenetics. *Trends in Plant Science*, *15*, 204–210.

Mutwil, M., Klie, S., Tohge, T., Giorgi, F. M., Wilkins, O., Campbell, M. M., et al. (2011). PlaNet: Combined sequence and expression comparisons across plant networks derived from seven species. *The Plant Cell*, *23*, 895–910.

Myburg, A. A., Grattapaglia, D., Tuskan, G. A., Hellsten, U., Hayes, R. D., Grimwood, J., et al. (2014). The genome of Eucalyptus grandis. *Nature*, *510*, 356–362.

Nakato, R., & Shirahige, K. (2017). Recent advances in ChIP-seq analysis: From quality management to whole-genome annotation. *Briefings in Bioinformatics, 18*, 279–290.

Neale, D. B., & Ingvarsson, P. K. (2008). Population, quantitative and comparative genomics of adaptation in forest trees. *Current Opinion in Plant Biology, 11*, 149–155.

Neale, D. B., & Kremer, A. (2011). Forest tree genomics: Growing resources and applications. *Nature Reviews. Genetics, 12*, 111–122.

Neale, D. B., Martínez-García, P. J., De La Torre, A. R., Montanari, S., & Wei, X.-X. (2017). Novel insights into tree biology and genome evolution as revealed through genomics. *Annual Review of Plant Biology, 68*, 457–483.

Neale, D. B., McGuire, P. E., Wheeler, N. C., Stevens, K. A., Crepeau, M. W., Cardeno, C., et al. (2017). The Douglas-fir genome sequence reveals specialization of the photosynthetic apparatus in Pinaceae. *G3 (Bethesda, Md.), 7*, 3157–3167.

Netotea, S., Sundell, D., Street, N. R., & Hvidsten, T. R. (2014). ComPlEx: Conservation and divergence of co-expression networks in A. thaliana, Populus and O. sativa. *BMC Genomics, 15*, 106.

Nowoshilow, S., Schloissnig, S., Fei, J.-F., Dahl, A., Pang, A. W. C., Pippel, M., et al. (2018). The axolotl genome and the evolution of key tissue formation regulators. *Nature, 554*, 50–55.

Nystedt, B. B., Street, N. R., Wetterbom, A., Zuccolo, A., Lin, Y.-C., Scofield, D. G., et al. (2013). The Norway spruce genome sequence and conifer genome evolution. *Nature, 497*, 579–584.

Obayashi, T., Aoki, Y., Tadaka, S., Kagaya, Y., & Kinoshita, K. (2017). ATTED-II in 2018: A plant Coexpression database based on investigation of statistical property of the mutual rank index. *Plant & Cell Physiology, 59*, 440.

Pajoro, A., Muiño, J. M., Angenent, G. C., & Kaufmann, K. (2018). Profiling nucleosome occupancy by MNase-seq: Experimental protocol and computational analysis. *Methods in molecular biology (Clifton, N.J.), 1675*, 167–181.

Parent, G. J., Raherison, E., Sena, J., & MacKay, J. J. (2015). Forest tree genomics: Review of Progress. *Advances in Botanical Research, 74*, 39–92.

Pass, D. A., Sornay, E., Marchbank, A., Crawford, M. R., Paszkiewicz, K., Kent, N. A., et al. (2017). Genome-wide chromatin mapping with size resolution reveals a dynamic sub-nucleosomal landscape in Arabidopsis. (O Mittelsten Scheid, Ed.). *PLoS Genetics, 13*, e1006988.

Patro, R., Duggal, G., Love, M. I., Irizarry, R. A., & Kingsford, C. (2017). Salmon provides fast and bias-aware quantification of transcript expression. *Nature Methods, 14*, 417–419.

Patro, R., Mount, S. M., & Kingsford, C. (2014). Sailfish enables alignment-free isoform quantification from RNA-seq reads using lightweight algorithms. *Nature Biotechnology, 32*, 462–464.

Paul, A., Jha, A., Bhardwaj, S., Singh, S., Shankar, R., & Kumar, S. (2015). RNA-seq-mediated transcriptome analysis of actively growing and winter dormant shoots identifies non-deciduous habit of evergreen tree tea during winters. *Scientific Reports, 4*, 5932.

Pinosio, S., Giacomello, S., Faivre-Rampant, P., Taylor, G., Jorge, V., Le Paslier, M. C., et al. (2016). Characterization of the poplar pan-genome by genome-wide identification of structural variation. *Molecular Biology and Evolution, 33*, 2706–2719.

Plomion, C., Bastien, C., Bogeat-Triboulot, M.-B., Bouffier, L., Déjardin, A., Duplessis, S., et al. (2016). Forest tree genomics: 10 achievements from the past 10 years and future prospects. *Annals of Forest Science, 73*, 77–103.

Porth, I., Klápště, J., McKown, A. D., La Mantia, J., Guy, R. D., Ingvarsson, P. K., et al. (2015). Evolutionary quantitative genomics of Populus trichocarpa. *PLoS One, 10*, e0142864.

Porth, I., Klapšte, J., Skyba, O., Hannemann, J., McKown, A. D., Guy, R. D., et al. (2013). Genome-wide association mapping for wood characteristics in Populus identifies an array of candidate single nucleotide polymorphisms. *The New Phytologist*, *200*, 710–726.

Price, M. N., & Arkin, A. P. (2017). PaperBLAST: Text mining papers for information about homologs. (MGI Langille, Ed.). *mSystems*, *2*, e00039-17.

Proost, S., Van Bel, M., Vaneechoutte, D., Van de Peer, Y., Inzé, D., Mueller-Roeber, B., et al. (2015). PLAZA 3.0: An access point for plant comparative genomics. *Nucleic Acids Research*, *43*, D974–D981.

Putnam, N. H., O'Connell, B. L., Stites, J. C., Rice, B. J., Blanchette, M., Calef, R., et al. (2016). Chromosome-scale shotgun assembly using an in vitro method for long-range linkage. *Genome Research*, *26*, 342–350.

R Core Team. (2013). *R: A language and environment for statistical computing*. the R Foundation for Statistical Computing.

Raj, S., Bräutigam, K., Hamanishi, E. T., Wilkins, O., Thomas, B. R., Schroeder, W., et al. (2011). Clone history shapes Populus drought responses. *Proceedings of the National Academy of Sciences of the United States of America*, *108*, 12521–12526.

Ratke, C., Terebieniec, B. K., Winestrand, S., Derba-Maceluch, M., Grahn, T., Schiffthaler, B., et al. (2018). Downregulating aspen xylan biosynthetic GT43 genes in developing wood stimulates growth via reprograming of the transcriptome. *The New Phytologist*, *219*, 230–245.

Reynoso, M. A., Pauluzzi, G. C., Kajala, K., Cabanlit, S., Velasco, J., Bazin, J., et al. (2018). Nuclear transcriptomes at high resolution using retooled INTACT. *Plant Physiology*, *176*, 270–281.

Roberts, A., Pimentel, H., Trapnell, C., & Pachter, L. (2011). Identification of novel transcripts in annotated genomes using RNA-Seq. *Bioinformatics*, *27*, 2325–2329.

Robinson, K. M., Delhomme, N., Mähler, N., Schiffthaler, B., Önskog, J., Albrectsen, B. R., et al. (2014). Populus tremula (European aspen) shows no evidence of sexual dimorphism. *BMC Plant Biology*, *14*, 276.

Rodgers-Melnick, E., Vera, D. L., Bass, H. W., & Buckler, E. S. (2016). Open chromatin reveals the functional maize genome. *Proceedings of the National Academy of Sciences of the United States of America*, *113*, E3177–E3184.

Sahlin, K., Street, N., Lundeberg, J., & Arvestad, L. (2012). Improved gap size estimation for scaffolding algorithms. *Bioinformatics*, *28*.

Schattner, P., Brooks, A. N., & Lowe, T. M. (2005). The tRNAscan-SE, snoscan and snoGPS web servers for the detection of tRNAs and snoRNAs. *Nucleic Acids Research*, *33*, W686–W689.

Shim, D., Ko, J.-H., Kim, W.-C., Wang, Q., Keathley, D. E., & Han, K.-H. (2014). A molecular framework for seasonal growth-dormancy regulation in perennial plants. *Horticulture Research*, *1*, 14059.

Simão, F. A., Waterhouse, R. M., Ioannidis, P., Kriventseva, E. V., & Zdobnov, E. M. (2015). BUSCO: Assessing genome assembly and annotation completeness with single-copy orthologs. *Bioinformatics (Oxford, England)*, *31*, 3210–3212.

Smit A, Hubley R, Green P. 1996. RepeatMasker Open-3.0. http://www.repeatmasker.org.

Sollars, E. S. A., & Buggs, R. J. A. (2018). Genome-wide epigenetic variation among ash trees differing in susceptibility to a fungal disease. *BMC Genomics*, *19*, 502.

Song, L., & Crawford, G. E. (2010). DNase-seq: A high-resolution technique for mapping active gene regulatory elements across the genome from mammalian cells. *Cold Spring Harbor Protocols*, *2010*. (pdb.prot5384-pdb.prot5384).

Song, Y., Ma, K., Ci, D., Zhang, Z., & Zhang, D. (2013). Sexual dimorphism floral microRNA profiling and target gene expression in andromonoecious poplar (Populus tomentosa) (R Wu, Ed.). *PLoS One*, *8*, e62681.

Srivastava, P. K., Moturu, T., Pandey, P., Baldwin, I. T., & Pandey, S. P. (2014). A comparison of performance of plant miRNA target prediction tools and the characterization of features for genome-wide target prediction. *BMC Genomics, 15*, 348.

Su, C., Wang, C., He, L., Yang, C., & Wang, Y. (2014). Shotgun bisulfite sequencing of the Betula platyphylla genome reveals the Tree's DNA methylation patterning. *International Journal of Molecular Sciences, 15*, 22874–22886.

Sundell, D., Mannapperuma, C., Netotea, S., Delhomme, N., Lin, Y.-C., Sjödin, A., et al. (2015). The plant genome integrative explorer resource: PlantGenIE.org. *New Phytologist, 208*, 1149–1156.

Sundell, D., Street, N. R., Kumar, M., Mellerowicz, E. J., Kucukoglu, M., Johnsson, C., et al. (2017). AspWood: High-spatial-resolution transcriptome profiles reveal uncharacterized modularity of wood formation in Populus tremula. *The Plant Cell, 29*, 1585–1604.

Thavamanikumar, S., Southerton, S., & Thumma, B. (2014). RNA-Seq using two populations reveals genes and alleles controlling wood traits and growth in Eucalyptus nitens. *PLoS One, 9*, e101104.

Tillich, M., Lehwark, P., Pellizzer, T., Ulbricht-Jones, E. S., Fischer, A., Bock, R., et al. (2017). GeSeq—Versatile and accurate annotation of organelle genomes. *Nucleic Acids Research, 45*, W6–W11.

Torales, S. L., Rivarola, M., Pomponio, M. F., Gonzalez, S., Acuña, C. V., Fernández, P., et al. (2013). De novo assembly and characterization of leaf transcriptome for the development of functional molecular markers of the extremophile multipurpose tree species Prosopis alba. *BMC Genomics, 14*, 705.

Torre, S., Tattini, M., Brunetti, C., Fineschi, S., Fini, A., Ferrini, F., et al. (2014). RNA-Seq analysis of quercus pubescens leaves: De Novo transcriptome assembly, annotation and functional markers development. (Z-J Liu, Ed.). *PLoS One, 9*, e112487.

Tuskan, G. A., Difazio, S., Jansson, S., Bohlmann, J., Grigoriev, I., Hellsten, U., et al. (2006). The genome of black cottonwood, Populus trichocarpa (Torr. & Gray). *Science (New York, N.Y.), 313*, 1596–1604.

Tylewicz, S., Petterle, A., Marttila, S., Miskolczi, P., Azeez, A., Singh, R. K., et al. (2018). Photoperiodic control of seasonal growth is mediated by ABA acting on cell-cell communication. *Science, 360*, 212–215.

Udall, J. A., & Dawe, R. K. (2018). Is it ordered correctly? Validating genome assemblies by optical mapping. *The Plant Cell, 30*, 7–14.

Ungaro, A., Pech, N., Martin, J.-F., McCairns, R. J. S., Mévy, J.-P., Chappaz, R., et al. (2017). Challenges and advances for transcriptome assembly in non-model species. (MFW te Pas, Ed.). *PLoS One, 12*, e0185020.

Van Bel, M., Proost, S., Wischnitzki, E., Movahedi, S., Scheerlinck, C., Van de Peer, Y., et al. (2011). Dissecting plant genomes with the PLAZA comparative genomics platform. *Plant Physiology, 158*, 590–600.

Veeckman, E., Ruttink, T., & Vandepoele, K. (2016). Are we there yet? Reliably estimating the completeness of plant genome sequences. *The Plant Cell, 28*, 1759–1768.

Vezzi, F., Narzisi, G., & Mishra, B. (2012). Feature-by-feature—Evaluating De Novo sequence assembly. (A Rzhetsky, Ed.). *PLoS One, 7*, e31002.

Vining, K. J., Pomraning, K. R., Wilhelm, L. J., Priest, H. D., Pellegrini, M., Mockler, T. C., et al. (2012). Dynamic DNA cytosine methylation in the Populus trichocarpa genome: Tissue-level variation and relationship to gene expression. *BMC Genomics, 13*, 27.

Wang, D., & Deal, R. B. (2015). Epigenome profiling of specific plant cell types using a streamlined INTACT protocol and ChIP-seq. *Methods in molecular biology (Clifton, N.J.)., 1284*, 3–25.

Wang, J., Ding, J., Tan, B., Robinson, K. M., Michelson, I. H., Johansson, A., et al. (2018). A major locus controls local adaptation and adaptive life history variation in a perennial plant. *Genome Biology*, *19*, 72.

Wang, Z., Gerstein, M., & Snyder, M. (2009). RNA-Seq: A revolutionary tool for transcriptomics. *Nature Reviews. Genetics*, *10*, 57–63.

Wang, J., Street, N. R., Scofield, D. G., & Ingvarsson, P. K. (2016a). Natural selection and recombination rate variation shape nucleotide polymorphism across the genomes of three related Populus species. *Genetics*, *202*, 1185–1200.

Wang, J., Street, N. R., Scofield, D. G., & Ingvarsson, P. K. (2016b). Variation in linked selection and recombination drive genomic divergence during allopatric speciation of European and American aspens. *Molecular Biology and Evolution*, *33*, 1754–1767.

Wardenaar, R., Liu, H., Colot, V., Colomé-Tatché, M., & Johannes, F. (2013). Evaluation of MeDIP-Chip in the context of whole-genome bisulfite sequencing (WGBS-Seq) in Arabidopsis. *Methods in molecular biology (Clifton, N.J.)*, *1067*, 203–224.

Warren, R. L., Keeling, C. I., Yuen, M. M., Raymond, A., Taylor, G. A., Vandervalk, B. P., et al. (2015). Improved white spruce (*Picea glauca*) genome assemblies and annotation of large gene families of conifer terpenoid and phenolic defense metabolism. *The Plant Journal*, *83*, 189–212.

Wegrzyn, J., Lee, J., Tearse, B., & Neale, D. (2008). TreeGenes: A forest tree genome database. *International Journal of Plant Genomics*, *2008*, 412875.

Wegrzyn, J. L., Liechty, J. D., Stevens, K. A., Wu, L.-S., Loopstra, C. A., Vasquez-Gross, H. A., et al. (2014). Unique features of the loblolly pine (Pinus taeda L.) megagenome revealed through sequence annotation. *Genetics*, *196*, 891–909.

Wolf, T., Kämmer, P., Brunke, S., & Linde, J. (2018). Two's company: Studying interspecies relationships with dual RNA-seq. *Current Opinion in Microbiology*, *42*, 7–12.

Wyman, S. K., Jansen, R. K., & Boore, J. L. (2004). Automatic annotation of organellar genomes with DOGMA. *Bioinformatics*, *20*, 3252–3255.

Yakovlev, I. A., Lee, Y., Rotter, B., Olsen, J. E., Skrøppa, T., Johnsen, Ø., et al. (2014). Temperature-dependent differential transcriptomes during formation of an epigenetic memory in Norway spruce embryogenesis. *Tree Genetics & Genomes*, *10*, 355–366.

Yandell, M., & Ence, D. (2012). A beginner's guide to eukaryotic genome annotation. *Nature Reviews Genetics*, *13*, 329–342.

Yong, W.-S., Hsu, F.-M., & Chen, P.-Y. (2016). Profiling genome-wide DNA methylation. *Epigenetics & Chromatin*, *9*, 26.

Zdobnov, E. M., & Apweiler, R. (2001). InterProScan—An integration platform for the signature-recognition methods in InterPro. *Bioinformatics (Oxford, England)*, *17*, 847–848.

Zeng, L., Kortschak, R. D., Raison, J. M., Bertozzi, T., & Adelson, D. L. (2018). Superior ab initio identification, annotation and characterisation of TEs and segmental duplications from genome assemblies. (R Kalendar, Ed.). *PLoS One*, *13*, e0193588.

Zhang, C., Zhang, B., Lin, L.-L., & Zhao, S. (2017). Evaluation and comparison of computational tools for RNA-seq isoform quantification. *BMC Genomics*, *18*, 583.

Zhang, W., Zhang, T., Wu, Y., & Jiang, J. (2014). Open chromatin in plant genomes. *Cytogenetic and Genome Research*, *143*, 18–27.

Zimin, A. V., Marçais, G., Puiu, D., Roberts, M., Salzberg, S. L., & Yorke, J. A. (2013). The MaSuRCA genome assembler. *Bioinformatics (Oxford, England)*, *29*, 2669–2677. btt476.

Zimin, A., Stevens, K. A., Crepeau, M. W., Holtz-Morris, A., Koriabine, M., Marçais, G., et al. (2014). Sequencing and assembly of the 22-gb loblolly pine genome. *Genetics*, *196*, 875–890.

Zuccolo, A., Scofield, D. G., De Paoli, E., & Morgante, M. (2015). The Ty1-copia LTR retroelement family PARTC is highly conserved in conifers over 200MY of evolution. *Gene*, *568*, 89–99.

CHAPTER TWO

Resources for conifer functional genomics at the omics era

Rafael A. Cañas*,†, Mª Belén Pascual, Fernando N. de la Torre, Concepción Ávila, Francisco M. Cánovas
Departamento de Biología Molecular y Bioquímica, Universidad de Málaga, Málaga, Spain
*Corresponding author: e-mail address: rcanas@uma.es

Contents

Abstract

Conifers are a group of woody plants that dominate vast forest ecosystems and represent a fundamental source of raw materials such as wood and other forest-derived products of great interest for humankind. Thus, understanding the molecular basis underlying the growth and development of these woody plants, as well as the response of these plants to environmental cues, is of paramount importance for the conservation of coniferous natural forests and the production of new tree lines with improved traits for sustainable forest plantations. Conifers have a series of characteristics that hinder the application of genetics and molecular biological approaches to generate data of high biological value. However, nowadays the biological sciences are dominated by omics techniques that have extensively increased the fundamental knowledge of conifer biology over the last decade. Novel resources are currently available for functional genomics research, and these resources are accessible to researchers all over the world. Recent advances in next-generation sequencing and assembly capacity have enabled whole-genome sequencing of several conifer species and have favoured the development of functional genomics studies in these non-model plant species. Currently, it is

† Senior author.

Advances in Botanical Research, Volume 89
ISSN 0065-2296
https://doi.org/10.1016/bs.abr.2018.11.002

possible to obtain large amounts of data to explore the transcriptome or genome of any species with a limited budget. However, functional genomics studies using genetics approaches and gene manipulation remain tremendously challenging, and the application of recent gene editing techniques in conifers is awaited.

1. Introduction

Conifers are a group of woody gymnosperms that appeared during the Carboniferous period, approximately 300 million years ago (Farjon & Filer, 2013). This group of plants is distributed across all continents except Antarctica, dominating the forest ecosystems found in the vast boreal regions of North America and Eurasia (Farjon & Filer, 2013). These boreal forests are considered one of the major global biomes covering around the 11% of the land surface on Earth (Essen, Ehnström, Ericsson, & Sjöberg, 1997; Lorenz & Lal, 2010). Conifers are also an important source of industrial raw materials, including resins and wood (Farjon, 2010). Conifers accounted for the 64.8% of the total global production of sawlogs and veneer logs and for the 50.9% of that of pulpwood in 2015 (Fig. 1; FAOSTAT, http://www.fao.org/faostat/en/?#home).

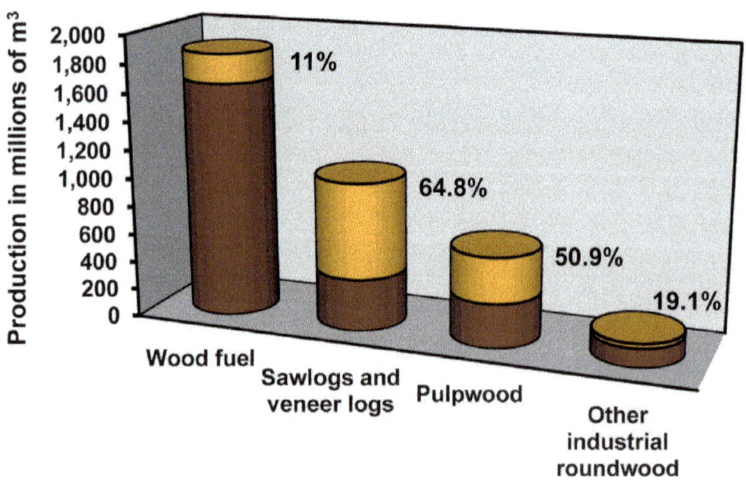

Fig. 1 World wood production in 2015 including the amount and percentage corresponding to conifers. Dark brown cylinders correspond to wood products from non-coniferous sources. Light brown cylinders correspond to wood products from coniferous sources. *Data from FAOSTAT (http://www.fao.org/faostat/en/).*

Over their evolutionary story, conifers have acquired a series of unique characteristics in response to interactions with their environments and to overcome biological challenges. There are very tall species, reaching up to 100 m, such as Douglas fir (*Pseudotsuga menziesii*) (Domec et al., 2008). Most conifers are long-living evergreens, including species with extreme longevity such as *Pinus longaeva* (some specimens of which are approximately 5000 years old) (Farjon & Filer, 2013) and *Fitzroya cupressoides* (some specimens of which are approximately 3600 years old) (Lara & Villalba, 1993). Due to the long-life spans of conifers, these trees are exposed to changing environmental conditions, mainly due to seasonal changes (Cañas et al., 2015; Cañas, de la Torre, Pascual, Avila, & Cánovas, 2016). Therefore, some conifers species have developed extraordinary adaptations, such as the capacity of the Siberian *Picea obovata* to survive at extremely low temperatures (Kjellsen, Shiryaeva, Schröder, & Strimbeck, 2010).

Conifers also have very large genomes, which range from 20 to 40 Gb (De La Torre et al., 2014; Mackay et al., 2012; Ritland, 2012) and include a large number of repeated sequences with long telomeres up to 57 kb in size (Aronen & Ryynänen, 2012; Flanary & Kletetschka, 2005; Neale et al., 2014), big introns (García-Gutiérrez, Cánovas, & Avila, 2018; Nystedt et al., 2013) and large amounts of transposable elements, mainly LTR (Long Terminal Repeat) retrotransposons, which may account for more than 70% of the genome (Cañas et al., 2017; Neale et al., 2014; Nystedt et al., 2013; Stevens et al., 2016).

In spite of the benefits of these biological characteristics of conifers, these characteristics have hindered conifer breeding, propagation and research, especially in the context of genetic manipulation (mutant libraries, production of transgenic plants, *etc.*). In some cases, these limitations have been overcome using large and costly infrastructures, which are available only to a small number of research groups. However, the emergence of "omics" techniques has partially palliated this problem at the research level. Of these techniques, next-generation sequencing (NGS) has allowed cheaper and faster sequencing of conifer genomes compared with classical reference models such as humans and several plant crop species, which would have been unbelievable in the recent past. Furthermore, these techniques have provided a way to explore gene function and perform functional genomics of conifers, including the characterization of the transcriptomes, proteomes and metabolomes but also the interactions and biological functions of their metabolic and regulatory pathways. An overview of recent advancements is presented in this chapter, which attempts to trace the current status of the resources that are available for functional genomics of conifers.

2. Resources

Nearly every current omics technique requires the development of a reference dataset, such as genome and transcriptome sequences or metabolite libraries. An example of such a dataset is a physical or virtual cDNA library containing enough transcriptomic information to identify cDNAs and design and synthesize oligonucleotides for microarray construction. When omics techniques first started to be used, the procurement of these sequence pools was very time consuming and costly. However, this situation has changed with the emergence of NGS platforms, allowing the rapid sequencing of transcriptomes and even entire genomes, which remains a real challenge in conifers due to the enormous genomes of these species (De La Torre et al., 2014). In conifer research, resources for omics techniques have greatly increased over the last decade, which has led to the increased research activity associated with some functional genomics methods.

The resources available for conifer functional genomics include genome and transcriptome sequence assemblies, microarrays and bioinformatic tools. Most of these resources are publicly available through online databases, which can also be considered as resources since they usually provide additional tools for the analysis and visualization of biological data.

2.1 Genome assemblies

Despite the large sizes and repetitive nature of conifer genomes, five genome assemblies have been published to date (Table 1): *Picea abies* (Nystedt et al., 2013), *Picea glauca* (Birol et al., 2013), *Pinus taeda* (Zimin et al., 2014), *Pinus lambertiana* (Stevens et al., 2016) and the first draft of the *P. menziesii* genome (Neale et al., 2017). In addition, the genome assembly of *Gnetum montanum*, a gnetophyte, has been recently published (Wan et al., 2018). Current phylogenomic analyses indicate that gnetophytes belong to the same clade as conifers (Li et al., 2017; Wickett et al., 2014).

Overall, the size and complexity of conifer genomes lead to these assemblies having too many contigs with very partitioned drafts. This characteristic along with the presence of very large introns in a considerable number of genes has also limited the bioinformatic identification of many genes. With the exception of *P. abies*, the number of high confidence genes is very

Table 1 Conifer genome assemblies.
Sequenced genomes

Species	Reference of last version	Online database	Genes[a]	Year of first version
Pseudotsuga menziesii	Neale et al. (2017)	http://treegenesdb.org	35,880	2017
Pinus lambertiana	Stevens et al. (2016)	http://treegenesdb.org	13,936	2016
Pinus taeda	Zimin et al. (2017)	http://treegenesdb.org	9024	2014
Picea glauca	Warren et al. (2015)	http://congenie.org	14,462	2013
Picea abies	Nystedt et al. (2013)	http://congenie.org	26,359	2013

[a]High confidence genes.

low in every assembly (Table 1). Each one of these limitations has led to new assembly strategies in order to obtain better draft genomes (Warren et al., 2015; Zimin et al., 2017). The use of sequencing platforms rendering long reads such as PacBio (https://www.pacb.com/) and Oxford Nanopore (https://nanoporetech.com/) in combination with Illumina could led to high-quality genome assemblies in conifers.

In functional genomics, genome assemblies are excellent tools to analyse different pools of data from NGS sequencing, which can include RNA-Seq and chromatin-derived samples (Alonso & Stepanova, 2015; Pevsner, 2015). However, there have been no reports of conifer genome assemblies being used for these purposes. In the case of RNA-Seq, the use of genome assemblies as a reference not only is useful to quantify gene expression but also can allow the identification of new transcripts derived from alternative splicing of genes (Steijger et al., 2013). Therefore, due to the limitations imposed by the quality of the constructed genome, reference transcriptomes and *de novo* transcriptome assemblies are much more adequate for RNA-Seq analyses (Steijger et al., 2013). In contrast, RNA-Seq data can also be used to aid the annotation of conifer genomes by identifying exons and introns from protein coding and non-coding genes.

For chromatin sequencing methods, read mapping with a genome assembly is essential for analysis. However, this process is dependent on the completeness and accuracy of the assembly. Fragmented and poorly annotated drafts are not adequate references for these analyses and lead to bias in sequence coverage (Meyer & Liu, 2014). Another problem associated with conifer genomes is the number of repeated sequences, accounting for almost 80% of the genome in some species such as *P. lambertiana* (Stevens et al., 2016). This phenomenon can make it challenging to use these genomes as references for read mapping due to the introduction of bias in the coverage of certain regions, which, in the case of conifers, could be very substantial.

An important utility of genome sequences is that these sequences can be used for the implementation of genome editing tools such as TALENs (Boch, 2011) or CRISPR/Cas9 (Charpentier & Doudna, 2013; Feng et al., 2013). The application of these tools requires precise knowledge of the target sequence (see Section 2.5). However, considering the difficulties in using forward and reverse genetics for functional validation in conifers, the use of genome assemblies in the future as a reference for certain methods such as ChIP-Seq (Johnson, Mortazavi, Myers, & Wold, 2007) and DAP-Seq (O'Malley et al., 2016) is crucial for developing functional genomic research in conifers. The identification of epigenomic regulation sites and transcription factor-binding sites in chromatin will be of great utility for conifer research in the near future. Several studies have paved the way by characterizing transcription factors associated with plant development (Alvarez et al., 2018, 2015; Bravo et al., 2017; Carlsbecker et al., 2013; Hedman, Zhu, von Arnold, & Sohlberg, 2013; Niu et al., 2016; Rupps, Raschke, Rümmler, Linke, & Zoglauer, 2016; Uddenberg, Akhter, Ramachandran, Sundström, & Carlsbecker, 2015; Uddenberg et al., 2013; Valledor, Pascual, Meijón, Escandón, & Cañal, 2015; Zhu, Moschou, Alvarez, Sohlberg, & Von Arnold, 2014; Zhu, Moschou, Alvarez, Sohlberg, & Arnold, 2016) and those associated with lignification and regulation of secondary metabolism (Bedon et al., 2010; Bedon, Grima-Pettenati, & Mackay, 2007; Bomal et al., 2008, 2014; Craven-Bartle, Pascual, Cánovas, & Ávila, 2013; Pascual, Canovas, & Avila, 2015; Pascual et al., 2016, 2018; Rueda-López, García-Gutiérrez, Cánovas, & Ávila, 2013). The application of these types of techniques will be extremely helpful in the functional characterization of regulatory networks involved in the control of essential processes in conifers. The elucidation of these regulatory networks will provide essential biological data to be employed in the development of new tree lines with improved traits. However, the currently available methods are insufficient, and it is necessary to improve the assemblies and annotations to implement accurate genome-wide analyses.

2.2 Transcriptome assemblies

Whole transcriptomes are fundamental tools to design expression microarrays, analyse RNA-Seq experiments, identify peptides in proteomic studies using mass spectrometric analyses, annotate genome assemblies and identify genetic markers such as single nucleotide polymorphisms (SNPs) and simple sequence repeats (SSRs) (Dufva, 2009; Lowe, Shirley, Bleackley, Dolan, & Shafee, 2017; Pevsner, 2015; Romero-Rodríguez, Pascual, Valledor, & Jorrín-Novo, 2014). However, each tissue and cell type has its own transcriptome, which depends on development, metabolic state, environmental conditions, *etc.* (Yang & Kim, 2015). Therefore, the transcriptome is dynamic and changes with the forms (alternative splicing and chemical modifications of N bases) and amounts of individual transcripts (Ozsolak & Milos, 2011). Thus, the whole transcriptome of a species is the set of different transcriptomes expressed in each of the cell types and under different conditions.

This biological complexity leads to technical difficulties in obtaining complete transcriptomic data. In addition, before the emergence of NGS technologies, the procurement of transcriptomic data was very laborious and costly. The first resources for the generation of large-scale transcriptomic data were expressed sequence tag (EST) databases (Adams et al., 1991). The need for physical libraries of cDNA in *Escherichia coli* and the individual Sanger sequencing of each clone limited the number of EST sequences and the assembly of transcriptomes (Lowe et al., 2017). In addition, the experimental procedures favoured the identification of transcripts coding for proteins, while sequences belonging to non-coding RNA (ncRNA) were not analysed because these sequences were considered sequencing errors or junk. Different EST databases were generated for conifer species for gene discovery and annotation (Table 2), and in addition, a substantial number of studies in conifers have developed ESTs for genetic marker identification (*e.g.* Bérubé et al., 2007; Perry & Bousquet, 1998; Rungis et al., 2004). The objectives and laboratory resources involved are indicative of the variety of efforts in EST studies, with projects ranging from 282 sequences in *Pinus pinea* (Alonso et al., 2007) to 206,875 sequences in *Picea* sp. (Ralph et al., 2008). Due to the lack of genomic information about conifers caused by the enormous sizes of the genomes of these species, the main objective of most of these EST studies was the construction of gene catalogues. Therefore, the research efforts to develop EST data can be considered the pioneering functional genomic studies of conifers. Additionally,

Table 2 Representative published EST data in conifer species. Sequences obtained by Sanger's method.

Article	Species	Tissues	Reads	Unigenes	Contings	Singletons
Allona et al. (1998)	*Pinus taeda*	Differentiating compression wood and side wood	1097	ND	ND	ND
Ujino-Ihara et al. (2000)	*Cryptomeria japonica*	Inner bark	2231	1398	353	1045
Kirst et al. (2003)	*Pinus taeda*	Xylem-forming tissues	59,797	20,377	8070	12,307
Ujino-Ihara, Taguchi, Yoshimura, and Tsumura (2003)	*Cryptomeria japonica*	Male and female cones	1210	1173	ND	ND
Pavy et al. (2005)	*Picea glauca*	Different tissues and treatments	49,101	16,578	9354	7224
Asiegbu, Nahalkova, and Li (2005)	*Pinus sylvestris*	Infected roots	966	656	147	509
Lorenz et al. (2006)	*Pinus taeda*	Water stressed roots	12,918	6765	2006	4759
Cairney et al. (2006)	*Pinus taeda*	Zygotic embryos	68,721	12,154	5274	6880
Yakovlev, Fossdal, Johnsen, Junttila, and Skrøppa (2006)	*Picea abies*	Bud burst	1585	673	270	403
Lee, Chan, Wang, and Chu (2006)	*Taiwania cryptomerioides*	Inner bark	973	303	144	159
Alonso et al. (2007)	*Pinus pinea*	Cultured cotyledons	282	245	24	221

Ralph et al. (2008)	*Picea* sp. (*P. sitchensis, P. glauca, P. glauca-engelmannii*)	Different tissues and treatments	206,875	46,745	19,941	26,804
Futamura et al. (2008)	*Cryptomeria japonica*	Male strobili	36,011	10,463	7686	2777
Li, Wu, Dillon, and Southerton (2009)	*Pinus radiata*	Xylem-forming tissues	5952	3304	952	2352
Chancerel et al. (2011)	*Pinus pinaster*	Different tissues and treatments	31,678	13,730	4483	9247
Rigault et al. (2011)[a]	*Picea glauca*	Different tissues and treatments	146,616[b]	27,720	ND	ND
Fernández-Pozo et al. (2011)	*Pinus pinaster*	Different tissues and treatments	19,160	10,843	2814	8029
Fernández-Pozo et al. (2011)	*Pinus sylvestris*	Different tissues and treatments	674	461	92	369
Villalobos et al. (2012)	*Pinus pinaster*	Differentiating compression wood and side wood	8429	ND	ND	ND
Mann, Wegrzyn, and Rajora (2013)	*Picea mariana*	Needles	4594	2731	497	2234

[a]Additional reads from NGS.
[b]In addition to EST from Pavy et al. (2005) and Ralph et al. (2008).
ND, Not determined.

the resulting EST data were used to design different cDNA expression microarrays for conifer species such as those developed for *P. taeda* (Van Zyl et al., 2002), *P. sitchensis* (Ralph et al., 2006) and *Pinus pinaster* (Villalobos et al., 2012) (see Section 2.3).

However, times have changed, and NGS allows rapid and inexpensive procurement of transcriptomic sequences in comparison with the old approaches. NGS makes it possible for individual research groups working in non-model species to generate appropriate transcriptomic resources. Although these data could be used to design new expression microarrays, this approach has become meaningless since most transcriptomic data are obtained from RNA-Seq experiments, accounting for transcript accumulation. Thus, the main uses of transcriptomic data are as follows: (i) as a reference to map sequenced reads in RNA-Seq experiments (*e.g.* Cañas et al., 2015), (ii) as a guide for genome annotations (*e.g.* Warren et al., 2015; Wegrzyn et al., 2014), (iii) to generate a virtual proteome for peptide identification in proteomic experiments (*e.g.* Romero-Rodríguez et al., 2014), (iv) as a reference for the design of gene/exome capture experiments (Lu et al., 2016; Neves, Davis, Barbazuk, & Kirst, 2013, 2014; Seoane-Zonjic et al., 2016; Suren et al., 2016) and (v) as a source of information about genetic markers such as microsatellites or SSRs (*e.g.* Canales et al., 2014; Hu et al., 2016) (Fig. 2).

In most cases, the results obtained from a single experiment are sufficient for the analyses (*e.g.* RNA-Seq). However, in some situations, complex transcriptomic assemblies that use data from a variety of tissues or conditions

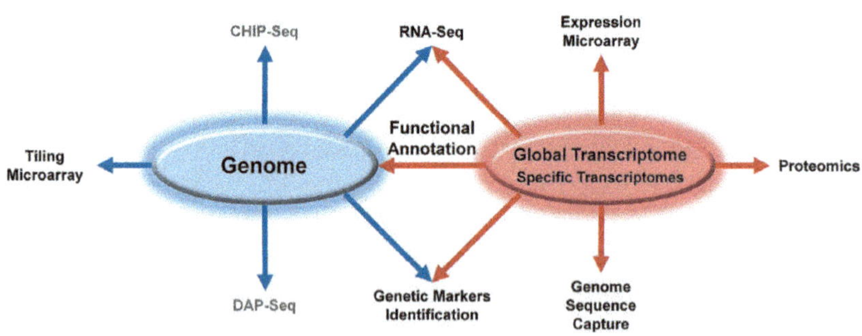

Fig. 2 Some of the main roles of genome and transcriptome assemblies as references in research. Techniques highlighted with blue lines are dependent on genome assemblies or sequences. Techniques highlighted with red lines are dependent on transcriptome assemblies or sequences. Techniques in grey still have not been developed in conifers due to fragmented genome assemblies. *ChIP-Seq*, chromatin immunoprecipitation sequencing; *DAP-Seq*, DNA affinity purification sequencing.

are required, for example, when using the transcriptome for genome annotation. In such cases, the reference transcriptome is used in an attempt to cover most of the transcripts expressed in a species, including splice variants and non-coding protein sequences (*e.g.* Canales et al., 2014; Lorenz et al., 2012; Rigault et al., 2011; Yeaman et al., 2014). In this regard, studies on microRNA (miRNA) are very limited, with some reports identifying and exploring miRNA expression under different physiological and developmental conditions (de Vega-Bartol et al., 2013; Fossdal et al., 2012; Liu & El-Kassaby, 2017; Morin et al., 2008; Niu et al., 2015; Qiu et al., 2015; Quinn, Iriyama, & Fernando, 2014; Wan et al., 2012; Xie et al., 2017; Zhang et al., 2013) or performing computational analyses in species with a genome draft, such as *P. taeda* (Quinn, Iriyama, & Fernando, 2015) and *P. glauca* (Yakovlev & Fossdal, 2017). Studies on long non-coding RNA (lncRNA) are nearly non-existent with the exception of the lncRNA catalogue for Norway spruce which was included in the first whole-genome assembly (Nystedt et al., 2013). Nevertheless, considering that a good genome assembly is necessary for classification of the identified lncRNA, it is understandable that lncRNAs have been poorly characterized in conifer species.

2.3 Expression microarrays

Expression microarray technology makes it possible to experimentally explore important portions of a transcriptome including miRNA by using specialized platforms. Expression microarrays are a hybridization-based technique that uses thousands of probes, cDNAs or oligonucleotides, corresponding to different transcripts fixed on a solid surface and hybridized with experimental samples labelled with fluorophores (the current state of the art is reviewed in Li, Sedighi, & Wang, 2016). In comparison with previous methods of gene expression analysis such as dot, slot and northern blots, the emergence of microarrays was an extraordinary advancement and provided the ability to evaluate the expression dynamics of thousands of transcripts simultaneously. Regarding expression studies in conifers, this methodology was used to complement and, finally, to overcome the transcript catalogue age of EST libraries. A significant, early study was published by Whetten, Sun, Zhang, and Sederoff (2001). Currently, in the NCBI's Gene Expression Omnibus (NCBI's GEO, https://www.ncbi.nlm.nih.gov/geo/) more than 30 microarray platforms for conifer species have been deposited since 2004 (Tables 3 and 4). This resource includes improved

Table 3 Gene expression microarrays available for conifers (*Pinidae* term in a platform search) at NCBI Gene Expression Omnibus (GEO).

GEO accession	Species	Type	Probe	Spot number	Year
No submitted[a]	*Pinus tabuliformis*	Commercial	*In situ* oligonucleotide	60,000	2016
GPL21977	*Pinus taeda*	Commercial	*In situ* oligonucleotide	180,880	2016
GPL22783	*Picea glauca*	Commercial	*In situ* oligonucleotide	177,321	2016
GPL19600	*Cryptomeria japonica*	Commercial	*In situ* oligonucleotide	71,984	2016
GPL21757	*Pinus sylvestris; Cenococcum geophilum*	Commercial	*In situ* oligonucleotide	43,926	2016
GPL19234	*Pinus pinaster*	Commercial	*In situ* oligonucleotide	42,500	2015
GPL19078	*Pinus sylvestris*	Commercial	*In situ* oligonucleotide	36,349	2015
GPL18419	*Pinus pinaster*	Non-commercial	Spotted cDNA	8208	2015
GPL18152	*Cryptomeria japonica*	Commercial	*In situ* oligonucleotide	62,892	2014
GPL18030	*Cryptomeria japonica*	Commercial	*In situ* oligonucleotide	75,191	2013
GPL18004	*Pinus massoniana*	Commercial	Spotted oligonucleotide	48,598	2013
GPL14845	*Cryptomeria japonica*	Commercial	*In situ* oligonucleotide	45,753	2013
GPL16286	*Pinus taeda*	Commercial	*In situ* oligonucleotide	180,880	2012
GPL15915	*Pinus taeda*	Non-commercial	Spotted cDNA	26,496	2012
GPL15526	*Pinus pinaster*	Non-commercial	Spotted cDNA	4223	2012
GPL15510	*Pinus pinaster*	Non-commercial	Spotted cDNA	2910	2012
GPL15033	*Picea glauca; Picea sitchensis*	Non-commercial	Spotted oligonucleotide	34,992	2011
GPL11184	*Pinus taeda*	Non-commercial	Spotted cDNA	26,496	2011

Table 3 Gene expression microarrays available for conifers (*Pinidae* term in a platform search) at NCBI Gene Expression Omnibus (GEO).—cont'd

GEO accession	Species	Type	Probe	Spot number	Year
GPL10754	*Picea glauca*	Non-commercial	Spotted cDNA	11,036	2010
GPL8302	*Pinus radiata*	Non-commercial	Spotted cDNA	18,432	2009
GPL6582	*Pinus pinaster*	Commercial	Spotted cDNA	24,288	2008
GPL5189	*Pinus taeda*	Non-commercial	Spotted cDNA	49,152	2007
GPL5423	*Picea*	Non-commercial	Spotted cDNA	25,344	2007
GPL5763	*Pinus taeda*	Non-commercial	Spotted cDNA	5376	2007
GPL2027	*Pinus taeda*; *Pinus elliottii* var. *elliottii*	Non-commercial	Spotted cDNA	8448	2006
GPL4039	*Pinus taeda*	Non-commercial	Spotted cDNA	6144	2006
GPL4657	*Pinus taeda*	Non-commercial	Spotted cDNA	5376	2006
GPL1880	*Pinus taeda*	Non-commercial	Spotted cDNA	4794	2005
GPL1014	*Pinus taeda*	Non-commercial	Spotted cDNA	9216	2004
GPL1508	*Pinus taeda*	Non-commercial	Spotted cDNA	9204	2004
No submitted[b]	*Pinus taeda*	Non-commercial	Spotted cDNA	3000	2002

[a]Presented at Niu et al. 2016.
[b]Presented at Whetten et al. (2001).

Table 4 miRNA expression microarrays available for plants including conifers (*Pinidae* term in a platform search) at NCBI Gene Expression Omnibus (GEO).

GEO accession	Species	Type	Probe	Spot number	Year
GPL17805	Different plant species including *P. glauca*, *P. sitchensis* and *P. taeda*	Commercial	*In situ* oligonucleotide	1512	2013
GPL9858	Different plant species including *P. taeda*	Commercial	*In situ* oligonucleotide	3968	2011
GPL9344	Different plant species including *P. taeda*	Commercial	*In situ* oligonucleotide	3968	2009
GPL9736	Different plant species including *P. taeda*	Commercial	*In situ* oligonucleotide	3540	2009
GPL9738	Different plant species including *P. taeda*	Commercial	*In situ* oligonucleotide	3281	2009
GPL9737	Different plant species including *P. taeda*	Commercial	*In situ* oligonucleotide	3122	2009

versions of the first microarrays for determined species and some versions that cover plant miRNA catalogues with sequences from certain conifer species.

However, the nature and characteristics of this technology had led to major limitations in the use of microarrays in research compared with RNA-Seq. Some of the main problems associated with microarray technology in comparison to RNA-Seq are lower sensitivity, lower dynamic range, hybridization bias and that microarray technology only yields information about known transcripts (Martin, Dehler, & Król, 2016). Therefore, the depth of the analysis depends on the number and the origin of the probes deposited over the microarray. In contrast, in RNA-Seq, the depth of the analysis depends on the number of sequenced reads in the experiment. Thus, in expression microarrays, it is not possible to cover both gene expression and the expression of alternative transcript forms that are not physically contained in the microarray. Therefore, the previous work to design and produce the microarrays determines the final quality of the data obtained. Microarrays containing spotted cDNAs as probes can be limited by the number of analysed transcripts and by transcriptome coverage (Martin et al., 2016). cDNAs must be cloned and physically preserved but also arranged in plates to be printed on the microarray surface, which is time consuming, costly and requires a lot of space. In this context, covering the whole transcriptome of a species can be a challenging objective. The transcriptome coverage of a microarray is, of course, strongly associated with the origins of the sequences (tissues, environmental, physiological and metabolic conditions). However, microarrays using synthesized oligonucleotides as probes allow more extensive coverage of the transcriptome. In this case, the sequences employed for oligonucleotide design come from the large transcriptomic databases that have become available since the advent of NGS (*e.g.* Cañas, Feito, et al., 2015; Raherison et al., 2012). The generation of microarrays has become more accessible making it possible to include a greater number of probes. In conifers, the platform with the higher number of spotted cDNA is GPL5189 for *P. taeda* with 49,152 probes (Table 3), while the oligonucleotide platform for a conifer species with the most probes has 180,880 probes (GPL21977 for *P. taeda*, Table 3), which is 3.7 times higher than the value obtained for GPL5189. Since 2012, concurrent with the rise of NGS, *in situ* oligonucleotide microarrays have been preferred to cDNA microarrays and are usually commercially manufactured by specialized companies such as Agilent or Affymetrix. NGS and RNA-Seq are increasingly replacing gene expression microarrays, although microarrays have demonstrated extraordinary utility over the last 15 years. Since 2012, there have been an increasing number of studies on conifers using RNA-Seq, and in the last few years, the number of studies using RNA-Seq has surpassed the number of publications using microarrays (Fig. 3). Currently, there is

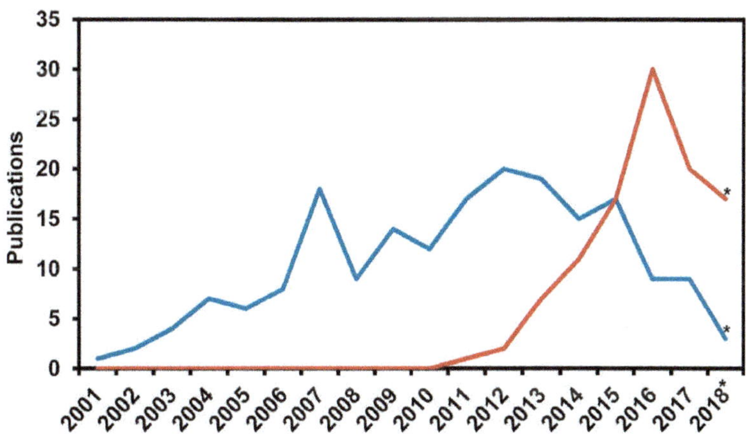

Fig. 3 Publications in conifers using microarrays and RNA-Seq. The publication numbers were estimated through a search in the Web of Science. Blue line shows microarray publications per year (search: "microarray AND conifer*"). Red line shows RNA-Seq publications per year (search: "RNA-Seq AND conifer*".). *Data accessed at 28th November 2018.

a clear trend towards performing transcriptomic analyses *via* NGS, which is mainly because the output of a large amount of data does not require previously determined information about the transcriptome. Each experiment is an opportunity to discover new transcript variants, including those with low expression levels. In the case of non-conventional model species such as conifers, the use of RNA-Seq is extremely advantageous because the transcriptome is obtained primarily from the assembly of sequencing data (see Section 2.2). Additionally, microarray hybridization can also lead to more false positives than the RNA-Seq procedures (Cañas, Feito, et al., 2015), which is mainly due to hybridization bias; the probe can hybridize with transcripts of different genes with extremely high identity (*e.g.* two paralog genes appearing in a recent genetic duplication) or does not differentiate between different transcripts of a single gene, mainly if spotted cDNA microarrays are used.

2.4 Online resources

New high-throughput techniques coupled with functional genomics produce large amounts of data requiring laborious analyses, resulting in most of the data remaining unseen. Therefore, access to data is an important issue in the omics era. The need to revisit the data and make new comparisons or visualizations has promoted the emergence of accessible online databases (Galperin, Fernández-Suárez, & Rigden, 2017) and web-servers (Benson, 2017). There are general web resources that can be employed for every species in biological research, for example, the three main databases containing nucleotide and protein sequences: NCBI's GenBank (Benson et al., 2017), DDBJ (Mashima et al., 2016) and ENA (Gibson et al., 2016). Whole-genome sequence assemblies of four conifer species have been deposited in NCBI's GenBank: *P. menziesii* (GCA_001517045.1); *P. glauca* (GCA_000411955.5, GCA_000966675.1 and GCA_001687225.1); *P. taeda* (GCA_000404065.3) and *P. lambertiana* (GCA_001447015.2). Although sequence information is basic, such resources may also store other types of functional genomic data and tools such as gene expression data from RNA-Seq and microarrays (GEO, Barrett et al., 2013), data regarding plant metabolic and regulatory pathways (Plant Reactome, Naithani et al., 2017) and mass-spectrometric data from proteomic analyses (ProteomeXchange, Vizcaíno et al., 2014).

In addition to these general web resources, there are several databases and web-servers containing specific information and tools that can be used for functional genomic studies of conifers (Table 5). Usually, these resources

Table 5 Online resources for conifer functional genomics.

Name	Data type	Species	Link
Dendrome/TreeGenes	Genome assemblies, transcriptomes	*P. taeda*, *P. glauca*, *P. sitchensis*, and other species	https://dendrome.ucdavis.edu/index.php/ http://tgwebprod.cam.uchc.edu/Drupal/
PineRefSeq	Genome assemblies	*P. taeda*, *P. lambertiana*, *P. menziesii*	http://pinegenome.org/pinerefseq/
ConGenIE	Genome assemblies, transcriptomes, gene expression, visualization and analysis tools	*P. abies*, *P. glauca*, *P. taeda* and other species	http://congenie.org/
NorWood	Wood gene expression network	*P. abies*	http://norwood.congenie.org/norwood-v1.0/
GymnoPLAZA	Transcriptomes	*P. taeda*, *P. abies* and other conifers	https://bioinformatics.psb.ugent.be/plaza/versions/gymno-plaza/
EVOLTREE	Genomic and EST sequences	*A. alba*, *P. abies*, *P. pinaster*, *P. sylvestris*	http://www.evoltree.eu/index.php
PineDB	Transcriptome	*P. taeda*	http://bioinfolab.muohio.edu/txid3352v1/index.php
Sugi Genome DB	Genome sequences Transcriptomes	*C. japonica*, *C. obtusa*	http://www.ffpri.affrc.go.jp/labs/cjgenome/index.html
SustainPineDB	Transcriptome	*P. pinaster*	http://www.scbi.uma.es/sustainpinedb/home_page

have been developed in the framework of research projects, and therefore most of these resources have either disappeared or have been integrated into other specific or general databases upon culmination of the project (*e.g.* Fernández-Pozo et al., 2011; Lippert, Yuen, & Bohlmann, 2009; Pavy et al., 2007). Among current online resources, Dendrome/TreeGenes is one of the older resources available (Wegrzyn, Lee, Tearse, & Neale, 2008). Dendrome/TreeGenes was developed under the framework of the Dendrome Project as a resource to integrate tools and tree-genomic data including data regarding species, publications, genetic maps, molecular markers, and genotypes as well as phenotypic data, genomic sequences, transcriptomes and expression data. Although the database was not conceived only for conifers, they constitute a fundamental part of the database. Of the 271 tree species in the transcriptome database 176 correspond to conifers; however, a substantial number of these sequences correspond to only a few species such as *P. taeda* and *P. sitchensis*. Another interesting web resource is PineRefSeq, which is closely associated with Dendrome/TreeGenes in terms of the project partners, primarily the University of California, Davis (UC Davis). Both databases provide FTP access to the genome assemblies and transcriptomes of three different conifers from North America (*P. taeda*, *P. menziesii* and *P. lambertiana*) including new unpublished versions.

Similar to Dendrome/TreeGenes, Conifer Genome Integrative Explorer (ConGenIE) is a web resource that integrates data about different conifer species, although this resource is mainly focused on *P. abies* as a part of the Norway spruce genome sequencing project (Sundell et al., 2015). This resource includes the genome assemblies of *P. abies*, *P. glauca* and *P. taeda*; transcriptomic sequences of these three species and different conifer and gymnosperm species; gene expression data from *P. abies;* and different visualization and analytic tools. ConGenIE is part of the Plant Genome Integrative Explorer (PlantGenIE) platform, which includes genomic resources for *Populus* (PopGenIE) and *Arabidopsis* (AtGenIE) (Sundell et al., 2015). An important characteristic of ConGenIE is the integration of this resource with additional web tools to analyse the gene expression data from *P. abies* or even other conifers such as *P. pinaster.* The Comparative Plant Expression tool (ComPlEx, http://complex.plantgenie.org/) allows the comparison and visualization of the gene co-expression networks of the three species in PlantGenIE (Netotea, Sundell, Street, & Hvidsten, 2014). Another example is the exImage tool for visualizing gene expression over organ and tissue pictograms in *P. abies* and more recently in *P. pinaster* (Cañas et al., 2017; http://v22.popgenie.org/microdisection/). A web resource associated with

ConGenIE is NorWood, which contains gene expression profiles and co-expression networks from high-spatial-resolution tissue sections of the cambial and woody tissues of *P. abies* (Jokipii-Lukkari et al., 2017). Integration of the data from NorWood with PlantGenIE data from *Arabidopsis* and *Populus* provides evolutionary insight into plant wood formation.

Another web resource that includes data from different species is GymnoPLAZA, which is part of a bigger platform for plants, PLAZA (Proost et al., 2015). GymnoPLAZA contains sequence and gene annotation data from different conifer species and other gymnosperms and angiosperms. This web resource also provides tools for comparative genomics. GymnoPLAZA is a good source of data for functional genomics, although the main aim of this resource is the facilitation of comparative and evolutionary genomics. Other resources that were not developed for functional genomics but contain useful information for such studies in conifers are the transcriptomic and genomic databases generated in the framework of the EVOLTREE project (http://www.evoltree.eu/). These databases contain data from conifer species such as *Abies alba, P. abies, P. pinaster, P. sylvestris* and *Podocarpus oleifolius* and also from angiosperm species of the genera *Fagus* and *Quercus*.

Alternatively, several web resources containing data about specific conifer species have been mainly developed to store the results generated by other projects and research groups. For example, PineDB at Miami University (Oxford, OH, USA) is a web database that contains extensive transcriptomic information for *P. taeda* (Table 5). Among this class of resources, a constant and meritorious effort is that of the Sugi Genome DB, which centralizes the genetic, genomic and transcriptomic data of the Asian cypresses; *Cryptomeria japonica* (*sugi*, Japanese cedar) and *Chamaecyparis obtusa* (*hinoki*, Japanese cypress) (Table 5). This resource contains information that has been generated since 1998 by the Japanese Forestry and Forest Products Research Institute. Finally, the SustainPineDB contains the transcriptomic sequence information of the western Mediterranean conifer *P. pinaster* (Canales et al., 2014). This database replaced an earlier resource, the EuroPineDB (Fernández-Pozo et al., 2011), and was developed in the framework of the European research project SustainPine (http://www.scbi.uma.es/sustainpine/).

2.5 Genetics for conifer functional genomics

Genetic techniques are powerful tools for elucidating gene functions. In model plant species such as *Arabidopsis thaliana*, these techniques have been widely used for the functional characterization of genes by taking advantage

Arabidopsis		Conifers
20-25 cm	**Plant size**	0.03-100 m
Herbaceous	**Plant type**	Woody
65,529	**Articles***	9,722
135 Mp	**Genome size**	4-36 Gb
High quality	**Genome assembly**	Fragmented
Annual	**Life time**	Perennial
6 weeks	**Life cycle**	Several years
≅ 300,000 lines	**Mutant collections**	-
Fast "floral dip"	**Transformation**	Somatic embryogenesis

Fig. 4 Characteristics of *Arabidopsis* vs. conifers. *Articles in the Web of Science from 2008 to 2017 (searches "*Arabidopsis thaliana*" and "Conifer").

of the biological characteristics of the model, including small plant and genome sizes, short lifespans and ease of transformation (Fig. 4). These advantages are complemented by the high number of labs and researchers working on this model, leading to a high potential for knowledge accumulation and to the development of new methods, mainly using reverse genetics, to study gene function (Provart et al., 2016). However, conifer species exhibit characteristics that make it challenging to apply certain genetic methods: large body sizes and long-life cycles (Farjon, 2010), large genome sizes (Ritland, 2012), high heterozygosity (Badenes, Fernández i Martí, Ríos, & Rubio-Cabetas, 2016) and difficulties in obtaining viable transgenic plants (Klimaszewska, Lachance, Pelletier, Lelu, & Seguin, 2001) (Fig. 4).

In this context, the genetic approaches used with conifers have been focussed on genotype-phenotype association, breeding, diversity and population structure more than on large-scale programs of mutant characterization or genetic engineering associated with functional genomics (Badenes et al., 2016; Plomion et al., 2016). The final aim of these studies is to assist and improve conifer breeding programs. Breeding programs for trees, including conifers, are large, expensive and long, requiring many years and resources to produce desired results (West, 2014). Many of these programs were started in the 1950s and have continued till the present (Pâques, 2013). Although these approaches are so powerful that important gains in the desired traits can be obtained in a few generations (West, 2014), there is a need in to accelerate these programs due to the long breeding cycles.

An alternative method to reach that goal is to use marker-assisted selection (MAS) of trees and quantitative trait loci (QTL) mapping. In recent years, the emergence of omics techniques has improved the identification and use of markers, mainly single nucleotide polymorphisms (SNPs), for gene mapping and genotyping. Transcriptome sequencing is widely employed, even in species rarely used in research, to identify thousands of SNPs (*e.g.* Liu et al., 2016; Wachowiak, Trivedi, Perry, & Cavers, 2015). The extent of genotyping has widened since the advent of SNP microarrays (*e.g.* Chancerel et al., 2013; Pavy et al., 2013, 2017), CGH microarrays for the identification of DNA copy number variations (CNVs) (Prunier, Caron, & MacKay, 2017) and exome capture sequencing, which also allows the identification of new markers (*e.g.* Lu et al., 2016; Neves et al., 2014; Pavy et al., 2016; Suren et al., 2016). Although SNPs in allelic variants can be considered mutations, most SNPs have no evident associations with a defined phenotype. In some cases, these relationships have been established allowing assignments of function to specific genes (*e.g.* Lu et al., 2017). A promising step forward in tree domestication seems to be the combined use of genomic selection (GS) of conifers with clonal propagation *via* somatic embryogenesis (Harfouche et al., 2012). This approach can substantially accelerate breeding programs. Unlike MAS, the fundamental principle of GS is the development of a prediction model generated from all available genetic markers and phenotypic data collected from a sample population. The aim of this model is to determine the genomic breeding value of progeny (Harfouche et al., 2012).

The use of genetics to search for or validate gene function is very limited in conifers. There are no mutant collections and extensive transformation programs are slow and expensive. In the case of mutant collections, screening for individual conifers with modified characteristics would be challenging due to their long-life cycles and large sizes of these trees, and the time and infrastructure required to complete the study would be tremendous. Thus, few studies have been carried out with mutant conifers, and usually, these studies have been conducted on trees with apparent phenotypes (Carlsbecker et al., 2013; Guo, Zheng, & Yang, 2014; Han, Shinohara, Kakubari, & Mukai, 2003; Hirao, Watanabe, Kurita, Kondo, & Takata, 2009; Ning, Lu, Liu, Yang, & Wang, 2013; Tsubomura, Kurita, & Watanabe, 2016; Uddenberg et al., 2013), although the best characterized conifer mutant, a mutant of the gene encoding the cinnamyl alcohol dehydrogenase enzyme (CAD; EC 1.1.1.195) in *P. taeda*, was identified in a genetic screen (MacKay et al., 1997; Ralph et al., 1997). NGS and other omics techniques represent a new opportunity to characterize conifer mutants and, in some cases, to identify the mutated gene. An excellent example of such a study is that performed on the Norway spruce mutant *acrocona*, in which a MADS box transcription factor was identified as the regulator of early cone setting using RNA-Seq (Uddenberg et al., 2013).

Gene function assignment by using spontaneous mutants can also be driven by gene silencing using RNA interference (RNAi) (Watanabe, 2011). Despite the challenges associated with transforming conifers, gene knockdown characterization has provided the data required to explore candidate genes and assign molecular and physiological functions to these genes (*e.g.* Maloney, Samuels, & Mansfield, 2012; Moschou et al., 2016; Zhu et al., 2014). Similar to the study of mutants, the combination of this approach with omics analyses is a powerful tool for functional genomics. Recently, a global gene expression analysis of maritime pine seedlings in which the *PpNAC1* transcription factor had been silenced allowed the elucidation of the role of this gene in wood formation and the relationships of this gene with other gene expression regulators involved in this process (Pascual et al., 2018). Notably, gene silencing has been very useful for studying the phenylpropanoid synthesis pathway in the context of lignification in conifers. This method has allowed the functional characterization of each member of the pathway, highlighting the roles of these members and quantifying their importance (Möller, Steward, Phillips, Flint, & Wagner, 2005; Trontin et al., 2014; Wadenbäck et al., 2008; Wagner et al., 2009; Wagner, Phillips, Narayan, Moody, & Geddes, 2005; Wagner et al., 2007, 2013, 2011).

Additionally, different studies have used genetic transformation to overexpress genes (*e.g.* Dalman et al., 2017; Mageroy et al., 2017; Möller, McDonald, Walter, & Harris, 2003; Nagel, Berasategui, Paetz, Gershenzon, & Schmidt, 2014; Uddenberg, Abrahamsson, & von Arnold, 2016) and to study the functions of gene promoters (Bedon, Levasseur, Grima-Pettenati, Seguin, & MacKay, 2009). Furthermore, to study gene function, genetic manipulation can also provide a biotechnological opportunity to generate better tree lines with improved characteristics. For example, the transformation of tracheary elements of *P. taeda* with genes encoding ferulate 5-hydroxylase (F5H) and caffeic acid O-methyltransferase (COMT) from *Liquidambar styraciflua* resulted in the production of lignin containing sinapyl alcohol, which is a type of monolignol not found in conifers making this wood more appropriated for biofuel production (Wagner et al., 2015).

Despite the great potential of these techniques, their use to study conifers is still scarce due to their unfavourable characteristics, mainly their long-life cycles. As a resource, to avoid or minimize this challenge, researchers have used cultured cell lines of conifers to perform genetic manipulations and analyse these cultures (*e.g.* Möller et al., 2005), developing somatic embryos (*e.g.* Klimaszewska, Hargreaves, Lelu-Walter, & Trontin, 2016; Pascual et al., 2018), and *in vitro* tissues as tracheary elements (*e.g.* Wagner et al., 2005, 2015). Additionally, transient gene expression in protoplasts has been employed as a successful alternative to study conifer gene promoters (Craven-Bartle et al., 2013; Gómez-Maldonado, Cánovas, & Avila, 2004; Gómez-Maldonado, Crespillo, Ávila, & Cánovas, 2001; Pascual et al., 2018; Rueda-López, Crespillo, Cánovas, & Avila, 2008). A widely used strategy to analyse conifer genes *via* genetic manipulation is overexpression in heterologous systems such as *Arabidopsis,* tobacco or poplar (*e.g.* Gao et al., 2013; Hills, Khan, & López-Juez, 2015; Katahata, Futamura, Igasaki, & Shinohara, 2014; Rueda-López, Cañas, Canales, Cánovas, & Ávila, 2015; Rueda-López et al., 2017; Rupps et al., 2016). Complementation of *Arabidopsis* mutants with the putative homologue of the conifer gene under study is a commonly used approach (*e.g.* Karlgren, Gyllenstrand, Källman, & Lagercrantz, 2013; Larsson, Sundström, Sitbon, & Von Arnold, 2012; Li et al., 2013; Maloney et al., 2012; Nilsson, Carlsbecker, Sundås-Larsson, & Vahala, 2007). Alternatively, complementation of *E. coli* (*e.g.* Hosoo, Kimura, Nanatani, & Uozumi, 2014) and yeast mutants (*e.g.* Castro-Rodríguez et al., 2016) when possible is extremely useful. Mutant complementation in heterologous systems provides functional information about the genes and the products of these genes (*e.g.* El-Azaz, de la Torre, Ávila, & Cánovas, 2016), while overexpression is more adequate as a biotechnological tool (*e.g.* Castro-Rodríguez et al., 2016).

One way to overcome the problems associated with the generation of mutant and transgenic conifers would be the use of genome editing techniques such as CRISPR/Cas9 technology (Charpentier & Doudna, 2013; Doudna & Charpentier, 2014; Feng et al., 2013). This method can potentially be used for directed mutation, modification or substitution of specific genes, facilitating genetic studies of gene function modification (loss, reduction, improvement or shift, Fig. 5) (Nonaka, Arai, Takayama, Matsukura, & Ezura, 2017; Wang et al., 2014) and differential gene expression (Park, Dempewolf, Zhang, & Wang, 2017; Shi et al., 2017). The opportunities are diverse, and this method has the advantage of being a targeting method. Unfortunately, despite the promise of this method, there are no published studies using CRISPR/Cas9 technology in conifers. Thus, CRISPR/Cas9 seems to hold great potential for conifer research.

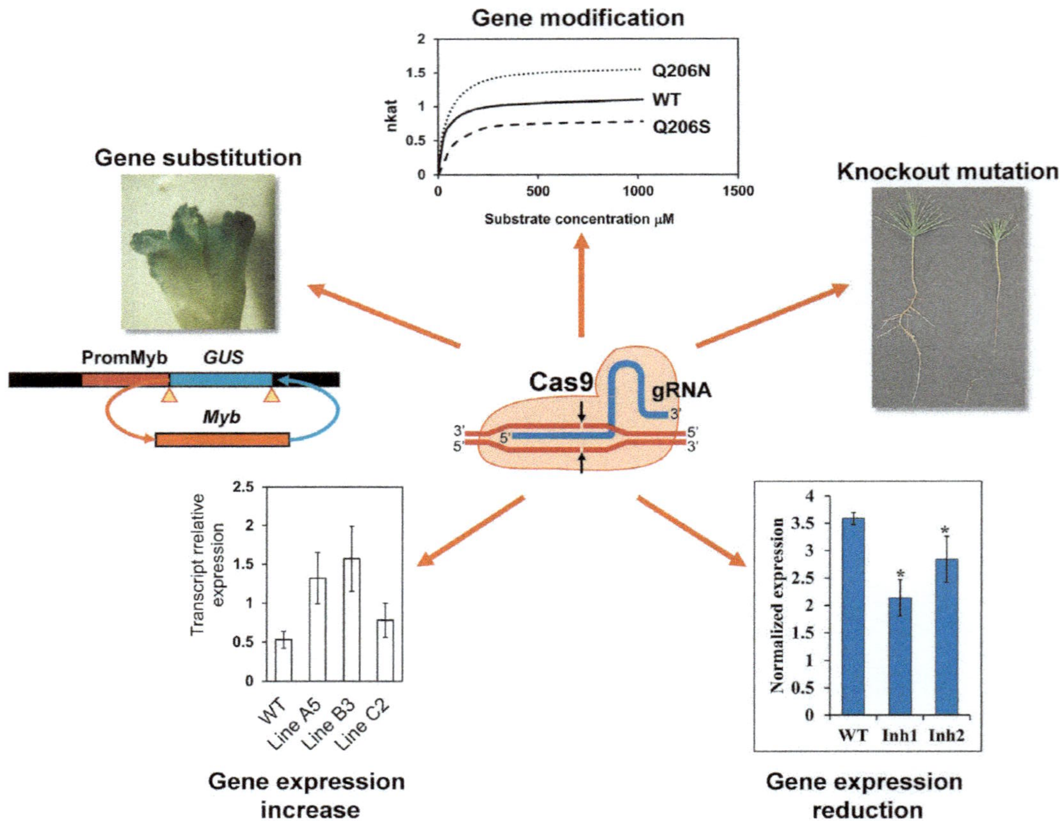

Fig. 5 Potential uses of gene edition methods (as CRISPR/CAS9 system) in conifers.

3. Conclusions and perspectives

Overall, functional genomic research in conifers has advanced enormously over the last 15–20 years following the development of omics techniques. Considering the main obstacle, which is the unfavourable characteristics of conifers as experimental models, the accumulation of data and the generation of scientific resources for conifer functional genomics are impressive achievements. Among these achievements, the assembly of the whole-genome sequences of several conifer species is the most relevant. Maybe, in the future, this achievement will be crucial for obtaining further information of biological and biotechnological value, although in order to reach that goal, improvement of the assemblies and annotations is mandatory. Current, and maybe future, techniques to study the regulation of gene expression and chromatin dynamics require high-quality whole-genome sequences.

At the genetic level, functional studies have been limited since the mutant screening and transgenesis are difficult to achieve. However, the application of genome editing techniques opens new perspectives in the field. When the possibility of manipulating the functions of conifer genes in a "surgical" way is realized, such studies will increase and profoundly change our knowledge about conifer biology. All these new resources are the "holy grail" towards obtaining trees with improved characteristics.

Acknowledgements

R.A.C. was supported by a grant from the Spanish *Ministerio de Economía y Competitividad* (BIO2015-73512-JIN; MINECO/AEI/FEDER, UE). Research work in the authors's laboratory was supported by grants from *Ministerio de Economía y Competitividad* (BIO2015–69285-R) and the European Union (FP7-KBBE-289841).

References

Adams, M. D., Kelley, J. M., Gocayne, J. D., Dubnick, M., Polymeropoulos, M. H., Xiao, H., et al. (1991). Complementary DNA sequencing: Expressed sequence tags and human genome project. *Science*, *252*, 1651–1656. https://doi.org/10.1126/science.2047873.

Allona, I., Quinn, M., Shoop, E., Swope, K., St Cyr, S., Carlis, J., et al. (1998). Analysis of xylem formation in pine by cDNA sequencing. *Proceedings of the National Academy of Sciences of the United States of America*, *95*, 9693–9698. https://doi.org/10.1073/pnas.95.16.9693.

Alonso, P., Cortizo, M., Cantón, F. R., Fernández, B., Rodríguez, A., Centeno, M. L., et al. (2007). Identification of genes differentially expressed during adventitious shoot induction in *Pinus pinea* cotyledons by subtractive hybridization and quantitative PCR. *Tree Physiology*, *27*, 1721–1730. https://doi.org/10.1093/treephys/27.12.1721.

Alonso, J. M., & Stepanova, A. N. (2015). Plant functional genomics. In J. M. Alonso & A. N. Stepanova (Eds.), *Vol. 1284. Plant functional genomics: Methods and protocols* (2nd ed.).New York NY: Springer New York. https://doi.org/10.1007/978-1-4939-2444-8.

Alvarez, J. M., Bueno, N., Cañas, R. A., Avila, C., Cánovas, F. M., & Ordás, R. J. (2018). Analysis of the WUSCHEL-RELATED HOMEOBOX gene family in *Pinus pinaster*: New insights into the gene family evolution. *Plant Physiology and Biochemistry, 123*, 304–318. https://doi.org/10.1016/j.plaphy.2017.12.031.

Alvarez, J. M., Sohlberg, J., Engström, P., Zhu, T., Englund, M., Moschou, P. N., et al. (2015). The WUSCHEL-RELATED HOMEOBOX 3 gene PaWOX3 regulates lateral organ formation in Norway spruce. *New Phytologist, 208*, 1078–1088. https://doi.org/10.1111/nph.13536.

Aronen, T., & Ryynänen, L. (2012). Variation in telomeric repeats of Scots pine (*Pinus sylvestris* L.). *Tree Genetics & Genomes, 8*, 267–275. https://doi.org/10.1007/s11295-011-0438-7.

Asiegbu, F. O., Nahalkova, J., & Li, G. (2005). Pathogen-inducible cDNAs from the interaction of the root rot fungus *Heterobasidion annosum* with Scots pine (*Pinus sylvestris* L.). *Plant Science, 168*, 365–372. https://doi.org/10.1016/j.plantsci.2004.08.010.

Badenes, M. L., Fernández i Martí, A., Ríos, G., & Rubio-Cabetas, M. J. (2016). Application of genomic technologies to the breeding of trees. *Frontiers in Genetics*, 7, 198. https://doi.org/10.3389/fgene.2016.00198.

Barrett, T., Wilhite, S. E., Ledoux, P., Evangelista, C., Kim, I. F., Tomashevsky, M., et al. (2013). NCBI GEO: Archive for functional genomics data sets—Update. *Nucleic Acids Research, 41*, 991–995. https://doi.org/10.1093/nar/gks1193.

Bedon, F., Bomal, C., Caron, S., Levasseur, C., Boyle, B., Mansfield, S. D., et al. (2010). Subgroup 4 R2R3-MYBs in conifer trees: Gene family expansion and contribution to the isoprenoid-and flavonoid-oriented responses. *Journal of Experimental Botany, 61*, 3847–3864. https://doi.org/10.1093/jxb/erq196.

Bedon, F., Grima-Pettenati, J., & Mackay, J. (2007). Conifer R2R3-MYB transcription factors: Sequence analyses and gene expression in wood-forming tissues of white spruce (*Picea glauca*). *BMC Plant Biology*, 7, 17. https://doi.org/10.1186/1471-2229-7-17.

Bedon, F., Levasseur, C., Grima-Pettenati, J., Seguin, A., & MacKay, J. (2009). Sequence analysis and functional characterization of the promoter of the *Picea glauca* cinnamyl alcohol dehydrogenase gene in transgenic white spruce plants. *Plant Cell Reports, 28*, 787–800. https://doi.org/10.1007/s00299-009-0688-0.

Benson, G. (2017). Editorial: The 15th annual nucleic acids research web server issue 2017. *Nucleic Acids Research, 45*, W1–W5. https://doi.org/10.1093/nar/gkx457.

Benson, D. A., Cavanaugh, M., Clark, K., Karsch-Mizrachi, I., Lipman, D. J., Ostell, J., et al. (2017). GenBank. *Nucleic Acids Research, 45*, D37–D42. https://doi.org/10.1093/nar/gkw1070.

Bérubé, Y., Zhuang, J., Rungis, D., Ralph, S., Bohlmann, J., & Ritland, K. (2007). Characterization of EST-SSRs in loblolly pine and spruce. *Tree Genetics & Genomes, 3*, 251–259. https://doi.org/10.1007/s11295-006-0061-1.

Birol, I., Raymond, A., Jackman, S. D., Pleasance, S., Coope, R., Taylor, G. A., et al. (2013). Assembling the 20 Gb white spruce (*Picea glauca*) genome from whole-genome shotgun sequencing data. *Bioinformatics, 29*, 1492–1497. https://doi.org/10.1093/bioinformatics/btt178.

Boch, J. (2011). TALEs of genome targeting. *Nature Biotechnology, 29*, 135–136. https://doi.org/ https://doi.org/10.1038/nbt.1767.

Bomal, C., Bedon, F., Caron, S., Mansfield, S. D., Levasseur, C., Cooke, J. E. K., et al. (2008). Involvement of *Pinus taeda* MYB1 and MYB8 in phenylpropanoid metabolism and secondary cell wall biogenesis: A comparative in planta analysis. *Journal of Experimental Botany, 59*, 3925–3939. https://doi.org/10.1093/jxb/ern234.

Bomal, C., Duval, I., Giguère, I., Fortin, E., Caron, S., Stewart, D., et al. (2014). Opposite action of R2R3-MYBs from different subgroups on key genes of the shikimate and monolignol pathways in spruce. *Journal of Experimental Botany*, *65*, 495–508. https://doi.org/10.1093/jxb/ert398.

Bravo, S., Bertín, A., Turner, A., Sepúlveda, F., Jopia, P., Parra, M. J., et al. (2017). Differences in DNA methylation, DNA structure and embryogenesis-related gene expression between embryogenic and non embryogenic lines of *Pinus radiata* D. don. *Plant Cell, Tissue and Organ Culture*, *130*, 521–529. https://doi.org/10.1007/s11240-017-1242-3.

Cairney, J., Zheng, L., Cowels, A., Hsiao, J., Zismann, V., Liu, J., et al. (2006). Expressed sequence tags from loblolly pine embryos reveal similarities with angiosperm embryogenesis. *Plant Molecular Biology*, *62*, 485–501. https://doi.org/10.1007/s11103-006-9035-9.

Canales, J., Bautista, R., Label, P., Gómez-Maldonado, J., Lesur, I., Fernández-Pozo, N., et al. (2014). *De novo* assembly of maritime pine transcriptome: Implications for forest breeding and biotechnology. *Plant Biotechnology Journal*, *12*, 286–299. https://doi.org/10.1111/pbi.12136.

Cañas, R. A., Canales, J., Muñoz-Hernández, C., Granados, J. M., Ávila, C., García-Martín, M. L., et al. (2015). Understanding developmental and adaptive cues in pine through metabolite profiling and co-expression network analysis. *Journal of Experimental Botany*, *66*, 3113–3127. https://doi.org/10.1093/jxb/erv118.

Cañas, R. A., de la Torre, F., Pascual, M., Avila, C., & Cánovas, F. (2016). Nitrogen economy and nitrogen environmental interactions in conifers. *Agronomy*, *6*, 26. https://doi.org/10.3390/agronomy6020026.

Cañas, R. A., Feito, I., Fuente-Maqueda, J., Avila, C., Majada, J., & Canovas, F. (2015). Transcriptome-wide analysis supports environmental adaptations of two *Pinus pinaster* populations from contrasting habitats. *BMC Genomics*, *16*, 909. https://doi.org/10.1186/s12864-015-2177-x.

Cañas, R. A., Pascual, M. B., Castro-Rodríguez, V., Ávila, C., Sterck, L., Van de Peer, Y., et al. (2017). The gene expression landscape of pine seedling tissues. *The Plant Journal*, *91*, 1064–1087. https://doi.org/10.1111/tpj.13617.

Carlsbecker, A., Sundström, J. F., Englund, M., Uddenberg, D., Izquierdo, L., Kvarnheden, A., et al. (2013). Molecular control of normal and *acrocona* mutant seed cone development in Norway spruce (*Picea abies*) and the evolution of conifer ovule-bearing organs. *New Phytologist*, *200*, 261–275. https://doi.org/10.1111/nph.12360.

Castro-Rodríguez, V., Assaf-Casals, I., Pérez-Tienda, J., Fan, X., Avila, C., Miller, A., et al. (2016). Deciphering the molecular basis of ammonium uptake and transport in maritime pine. *Plant, Cell & Environment*, *39*, 1669–1682. https://doi.org/10.1111/pce.12692.

Castro-Rodríguez, V., García-Gutiérrez, A., Canales, J., Cañas, R. A., Kirby, E. G., Avila, C., et al. (2016). Poplar trees for phytoremediation of high levels of nitrate and applications in bioenergy. *Plant Biotechnology Journal*, *14*, 299–312. https://doi.org/10.1111/pbi.12384.

Chancerel, E., Lamy, J. B., Lesur, I., Noirot, C., Klopp, C., Ehrenmann, F., et al. (2013). High-density linkage mapping in a pine tree reveals a genomic region associated with inbreeding depression and provides clues to the extent and distribution of meiotic recombination. *BMC Biology*, *11*, 50. https://doi.org/10.1186/1741-7007-11-50.

Chancerel, E., Lepoittevin, C., Le Provost, G., Lin, Y., Jaramillo-correa, J. P., Eckert, A. J., et al. (2011). Development and implementation of a highly multiplexed SNP array for genetic mapping in maritime pine and comparative mapping with loblolly pine. *BMC Genomics*, *12*, 368. https://doi.org/10.1186/1471-2164-12-368.

Charpentier, E., & Doudna, J. A. (2013). Biotechnology: Rewriting a genome. *Nature*, *495*, 50–51. https://doi.org/10.1038/495050a.

Craven-Bartle, B., Pascual, M. B., Cánovas, F. M., & Ávila, C. (2013). A Myb transcription factor regulates genes of the phenylalanine pathway in maritime pine. *Plant Journal*, *74*, 755–766. https://doi.org/10.1111/tpj.12158.

Dalman, K., Wind, J. J., Nemesio-Gorriz, M., Hammerbacher, A., Lundén, K., Ezcurra, I., et al. (2017). Overexpression of *PaNAC03*, a stress induced NAC gene family transcription factor in Norway spruce leads to reduced flavonol biosynthesis and aberrant embryo development. *BMC Plant Biology*, *17*, 6. https://doi.org/10.1186/s12870-016-0952-8.

De La Torre, A. R., Birol, I., Bousquet, J., Ingvarsson, P. K., Jansson, S., Jones, S. J. M., et al. (2014). Insights into conifer giga-genomes. *Plant Physiology*, *166*, 1724–1732. https://doi.org/10.1104/pp.114.248708.

de Vega-Bartol, J. J., Simões, M., Lorenz, W., Rodrigues, A. S., Alba, R., Dean, J. F. D., et al. (2013). Transcriptomic analysis highlights epigenetic and transcriptional regulation during zygotic embryo development of *Pinus pinaster*. *BMC Plant Biology*, *13*, 123. https://doi.org/10.1186/1471-2229-13-123.

Domec, J. C., Lachenbruch, B., Meinzer, F. C., Woodruff, D. R., Warren, J. M., & McCulloh, K. A. (2008). Maximum height in a conifer is associated with conflicting requirements for xylem design. *Proceedings of the National Academy of Sciences of the United States of America*, *105*, 12069–12074. https://doi.org/10.1073/pnas.0710418105.

Doudna, J. A., & Charpentier, E. (2014). The new frontier of genome engineering with CRISPR-Cas9. *Science*, *346*, 1258096. https://doi.org/10.1126/science.1258096.

Dufva, M. (2009). Introduction to microarray technology. In M. Dufva (Ed.), *DNA microarrays for biomedical research: Methods and protocols* (pp. 1–22). Totowa, NJ: Humana Press. https://doi.org/10.1007/978-1-59745-538-1_1.

El-Azaz, J., de la Torre, F., Ávila, C., & Cánovas, F. M. (2016). Identification of a small protein domain present in all plant lineages that confers high prephenate dehydratase activity. *Plant Journal*, *87*, 215–229. https://doi.org/10.1111/tpj.13195.

Essen, P. A., Ehnström, B., Ericsson, L., & Sjöberg, K. (1997). Boreal forest. *Ecological Bulletins*, *46*, 16–47. http://www.jstor.org/stable/20113207.

Farjon, A. (2010). *A handbook of the world's conifers, Vol. 2*. Leiden-Boston: Brill. https://doi.org/10.1163/9789047430629.

Farjon, A., & Filer, D. (2013). *An atlas of the world's conifers: An analysis of their distribution, biogeography, diversity and conservation status*. Leiden-Boston: Brill. https://doi.org/10.1017/CBO9781107415324.004.

Feng, Z., Zhang, B., Ding, W., Liu, X., Yang, D. L., Wei, P., et al. (2013). Efficient genome editing in plants using a CRISPR/Cas system. *Cell Research*, *23*, 1229–1232. https://doi.org/10.1038/cr.2013.114.

Fernández-Pozo, N., Canales, J., Guerrero-Fernández, D., Villalobos, D. P., Díaz-Moreno, S. M., Bautista, R., et al. (2011). EuroPineDB: A high-coverage web database for maritime pine transcriptome. *BMC Genomics*, *12*, 366. https://doi.org/10.1186/1471-2164-12-366.

Flanary, B. E., & Kletetschka, G. (2005). Analysis of telomere length and telomerase activity in tree species of various life-spans, and with age in the bristlecone pine *Pinus longaeva*. *Biogerontology*, *6*, 101–111. https://doi.org/10.1007/s10522-005-3484-4.

Fossdal, C. G., Yaqoob, N., Krokene, P., Kvaalen, H., Solheim, H., & Yakovlev, I. A. (2012). Local and systemic changes in expression of resistance genes, *nb-lrr* genes and their putative microRNAs in Norway spruce after wounding and inoculation with the pathogen *Ceratocystis polonica*. *BMC Plant Biology*, *12*, 105. https://doi.org/10.1186/1471-2229-12-105.

Futamura, N., Totoki, Y., Toyoda, A., Igasaki, T., Nanjo, T., Seki, M., et al. (2008). Characterization of expressed sequence tags from a full-length enriched cDNA library of *Cryptomeria japonica* male strobili. *BMC Genomics*, *9*, 383. https://doi.org/10.1186/1471-2164-9-383.

Galperin, M. Y., Fernández-Suárez, X. M., & Rigden, D. J. (2017). The 24th annual nucleic acids research database issue: A look back and upcoming changes. *Nucleic Acids Research*, *45*, D1–D11. https://doi.org/10.1093/nar/gkw1188.

Gao, Y., Bian, L., Shi, J., Xu, J., Xi, M., & Wang, G. (2013). Expression of a conifer *COBRA*-like gene *ClCOBL1* from Chinese fir (*Cunninghamia lanceolata*) alters the leaf architecture in tobacco. *Plant Physiology and Biochemistry*, *70*, 483–491. https://doi.org/10.1016/j.plaphy.2013.06.013.

García-Gutiérrez, A., Cánovas, F. M., & Avila, C. (2018). Glutamate synthases from conifers: Gene structure and phylogenetic studies. *BMC Genomics*, *19*, 65. https://doi.org/10.1186/s12864-018-4454-y.

Gibson, R., Alako, B., Amid, C., Cerdeño-Tárraga, A., Cleland, I., Goodgame, N., et al. (2016). Biocuration of functional annotation at the European nucleotide archive. *Nucleic Acids Research*, *44*, D58–D66. https://doi.org/10.1093/nar/gkv1311.

Gómez-Maldonado, J., Cánovas, F. M., & Avila, C. (2004). Molecular analysis of the 5′-upstream region of a gibberellin-inducible cytosolic glutamine synthetase gene (GS1b) expressed in pine vascular tissue. *Planta*, *218*, 1036–1045. https://doi.org/10.1007/s00425-003-1185-2.

Gómez-Maldonado, J., Crespillo, R., Ávila, C., & Cánovas, F. M. (2001). Efficient preparation of maritime pine (*Pinus pinaster*) protoplasts suitable for transgene expression analysis. *Plant Molecular Biology Reporter*, *19*, 361–366. https://doi.org/10.1007/BF02772834.

Guo, A., Zheng, C. X., & Yang, Y. Y. (2014). Differential expression of *SLOW WALKER2* homologue in ovules of female sterile mutant and fertile clone of *Pinus tabulaeformis*. *Russian Journal of Developmental Biology*, *45*, 78–84. https://doi.org/10.1134/S1062360414020052.

Han, Q., Shinohara, K., Kakubari, Y., & Mukai, Y. (2003). Photoprotective role of rhodoxanthin during cold acclimation in *Cryptomeria japonica*. *Plant, Cell and Environment*, *26*, 715–723. https://doi.org/10.1046/j.1365-3040.2003.01008.x.

Harfouche, A., Meilan, R., Kirst, M., Morgante, M., Boerjan, W., Sabatti, M., et al. (2012). Accelerating the domestication of forest trees in a changing world. *Trends in Plant Science*, *17*, 64–72. https://doi.org/10.1016/j.tplants.2011.11.005.

Hedman, H., Zhu, T., von Arnold, S., & Sohlberg, J. J. (2013). Analysis of the WUSCHEL-RELATED HOMEOBOX gene family in the conifer *Picea abies* reveals extensive conservation as well as dynamic patterns. *BMC Plant Biology*, *13*, 89. https://doi.org/10.1186/1471-2229-13-89.

Hills, A. C., Khan, S., & López-Juez, E. (2015). Chloroplast biogenesis-associated nuclear genes: Control by plastid signals evolved prior to their regulation as part of photomorphogenesis. *Frontiers in Plant Science*, *6*, 1078. https://doi.org/10.3389/fpls.2015.01078.

Hirao, T., Watanabe, A., Kurita, M., Kondo, T., & Takata, K. (2009). A frameshift mutation of the chloroplast *mat*K coding region is associated with chlorophyll deficiency in the *Cryptomeria japonica* virescent mutant *Wogon*-Sugi. *Current Genetics*, *55*, 311–321. https://doi.org/10.1007/s00294-009-0247-9.

Hosoo, Y., Kimura, Y., Nanatani, K., & Uozumi, N. (2014). Molecular cloning and expression analysis of a gene encoding KUP/HAK/KT-type potassium uptake transporter from *Cryptomeria japonica*. *Trees*, *28*, 1527–1537. https://doi.org/10.1007/s00468-014-1059-1.

Hu, X. G., Liu, H., Jin, Y. Q., Sun, Y. Q., Li, Y., Zhao, W., et al. (2016). *De novo* transcriptome assembly and characterization for the widespread and stress-tolerant conifer *Platycladus orientalis*. *PLoS One*, *11*, e0148985. https://doi.org/10.1371/journal.pone.0148985.

Johnson, D. S., Mortazavi, A., Myers, R. M., & Wold, B. (2007). Genome-wide mapping of in vivo protein-DNA interactions. *Science*, *316*, 1497–1502. https://doi.org/10.1126/science.1141319.

Jokipii-Lukkari, S., Sundell, D., Nilsson, O., Hvidsten, T. R., Street, N. R., & Tuominen, H. (2017). NorWood: A gene expression resource for evo-devo studies of conifer wood development. *New Phytologist, 216*, 482–494. https://doi.org/10.1111/nph.14458.

Karlgren, A., Gyllenstrand, N., Källman, T., & Lagercrantz, U. (2013). Conserved function of core clock proteins in the gymnosperm Norway spruce (*Picea abies* L. Karst). *PLoS One, 8*, e60110. https://doi.org/10.1371/journal.pone.0060110.

Katahata, S. I., Futamura, N., Igasaki, T., & Shinohara, K. (2014). Functional analysis of *SOC1*-like and *AGL6*-like MADS-box genes of the gymnosperm *Cryptomeria japonica*. *Tree Genetics & Genomes, 10*, 317–327. https://doi.org/10.1007/s11295-013-0686-9.

Kirst, M., Johnson, A. F., Baucom, C., Ulrich, E., Hubbard, K., Staggs, R., et al. (2003). Apparent homology of expressed genes from wood-forming tissues of loblolly pine (*Pinus taeda* L.) with *Arabidopsis thaliana*. *Proceedings of the National Academy of Sciences of the United States of America, 100*, 7383–7388. https://doi.org/10.1073/pnas.1132171100.

Kjellsen, T. D., Shiryaeva, L., Schröder, W. P., & Strimbeck, G. R. (2010). Proteomics of extreme freezing tolerance in Siberian spruce (*Picea obovata*). *Journal of Proteomics, 73*, 965–975. https://doi.org/10.1016/j.jprot.2009.12.010.

Klimaszewska, K., Hargreaves, C., Lelu-Walter, M. A., & Trontin, J. F. (2016). Advances in conifer somatic embryogenesis since year 2000. *Methods in Molecular Biology, 1359*, 131–166. https://doi.org/10.1007/978-1-4939-3061-6_7.

Klimaszewska, K., Lachance, D., Pelletier, G., Lelu, M. A., & Seguin, A. (2001). Regeneration of transgenic *Picea Glauca, P. Mariana*, and *P. Abies* after cocultivation of embryogenic tissue with *Agrobacterium tumefaciens*. *In Vitro Cellular & Developmental Biology—Plant, 37*, 748–755. https://doi.org/10.1079/IVP2001230.

Lara, A., & Villalba, R. (1993). A 3620-year temperature record from *Fitzroya cupressoides* tree rings in southern South America. *Science, 260*, 1104–1106. https://doi.org/10.1126/science.260.5111.1104.

Larsson, E., Sundström, J. F., Sitbon, F., & Von Arnold, S. (2012). Expression of *PaNAC01*, a *Picea abies CUP-SHAPED COTYLEDON* orthologue, is regulated by polar auxin transport and associated with differentiation of the shoot apical meristem and formation of separated cotyledons. *Annals of Botany, 110*, 923–934. https://doi.org/10.1093/aob/mcS151.

Lee, C. H., Chan, M. H., Wang, Y. N., & Chu, F. H. (2006). Gene investigation into the inner bark of Taiwania (*Taiwania cryptomerioides*). *Botanical Studies, 47*, 111–118.

Li, Z., De La Torre, A., Sterck, L., Cánovas, F. M., Avila, C., Merino, I., et al. (2017). Single-copy genes as molecular markers for phylogenomic studies in seed plants. *Genome Biology and Evolution, 9*, 1130–1147. https://doi.org/10.1093/gbe/evx070.

Li, P. C. H., Sedighi, A., & Wang, L. (2016). In P. C. H. Li, A. Sedighi, & L. Wang (Eds.), *Vol. 1368. Microarray Technology*. New York, NY: Springer New York. https://doi.org/10.1007/978-1-4939-3136-1.

Li, X., Wu, H. X., Dillon, S. K., & Southerton, S. G. (2009). Generation and analysis of expressed sequence tags from six developing xylem libraries in *Pinus radiata* D. Don. *BMC Genomics, 10*, 41. https://doi.org/10.1186/1471-2164-10-41.

Li, L., Yu, Y., Wei, J., Huang, G., Zhang, D., Liu, Y., et al. (2013). Homologous HAP5 subunit from *Picea wilsonii* improved tolerance to salt and decreased sensitivity to ABA in transformed *Arabidopsis*. *Planta, 238*, 345–356. https://doi.org/10.1007/s00425-013-1894-0.

Lippert, D., Yuen, M., & Bohlmann, J. (2009). Spruce proteome DB: A resource for conifer proteomics research. *Tree Genetics and Genomes, 5*, 723–727. https://doi.org/10.1007/s11295-009-0220-2.

Liu, Y., & El-Kassaby, Y. A. (2017). Landscape of fluid sets of hairpin-derived 21-/24-nt-long small RNAs at seed set uncovers special epigenetic features in *Picea glauca*. *Genome Biology and Evolution*, *9*, 82–92. https://doi.org/10.1093/gbe/evw283.

Liu, J. J., Schoettle, A. W., Sniezko, R. A., Sturrock, R. N., Zamany, A., Williams, H., et al. (2016). Genetic mapping of *Pinus flexilis* major gene (Cr4) for resistance to white pine blister rust using transcriptome-based SNP genotyping. *BMC Genomics*, *17*, 753. https://doi.org/10.1186/s12864-016-3079-2.

Lorenz, W. W., Ayyampalayam, S., Bordeaux, J. M., Howe, G. T., Jermstad, K. D., Neale, D. B., et al. (2012). Conifer DBMagic: A database housing multiple *de novo* transcriptome assemblies for 12 diverse conifer species. *Tree Genetics & Genomes*, *8*, 1477–1485. https://doi.org/10.1007/s11295-012-0547-y.

Lorenz, K., & Lal, R. (2010). *Carbon sequestration in forest ecosystems*. Dordrecht: Springer Netherlands. https://doi.org/10.1007/978-90-481-3266-9.

Lorenz, W. W., Sun, F., Liang, C., Kolychev, D., Wang, H., Zhao, X., et al. (2006). Water stress-responsive genes in loblolly pine (*Pinus taeda*) roots identified by analyses of expressed sequence tag libraries. *Tree Physiology*, *26*, 1–16. https://doi.org/10.1093/treephys/26.1.1.

Lowe, R., Shirley, N., Bleackley, M., Dolan, S., & Shafee, T. (2017). Transcriptomics technologies. *PLoS Computational Biology*, *13*, e1005457. https://doi.org/10.1371/journal.pcbi.1005457.

Lu, M., Krutovsky, K. V., Nelson, C. D., Koralewski, T. E., Byram, T. D., & Loopstra, C. A. (2016). Exome genotyping, linkage disequilibrium and population structure in loblolly pine (*Pinus taeda* L.). *BMC Genomics*, *17*, 730. https://doi.org/10.1186/s12864-016-3081-8.

Lu, M., Krutovsky, K. V., Nelson, C. D., West, J. B., Reilly, N. A., & Loopstra, C. A. (2017). Association genetics of growth and adaptive traits in loblolly pine (*Pinus taeda* L.) using whole-exome-discovered polymorphisms. *Tree Genetics & Genomes*, *13*, 57. https://doi.org/10.1007/s11295-017-1140-1.

Mackay, J., Dean, J. F. D., Plomion, C., Peterson, D. G., Cánovas, F. M., Pavy, N., et al. (2012). Towards decoding the conifer giga-genome. *Plant Molecular Biology*, *80*, 555–569. https://doi.org/10.1007/s11103-012-9961-7.

MacKay, J. J., O'Malley, D. M., Presnell, T., Booker, F. L., Campbell, M. M., Whetten, R. W., et al. (1997). Inheritance, gene expression, and lignin characterization in a mutant pine deficient in cinnamyl alcohol dehydrogenase. *Proceedings of the National Academy of Sciences of the United States of America*, *94*, 8255–8260.

Mageroy, M. H., Lachance, D., Jancsik, S., Parent, G., Séguin, A., Mackay, J., et al. (2017). *In vivo* function of *Pgβglu-1* in the release of acetophenones in white spruce. *PeerJ*, *5*, e3535. https://doi.org/10.7717/peerj.3535.

Maloney, V. J., Samuels, A. L., & Mansfield, S. D. (2012). The endo-1,4-β-glucanase *Korrigan* exhibits functional conservation between gymnosperms and angiosperms and is required for proper cell wall formation in gymnosperms. *New Phytologist*, *193*, 1076–1087. https://doi.org/10.1111/j.1469-8137.2011.03998.x.

Mann, I. K., Wegrzyn, J. L., & Rajora, O. P. (2013). Generation, functional annotation and comparative analysis of black spruce (*Picea mariana*) ESTs: An important conifer genomic resource. *BMC Genomics*, *14*, 702. https://doi.org/10.1186/1471-2164-14-702.

Martin, S. A. M., Dehler, C. E., & Król, E. (2016). Transcriptomic responses in the fish intestine. *Developmental and Comparative Immunology*, *64*, 103–117. https://doi.org/10.1016/j.dci.2016.03.014.

Mashima, J., Kodama, Y., Kosuge, T., Fujisawa, T., Katayama, T., Nagasaki, H., et al. (2016). DNA data bank of Japan (DDBJ) progress report. *Nucleic Acids Research*, *44*, D51–D57. https://doi.org/10.1093/nar/gkv1105.

Meyer, C. A., & Liu, X. S. (2014). Identifying and mitigating bias in next-generation sequencing methods for chromatin biology. *Nature Reviews. Genetics*, *15*, 709–721. https://doi.org/10.1038/nrg3788.

Möller, R., McDonald, A. G., Walter, C., & Harris, P. J. (2003). Cell differentiation, secondary cell-wall formation and transformation of callus tissue of *Pinus radiata* D. Don. *Planta*, *217*, 736–747. https://doi.org/10.1007/s00425-003-1053-0.

Möller, R., Steward, D., Phillips, L., Flint, H., & Wagner, A. (2005). Gene silencing of cinnamyl alcohol dehydrogenase in *Pinus radiata* callus cultures. *Plant Physiology and Biochemistry*, *43*, 1061–1066. https://doi.org/10.1016/j.plaphy.2005.11.001.

Morin, R. D., Aksay, G., Dolgosheina, E., Ebhardt, H. A., Magrini, V., Mardis, E. R., et al. (2008). Comparative analysis of the small RNA transcriptomes of *Pinus contorta* and *Oryza sativa*. *Genome Research*, *18*, 571–584. https://doi.org/ https://doi.org/10.1101/gr.6897308.

Moschou, P. N., Savenkov, E. I., Minina, E. A., Fukada, K., Reza, S. H., Gutierrez-Beltran, E., et al. (2016). *EXTRA SPINDLE POLES* (Separase) controls anisotropic cell expansion in Norway spruce (*Picea abies*) embryos independently of its role in anaphase progression. *New Phytologist*, *212*, 232–243. https://doi.org/10.1111/nph.14012.

Nagel, R., Berasategui, A., Paetz, C., Gershenzon, J., & Schmidt, A. (2014). Overexpression of an isoprenyl diphosphate synthase in spruce leads to unexpected terpene diversion products that function in plant defense. *Plant Physiology*, *164*, 555–569. https://doi.org/10.1104/pp.113.228940.

Naithani, S., Preece, J., D'Eustachio, P., Gupta, P., Amarasinghe, V., Dharmawardhana, P. D., et al. (2017). Plant reactome: A resource for plant pathways and comparative analysis. *Nucleic Acids Research*, *45*, D1029–D1039. https://doi.org/10.1093/nar/gkw932.

Neale, D. B., McGuire, P. E., Wheeler, N. C., Stevens, K. A., Crepeau, M. W., Cardeno, C., et al. (2017). The Douglas-fir genome sequence reveals specialization of the photosynthetic apparatus in pinaceae. *G3 (Bethesda)*, 7, 3157–3167. https://doi.org/10.1534/g3.117.300078.

Neale, D. B., Wegrzyn, J. L., Stevens, K. A., Zimin, A. V., Puiu, D., Crepeau, M. W., et al. (2014). Decoding the massive genome of loblolly pine using haploid DNA and novel assembly strategies. *Genome Biology*, *15*, R59. https://doi.org/10.1186/gb-2014-15-3-r59.

Netotea, S., Sundell, D., Street, N. R., & Hvidsten, T. R. (2014). ComPlEx: Conservation and divergence of co-expression networks in *A. thaliana*, *Populus* and *O. sativa*. *BMC Genomics*, *15*, 106. https://doi.org/10.1186/1471-2164-15-106.

Neves, L. G., Davis, J. M., Barbazuk, W. B., & Kirst, M. (2013). Whole-exome targeted sequencing of the uncharacterized pine genome. *The Plant Journal*, *75*, 146–156. https://doi.org/10.1111/tpj.12193.

Neves, L. G., Davis, J. M., Barbazuk, W. B., & Kirst, M. (2014). A high-density gene map of loblolly pine (*Pinus taeda* L.) based on exome sequence capture genotyping. *G3 (Bethesda)*, *4*, 29–37. https://doi.org/10.1534/g3.113.008714.

Nilsson, L., Carlsbecker, A., Sundås-Larsson, A., & Vahala, T. (2007). *APETALA2* like genes from *Picea abies* show functional similarities to their *Arabidopsis* homologues. *Planta*, *225*, 589–602. https://doi.org/10.1007/s00425-006-0374-1.

Ning, D. L., Lu, T. C., Liu, G. F., Yang, C. P., & Wang, B. C. (2013). Proteomic analysis points to a role for RAD23 in apical dominance in *Pinus sylvestris* var. *Mongolica*. *Plant Molecular Biology Reporter*, *31*, 1283–1292. https://doi.org/10.1007/s11105-013-0593-0.

Niu, S. H., Liu, C., Yuan, H.-W., Li, P., Li, Y., & Li, W. (2015). Identification and expression profiles of sRNAs and their biogenesis and action-related genes in male and female cones of *Pinus tabuliformis*. *BMC Genomics*, *16*, 693. https://doi.org/10.1186/s12864-015-1885-6.

Niu, S., Yuan, H., Sun, X., Porth, I., Li, Y., El-Kassaby, Y. A., et al. (2016). A transcriptomics investigation into pine reproductive organ development. *New Phytologist, 209*, 1278–1289. https://doi.org/10.1111/nph.13680.

Nonaka, S., Arai, C., Takayama, M., Matsukura, C., & Ezura, H. (2017). Efficient increase of γ-aminobutyric acid (GABA) content in tomato fruits by targeted mutagenesis. *Scientific Reports*, 7, 7057. https://doi.org/10.1038/s41598-017-06400-y.

Nystedt, B., Street, N. R., Wetterbom, A., Zuccolo, A., Lin, Y. C., Scofield, D. G., et al. (2013). The Norway spruce genome sequence and conifer genome evolution. *Nature, 497*, 579–584. https://doi.org/10.1038/nature12211.

O'Malley, R. C., Huang, S. C., Song, L., Lewsey, M. G., Bartlett, A., Nery, J. R., et al. (2016). Cistrome and epicistrome features shape the regulatory DNA landscape. *Cell, 165*, 1280–1292. https://doi.org/10.1016/j.cell.2016.04.038.

Ozsolak, F., & Milos, P. M. (2011). RNA sequencing: Advances, challenges and opportunities. *Nature Reviews. Genetics, 12*, 87–98. https://doi.org/10.1038/nrg2934.

Pâques, L. E. (2013). In L. E. Pâques (Ed.), *Vol. 25. Forest tree breeding in Europe*. Dordrecht: Springer Netherlands. https://doi.org/10.1007/978-94-007-6146-9.

Park, J. J., Dempewolf, E., Zhang, W., & Wang, Z. Y. (2017). RNA-guided transcriptional activation via CRISPR/dCas9 mimics overexpression phenotypes in *Arabidopsis*. *PLoS One, 12*, e0179410 https://doi.org/10.1371/journal.pone.0179410.

Pascual, M., Canovas, F., & Avila, C. (2015). The NAC transcription factor family in maritime pine (*Pinus pinaster*): Molecular regulation of two genes involved in stress responses. *BMC Plant Biology, 15*, 254. https://doi.org/10.1186/s12870-015-0640-0.

Pascual, M. B., El-Azaz, J., de la Torre, F. N., Cañas, R. A., Avila, C., & Cánovas, F. M. (2016). Biosynthesis and metabolic fate of phenylalanine in conifers. *Frontiers in Plant Science*, 7, 1030. https://doi.org/10.3389/fpls.2016.01030.

Pascual, M. B., Llebrés, M. T., Craven-Bartle, B., Cañas, R. A., Cánovas, F. M., & Ávila, C. (2018). *PpNAC1*, a main regulator of phenylalanine biosynthesis and utilization in maritime pine. *Plant Biotechnology Journal, 16*, 1094–1104. https://doi.org/10.1111/pbi.12854.

Pavy, N., Gagnon, F., Deschênes, A., Boyle, B., Beaulieu, J., & Bousquet, J. (2016). Development of highly reliable in silico SNP resource and genotyping assay from exome capture and sequencing: An example from black spruce (*Picea mariana*). *Molecular Ecology Resources, 16*, 588–598. https://doi.org/10.1111/1755-0998.12468.

Pavy, N., Gagnon, F., Rigault, P., Blais, S., Deschênes, A., Boyle, B., et al. (2013). Development of high-density SNP genotyping arrays for white spruce (*Picea glauca*) and transferability to subtropical and nordic congeners. *Molecular Ecology Resources, 13*, 324–336. https://doi.org/10.1111/1755-0998.12062.

Pavy, N., Johnson, J. J., Crow, J. A., Paule, C., Kunau, T., MacKay, J., et al. (2007). ForestTreeDB: A database dedicated to the mining of tree transcriptomes. *Nucleic Acids Research, 35*(Suppl. 1), 888–894. https://doi.org/10.1093/nar/gkl882.

Pavy, N., Lamothe, M., Pelgas, B., Gagnon, F., Birol, I., Bohlmann, J., et al. (2017). A high-resolution reference genetic map positioning 8.8 K genes for the conifer white spruce: Structural genomics implications and correspondence with physical distance. *Plant Journal, 90*, 189–203. https://doi.org/10.1111/tpj.13478.

Pavy, N., Paule, C., Parsons, L., Crow, J. a., Morency, M.-J., Cooke, J., et al. (2005). Generation, annotation, analysis and database integration of 16,500 white spruce EST clusters. *BMC Genomics, 6*, 144. https://doi.org/10.1186/1471-2164-6-144.

Perry, D. J., & Bousquet, J. (1998). Sequence-tagged-site (STS) markers of arbitrary genes: Development, characterization and analysis of linkage in black spruce. *Genetics, 149*, 1089–1098.

Pevsner, J. (2015). *Bioinformatics and functional genomics. Vol. 1* (3rd ed.). Chichester, UK: John Wiley & Sons Inc. Retrieved from http://eu.wiley.com/WileyCDA/WileyTitle/productCd-1118581784.html#.

Plomion, C., Bastien, C., Bogeat-Triboulot, M. B., Bouffier, L., Déjardin, A., Duplessis, S., et al. (2016). Forest tree genomics: 10 Achievements from the past 10 years and future prospects. *Annals of Forest Science*, *73*, 77–103. https://doi.org/10.1007/s13595-015-0488-3.

Proost, S., Van Bel, M., Vaneechoutte, D., Van De Peer, Y., Inzé, D., Mueller-Roeber, B., et al. (2015). PLAZA 3.0: An access point for plant comparative genomics. *Nucleic Acids Research*, *43*, D974–D981. https://doi.org/10.1093/nar/gku986.

Provart, N. J., Alonso, J., Assmann, S. M., Bergmann, D., Brady, S. M., Brkljacic, J., et al. (2016). 50 Years of *Arabidopsis* research: Highlights and future directions. *New Phytologist*, *209*, 921–944. https://doi.org/10.1111/nph.13687.

Prunier, J., Caron, S., & MacKay, J. (2017). CNVs into the wild: Screening the genomes of conifer trees (*Picea* spp.) reveals fewer gene copy number variations in hybrids and links to adaptation. *BMC Genomics*, *18*, 97. https://doi.org/10.1186/s12864-016-3458-8.

Qiu, Z., Li, X., Zhao, Y., Zhang, M., Wan, Y., Cao, D., et al. (2015). Genome-wide analysis reveals dynamic changes in expression of microRNAs during vascular cambium development in Chinese fir, *Cunninghamia lanceolata*. *Journal of Experimental Botany*, *66*, 3041–3054. https://doi.org/10.1093/jxb/erv103.

Quinn, C. R., Iriyama, R., & Fernando, D. D. (2014). Expression patterns of conserved microRNAs in the male gametophyte of loblolly pine (*Pinus taeda*). *Plant Reproduction*, *27*, 69–78. https://doi.org/10.1007/s00497-014-0241-3.

Quinn, C. R., Iriyama, R., & Fernando, D. D. (2015). Computational predictions and expression patterns of conserved microRNAs in loblolly pine (*Pinus taeda*). *Tree Genetics & Genomes*, *11*, 806. https://doi.org/10.1007/s11295-014-0806-1.

Raherison, E., Rigault, P., Caron, S., Poulin, P. L., Boyle, B., Verta, J. P., et al. (2012). Transcriptome profiling in conifers and the PiceaGenExpress database show patterns of diversification within gene families and interspecific conservation in vascular gene expression. *BMC Genomics*, *13*, 434. https://doi.org/10.1186/1471-2164-13-434.

Ralph, S. G., Chun, H., Kolosova, N., Cooper, D., Oddy, C., Ritland, C. E., et al. (2008). A conifer genomics resource of 200,000 spruce (*Picea* spp.) ESTs and 6,464 high-quality, sequence-finished full-length cDNAs for Sitka spruce (*Picea sitchensis*). *BMC Genomics*, *9*, 484. https://doi.org/10.1186/1471-2164-9-484.

Ralph, J., MacKay, J., Hatfield, R. D., O'Malley, D. M., Whetten, R. W., & Sederoff, R. R. (1997). Abnormal lignin in a loblolly pine mutant. *Science*, *277*, 235–239. https://doi.org/10.1126/science.277.5323.235.

Ralph, S. G., Yueh, H., Friedmann, M., Aeschliman, D., Zeznik, J. A., Nelson, C. C., et al. (2006). Conifer defence against insects: Microarray gene expression profiling of Sitka spruce (*Picea sitchensis*) induced by mechanical wounding or feeding by spruce budworms (*Choristoneura occidentalis*) or white pine weevils (*Pissodes strobi*) reveals large-scale. *Plant, Cell and Environment*, *29*, 1545–1570. https://doi.org/10.1111/j.1365-3040.2006.01532.x.

Rigault, P., Boyle, B., Lepage, P., Cooke, J. E. K., Bousquet, J., & MacKay, J. J. (2011). A white spruce gene catalog for conifer genome analyses. *Plant Physiology*, *157*, 14–28. https://doi.org/10.1104/pp.111.179663.

Ritland, K. (2012). Genomics of a phylum distant from flowering plants: Conifers. *Tree Genetics and Genomes*, *8*, 573–582. https://doi.org/10.1007/s11295-012-0497-4.

Romero-Rodríguez, M. C., Pascual, J., Valledor, L., & Jorrín-Novo, J. (2014). Improving the quality of protein identification in non-model species. Characterization of *Quercus ilex* seed and *Pinus radiata* needle proteomes by using SEQUEST and custom databases. *Journal of Proteomics*, *105*, 85–91. https://doi.org/10.1016/j.jprot.2014.01.027.

Rueda-López, M., Cañas, R. A., Canales, J., Cánovas, F. M., & Ávila, C. (2015). The overexpression of the pine transcription factor PpDof5 in *Arabidopsis* leads to increased lignin content and affects carbon and nitrogen metabolism. *Physiologia Plantarum*, *155*, 369–383. https://doi.org/10.1111/ppl.12381.

Rueda-López, M., Crespillo, R., Cánovas, F. M., & Avila, C. (2008). Differential regulation of two glutamine synthetase genes by a single Dof transcription factor. *Plant Journal, 56*, 73–85. https://doi.org/10.1111/j.1365-313X.2008.03573.x.

Rueda-López, M., García-Gutiérrez, A., Cánovas, F. M., & Ávila, C. (2013). The family of Dof transcription factors in pine. *Trees—Structure and Function, 27*, 1547–1557. https://doi.org/10.1007/s00468-013-0903-z.

Rueda-López, M., Pascual, M. B., Pallero, M., Henao, L. M., Lasa, B., Jauregui, I., et al. (2017). Overexpression of a pine Dof transcription factor in hybrid poplars: A comparative study in trees growing under controlled and natural conditions. *PLoS One, 12*, e0174748. https://doi.org/10.1371/journal.pone.0174748.

Rungis, D., Bérubé, Y., Zhang, J., Ralph, S., Ritland, C. E., Ellis, B. E., et al. (2004). Robust simple sequence repeat markers for spruce (*Picea* spp.) from expressed sequence tags. *Theoretical and Applied Genetics, 109*, 1283–1294. https://doi.org/10.1007/s00122-004-1742-5.

Rupps, A., Raschke, J., Rümmler, M., Linke, B., & Zoglauer, K. (2016). Identification of putative homologs of *Larix decidua* to BABYBOOM (BBM), LEAFY COTYLEDON1 (LEC1), WUSCHEL-related HOMEOBOX2 (WOX2) and SOMATIC EMBRYOGENESIS RECEPTOR-like KINASE (SERK) during somatic embryogenesis. *Planta, 243*, 473–488. https://doi.org/10.1007/s00425-015-2409-y.

Seoane-Zonjic, P., Cañas, R. A., Bautista, R., Gómez-Maldonado, J., Arrillaga, I., Fernández-Pozo, N., et al. (2016). Establishing gene models from the *Pinus pinaster* genome using gene capture and BAC sequencing. *BMC Genomics, 17*, 148. https://doi.org/10.1186/s12864-016-2490-z.

Shi, J., Gao, H., Wang, H., Lafitte, H. R., Archibald, R. L., Yang, M., et al. (2017). ARGOS8 variants generated by CRISPR-Cas9 improve maize grain yield under field drought stress conditions. *Plant Biotechnology Journal, 15*, 207–216. https://doi.org/10.1111/pbi.12603.

Steijger, T., Abril, J. F., Engström, P. G., Kokocinski, F., Abril, J. F., Akerman, M., et al. (2013). Assessment of transcript reconstruction methods for RNA-seq. *Nature Methods, 10*, 1177–1184. https://doi.org/10.1038/nmeth.2714.

Stevens, K. A., Wegrzyn, J. L., Zimin, A., Puiu, D., Crepeau, M., Cardeno, C., et al. (2016). Sequence of the sugar pine megagenome. *Genetics, 204*, 1613–1626. https://doi.org/10.1534/genetics.116.193227.

Sundell, D., Mannapperuma, C., Netotea, S., Delhomme, N., Lin, Y. C., Sj??din, A., et al. (2015). The plant genome integrative explorer resource: PlantGenIE.org. *New Phytologist, 208*, 1149–1156. https://doi.org/10.1111/nph.13557.

Suren, H., Hodgins, K. A., Yeaman, S., Nurkowski, K. A., Smets, P., Rieseberg, L. H., et al. (2016). Exome capture from the spruce and pine giga-genomes. *Molecular Ecology Resources, 16*, 1136–1146. https://doi.org/10.1111/1755-0998.12570.

Trontin, J.-F., Teyssier, C., Avila, C., Debille, S., Le Matté, C., Lesage-Descauses, M., et al. (2014). Molecular phenotyping of Maritime pine somatic plants transformed with an RNAi construct targeting cinnamyl alcohol. In *Proceedings of the third international conference of the IUFRO unit 2.09.02: Somatic embryogenesis and other Vegetative propagation technologies* (p. 140). Woody Plant Production Integrating Genetic and Vegetative Propagation Technologies. https://doi.org/10.13140/RG.2.1.1036.7767.

Tsubomura, M., Kurita, M., & Watanabe, A. (2016). Determination of male strobilus developmental stages by cytological and gene expression analyses in Japanese cedar (*Cryptomeria japonica*). *Tree Physiology, 36*, 653–666. https://doi.org/10.1093/treephys/tpw001.

Uddenberg, D., Abrahamsson, M., & von Arnold, S. (2016). Overexpression of *PaHAP3A* stimulates differentiation of ectopic embryos from maturing somatic embryos of Norway spruce. *Tree Genetics & Genomes, 12*, 18. https://doi.org/10.1007/s11295-016-0974-2.

Uddenberg, D., Akhter, S., Ramachandran, P., Sundström, J. F., & Carlsbecker, A. (2015). Sequenced genomes and rapidly emerging technologies pave the way for conifer evolutionary developmental biology. *Frontiers in Plant Science, 6*, 970. https://doi.org/10.3389/fpls.2015.00970.

Uddenberg, D., Reimegård, J., Clapham, D., Almqvist, C., von Arnold, S., Emanuelsson, O., et al. (2013). Early cone setting in *Picea abies acrocona* is associated with increased transcriptional activity of a MADS box transcription factor. *Plant Physiology, 161*, 813–823. https://doi.org/10.1104/pp.112.207746.

Ujino-Ihara, T., Taguchi, Y., Yoshimura, K., & Tsumura, Y. (2003). Analysis of expressed sequence tags derived from developing seed and pollen cones of *Cryptomeria japonica*. *Plant Biology, 5*, 600–607. https://doi.org/10.1055/s-2003-44690.

Ujino-Ihara, T., Yoshimura, K., Ugawa, Y., Yoshimaru, H., Nagasaka, K., & Tsumura, Y. (2000). Expression analysis of ESTs derived from the inner bark of *Cryptomeria japonica*. *Plant Molecular Biology, 43*, 451–457. https://doi.org/10.1023/A:1006492103063.

Valledor, L., Pascual, J., Meijón, M., Escandón, M., & Cañal, M. J. (2015). Conserved epigenetic mechanisms could play a key role in regulation of photosynthesis and development-related genes during needle development of *Pinus radiata*. *PLoS One, 10*, e0126405. https://doi.org/10.1371/journal.pone.0126405.

Van Zyl, L., von Arnold, S., Chen, Y., Egertsdotter, U., Mackay, J., Sederoff, R., et al. (2002). Heterologous array analysis in Pinaceae: Hybridization of *Pinus taeda* cDNA arrays with cDNA from needles and embryogenic cultures of *P. taeda*, *P. sylvestris* or *Picea abies*. *Comparative and Functional Genomics, 3*, 306–318. https://doi.org/10.1002/cfg.199.

Villalobos, D. P., Díaz-Moreno, S. M., Said, E. S. S., Cañas, R. A., Osuna, D., Van Kerckhoven, S. H. E., et al. (2012). Reprogramming of gene expression during compression wood formation in pine: Coordinated modulation of S-adenosylmethionine, lignin and lignan related genes. *BMC Plant Biology, 12*, 100. https://doi.org/10.1186/1471-2229-12-100.

Vizcaíno, J. A., Deutsch, E. W., Wang, R., Csordas, A., Reisinger, F., Ríos, D., et al. (2014). ProteomeXchange provides globally coordinated proteomics data submission and dissemination. *Nature Biotechnology, 32*, 223–226. https://doi.org/10.1038/nbt.2839.

Wachowiak, W., Trivedi, U., Perry, A., & Cavers, S. (2015). Comparative transcriptomics of a complex of four European pine species. *BMC Genomics, 16*, 234. https://doi.org/10.1186/s12864-015-1401-z.

Wadenbäck, J., Von Arnold, S., Egertsdotter, U., Walter, M. H., Grima-Pettenati, J., Goffner, D., et al. (2008). Lignin biosynthesis in transgenic Norway spruce plants harboring an antisense construct for cinnamoyl CoA reductase (CCR). *Transgenic Research, 17*, 379–392. https://doi.org/10.1007/s11248-007-9113-z.

Wagner, A., Donaldson, L., Kim, H., Phillips, L., Flint, H., Steward, D., et al. (2009). Suppression of 4-coumarate-CoA ligase in the coniferous gymnosperm *Pinus radiata*. *Plant Physiology, 149*, 370–383. https://doi.org/10.1104/pp.108.125765.

Wagner, A., Phillips, L., Narayan, R. D., Moody, J. M., & Geddes, B. (2005). Gene silencing studies in the gymnosperm species *Pinus radiata*. *Plant Cell Reports, 24*, 95–102. https://doi.org/10.1007/s00299-004-0911-y.

Wagner, A., Ralph, J., Akiyama, T., Flint, H., Phillips, L., Torr, K., et al. (2007). Exploring lignification in conifers by silencing hydroxycinnamoyl-CoA:shikimate hydroxycinnamoyltransferase in *Pinus radiata*. *Proceedings of the National Academy of Sciences of the United States of America, 104*, 11856–11861. https://doi.org/10.1073/pnas.0701428104.

Wagner, A., Tobimatsu, Y., Goeminne, G., Phillips, L., Flint, H., Steward, D., et al. (2013). Suppression of *CCR* impacts metabolite profile and cell wall composition in *Pinus radiata* tracheary elements. *Plant Molecular Biology, 81*, 105–117. https://doi.org/10.1007/s11103-012-9985-z.

Wagner, A., Tobimatsu, Y., Phillips, L., Flint, H., Geddes, B., Lu, F., et al. (2015). Syringyl lignin production in conifers: Proof of concept in a Pine tracheary element system. *Proceedings of the National Academy of Sciences of the United States of America, 112*, 6218–6223. https://doi.org/10.1073/pnas.1411926112.

Wagner, A., Tobimatsu, Y., Phillips, L., Flint, H., Torr, K., Donaldson, L., et al. (2011). *CCoAOMT* suppression modifies lignin composition in *Pinus radiata*. *Plant Journal, 67*, 119–129. https://doi.org/10.1111/j.1365-313X.2011.04580.x.

Wan, T., Liu, Z. M., Li, L. F., Leitch, A. R., Leitch, I. J., Lohaus, R., et al. (2018). A genome for gnetophytes and early evolution of seed plants. *Nature Plants, 4*, 82–89. https://doi.org/10.1038/s41477-017-0097-2.

Wan, L. C., Zhang, H., Lu, S., Zhang, L., Qiu, Z., Zhao, Y., et al. (2012). Transcriptome-wide identification and characterization of miRNAs from *Pinus densata*. *BMC Genomics, 13*, 132. https://doi.org/10.1186/1471-2164-13-132.

Wang, Y., Cheng, X., Shan, Q., Zhang, Y., Liu, J., Gao, C., et al. (2014). Simultaneous editing of three homoeoalleles in hexaploid bread wheat confers heritable resistance to powdery mildew. *Nature Biotechnology, 32*, 947–951. https://doi.org/10.1038/nbt.2969.

Warren, R. L., Keeling, C. I., Yuen, M. M. S., Raymond, A., Taylor, G. A., Vandervalk, B. P., et al. (2015). Improved white spruce (*Picea glauca*) genome assemblies and annotation of large gene families of conifer terpenoid and phenolic defense metabolism. *Plant Journal, 83*, 189–212. https://doi.org/10.1111/tpj.12886.

Watanabe, Y. (2011). Overview of plant RNAi. In H. Kodama & A. Komamine (Eds.), *Vol. 744. Methods in molecular biology (Clifton, N.J.)*: (pp. 1–11). Totowa, NJ: Humana Press. https://doi.org/10.1007/978-1-61779-123-9_1.

Wegrzyn, J. L., Lee, J. M., Tearse, B. R., & Neale, D. B. (2008). TreeGenes: A forest tree genome database. *International Journal of Plant Genomics, 2008*, 412875. https://doi.org/10.1155/2008/412875.

Wegrzyn, J. L., Liechty, J. D., Stevens, K. A., Wu, L. S., Loopstra, C. A., Vasquez-Gross,-H. A., et al. (2014). Unique features of the loblolly pine (*Pinus taeda* L.) megagenome revealed through sequence annotation. *Genetics, 196*, 891–909. https://doi.org/10.1534/genetics.113.159996.

West, P. W. (2014). *Growing plantation forests. Growing plantation forests (second)*. New Delhi, India: Springer International Publishing. https://doi.org/10.1007/978-3-319-01827-0.

Whetten, R., Sun, Y. H., Zhang, Y., & Sederoff, R. (2001). Functional genomics and cell wall biosynthesis in loblolly pine. *Plant Molecular Biology, 47*, 275–291. https://doi.org/10.1023/A:1010652003395.

Wickett, N. J., Mirarab, S., Nguyen, N., Warnow, T., Carpenter, E., Matasci, N., et al. (2014). Phylotranscriptomic analysis of the origin and early diversification of land plants. *Proceedings of the National Academy of Sciences of the United States of America, 111*, E4859–E4868. https://doi.org/10.1073/pnas.1323926111.

Xie, W., Huang, A., Li, H., Feng, L., Zhang, F., & Guo, W. (2017). Identification and comparative analysis of microRNAs in *Pinus massoniana* infected by *Bursaphelenchus xylophilus*. *Plant Growth Regulation, 83*, 223–232. https://doi.org/10.1007/s10725-016-0221-8.

Yakovlev, I. A., & Fossdal, C. G. (2017). *In silico* analysis of small RNAs suggest roles for novel and conserved miRNAs in the formation of epigenetic memory in somatic embryos of Norway spruce. *Frontiers in Physiology, 8*, 674. https://doi.org/10.3389/fphys.2017.00674.

Yakovlev, I. A., Fossdal, C. G., Johnsen, Ø., Junttila, O., & Skrøppa, T. (2006). Analysis of gene expression during bud burst initiation in Norway spruce via ESTs from subtracted cDNA libraries. *Tree Genetics & Genomes, 2*, 39–52. https://doi.org/10.1007/s11295-005-0031-z.

Yang, I. S., & Kim, S. (2015). Analysis of whole transcriptome sequencing data: Workflow and software. *Genomics & Informatics*, *13*, 119–125. https://doi.org/10.5808/GI.2015.13.4.119.

Yeaman, S., Hodgins, K. A., Suren, H., Nurkowski, K. A., Rieseberg, L. H., Holliday, J. A., et al. (2014). Conservation and divergence of gene expression plasticity following c. 140 million years of evolution in lodgepole pine (*Pinus contorta*) and interior spruce (*Picea glauca* × *Picea engelmannii*). *The New Phytologist*, *203*, 578–591. https://doi.org/10.1111/nph.12819.

Zhang, J., Wu, T., Li, L., Han, S., Li, X., Zhang, S., et al. (2013). Dynamic expression of small RNA populations in larch (*Larix leptolepis*). *Planta*, *237*, 89–101. https://doi.org/10.1007/s00425-012-1753-4.

Zhu, T., Moschou, P. N., Alvarez, J. M., Sohlberg, J. J., & Arnold, S. (2016). *WUSCHEL-RELATED HOMEOBOX 2* is important for protoderm and suspensor development in the gymnosperm Norway spruce. *BMC Plant Biology*, *16*, 19. https://doi.org/10.1186/s12870-016-0706-7.

Zhu, T., Moschou, P. N., Alvarez, J. M., Sohlberg, J. J., & Von Arnold, S. (2014). *WUSCHEL-RELATED HOMEOBOX 8/9* is important for proper embryo patterning in the gymnosperm Norway spruce. *Journal of Experimental Botany*, *65*, 6543–6552. https://doi.org/10.1093/jxb/eru371.

Zimin, A., Stevens, K. A., Crepeau, M. W., Holtz-Morris, A., Koriabine, M., Marçais, G., et al. (2014). Sequencing and assembly of the 22-Gb loblolly pine genome. *Genetics*, *196*, 875–890. https://doi.org/10.1534/genetics.113.159715.

Zimin, A. V., Stevens, K. A., Crepeau, M. W., Puiu, D., Wegrzyn, J. L., Yorke, J. A., et al. (2017). An improved assembly of the loblolly pine mega-genome using long-read single-molecule sequencing. *GigaScience*, *6*, 1–4. https://doi.org/10.1093/gigascience/giw016.

CHAPTER THREE

The ectomycorrhizal contribution to tree nutrition

Adeline Becquer[a,b], Carmen Guerrero-Galán[c,†], Janice L. Eibensteiner[d], Gabriella Houdinet[c], Heike Bücking[d], Sabine D. Zimmermann[c], Kevin Garcia[e,*]

[a]CSIRO Agriculture, Canberra, ACT, Australia
[b]Eco&Sols, Univ. Montpellier, CIRAD, INRA, IRD, Montpellier SupAgro, Montpellier, France
[c]BPMP, Univ Montpellier, CNRS, INRA, SupAgro, Montpellier, France
[d]Biology and Microbiology Department, South Dakota State University, McFadden Biostress, Brookings, SD, United States
[e]Department of Crop and Soil Sciences, North Carolina State University, Raleigh, NC, United States
*Corresponding author: e-mail address: kgarcia2@ncsu.edu

Contents

† Present address: Centro de Biotecnología y Genómica de Plantas, Instituto Nacional de Investigación y Tecnología Agraria y Alimentación (INIA), Universidad Politécnica de Madrid (UPM), 28223 Pozuelo de Alarcón, Spain

Advances in Botanical Research, Volume 89
ISSN 0065-2296
https://doi.org/10.1016/bs.abr.2018.11.003

Abstract

Trees can be associated with dozens of fungi helping them to acquire resources from forest soils. The most widespread mutualistic association in boreal and temperate forests is the ectomycorrhizal symbiosis. This symbiosis involves mushroom-forming fungi of basidiomycota, ascomycota, and some zygomycota clades and the roots of woody plant species, including oaks, poplars or pines. Although the impact of this association on ecosystem production and tree nutrition is investigated for about a century, our understanding on the molecular mechanisms that control water and nutrient fluxes between plant and fungal partners is still limited. Here, we review the recent knowledge on the ectomycorrhizal contribution to tree nutrition. We specifically highlight the molecular mechanisms driving the acquisition, translocation and release of water and nutrients in ectomycorrhizal systems. We particularly focus on the transport of macronutrients, including nitrogen, phosphorus, potassium, sulphur and calcium, micronutrients, and water by the symbiotic partner. We also provide background on the evolution, diversity, and importance of this symbiosis, identify knowledge gaps, and propose future research directions.

1. Introduction

The hydro-mineral nutrition of trees from boreal and temperate forests greatly depends on the mycelium of mycorrhizal fungi growing in the soil. These fungi can explore with their extra-radical mycelium a larger volume of soil than roots, take up water and nutrients from the soil, and translocate them to the intraradical mycelium inside of tree roots (Smith & Read, 2008). Among the different types of mycorrhizal associations, the ectomycorrhizal (ECM) symbiosis is the most important for forests of the northern hemisphere. ECM fungi are evolutionary closely related to saprotrophic fungi and mainly belong to the basidiomycota clade, but also include some ascomycota and zygomycota. Only 3–5% of land plants form this interaction, but ECM fungi play a critical role for forest ecosystems because they colonise up to 95% of tree short roots (Martin et al., 2001) and actively participate in nutrient cycling and carbon sequestration (Clemmensen et al., 2013; Marmeisse & Girlanda, 2016). Consequently, ECM associations greatly influence many forest ecosystems across temperate, boreal and some subtropical regions all around the globe (Finlay, 2008; Smith & Read, 2008). The majority of woody plants with high economic value are highly dependent on the ECM symbiosis, making a better understanding of these interactions critical for the establishment of a successful commercial agroforestry.

Plant and fungal tissues form ectomycorrhizas and mutually benefit from this interaction through the exchange of "goods". The fungus provides the plant with water, macronutrients, and micronutrients, whereas the host transfers photosynthetically fixed carbon to its mycobionts. Although this mutualistic exchange of resources has been extensively described over the past decades, key information is still largely unknown. For example, we have only little information about (i) the proteins governing the acquisition of resources by the mycelium, and their release from the fungal cells to the plant roots, (ii) the functional characterisation and regulation of these transport proteins, and (iii) the molecular mechanisms driving the translocation of water and nutrients from extra-radical hyphae exploring the soil to intraradical mycelium in tree roots. Even though an increasing amount of genomic (Branco et al., 2015; Kohler et al., 2015; Martin et al., 2008, 2010; Martino et al., 2018; Mujic et al., 2017; Peter et al., 2016), transcriptomic (Doré et al., 2017; Hacquard et al., 2013), proteomic (Doré et al., 2017, 2015; Liang, Chen, Tang, & Shen, 2007) and metabolomic (Larsen et al., 2011) data is available, the validation of candidate players participating in resource exchange is hindered by the difficulty to genetically transform fungal and tree partners and the limited molecular tools that are currently available.

Here, we summarise the most recent knowledge on the diversity of this association and the transport of macronutrients (nitrogen, phosphorus, potassium, sulphur and calcium), micronutrients (primarily zinc), and water from the soil to the colonised tree roots. Aiming at highlighting the impact of ECM fungi on tree nutrition, we decided not to cover the transport of carbon from the plant to the fungal partners, since it was recently reviewed in a number of publications (Casieri et al., 2013; Courty, Doidy, Garcia, Wipf, & Zimmermann, 2016; Garcia, Doidy, Zimmermann, Wipf, & Courty, 2016). We particularly emphasise the molecular players of fungal and plant partners and the mechanisms that drive the hydro-mineral transport from soil to roots and propose research directions and future challenges in the study of ECM tree nutrition.

2. Evolution and diversity of ectomycorrhizas

2.1 Saprophytic to ectomycorrhizal fungal continuum

Fossil and molecular clock analyses suggest that ECM associations evolved about 100–200 million years ago (Heijden, Martin, Selosse, & Sanders, 2015; Kohler et al., 2015; Strullu-Derrien, Selosse, Kenrick, & Martin, 2018; Tedersoo, May, & Smith, 2010). Evidence from molecular, palaeographical, and palaeontological studies suggest that the evolution of ECM fungi approximately 180 million years ago coincided with (i) the origin of the Pinaceae lineage, which may represent the first group of ECM hosts, and (ii) the rapid diversification of angiosperms in the Jurassic and early Cretaceous period (Brundrett, 2002; Moyersoen, 2006; Strullu-

Derrien et al., 2018). Conversely, it is likely that the ability to interact with ECM fungi was a key step for the emergence of new host tree species and for the structure of ancient ecosystems (Quirk, Andrews, Leake, Banwart, & Beerling, 2014). Phylogenetic and phylogenomic studies revealed that ECM fungi have evolved independently 78–82 times from free-living saprotrophs in multiple fungal lineages (Eastwood et al., 2011; Floudas et al., 2012; Hibbett, Gilbert, & Donoghue, 2000; James et al., 2006; Tedersoo & Smith, 2017).

The shift from a saprophytic to a more mutualistic lifestyle was the result of a convergent loss of genes involved in the degradation of lignocellulose, including plant cell wall-degrading enzymes (PCWDEs), class II peroxidases or lignin oxidoreductases (Garcia & Ané, 2016; Kohler et al., 2015; Pellegrin, Morin, Martin, & Veneault-Fourrey, 2015; Tedersoo et al., 2010; Wolfe, Tulloss, & Pringle, 2012). Kohler et al. (2015) demonstrated that saprotrophic fungi, white-rot and litter decomposers, display on average 133 genes for PCWDEs, whereas brown-rot species have 81 genes and ECM species only 62 genes. Consequently, ECM fungi are generally less capable of utilising complex carbohydrates from forest soils, increasing their dependence for host plant photo-assimilates as a source of carbon (Section 3.1).

In addition, transcriptional profiling also revealed that the transition towards an ECM lifestyle resulted from the emergence of genes specifically expressed and/or regulated during the interaction with a host species. For example, Doré et al. (2017) reported that 3.6% of the genes from the ECM fungus *Hebeloma cylindrosporum* were significantly up-regulated during the symbiotic association with *Pinus pinaster*. Among them, many genes encode fungal effectors putatively involved in the molecular cross-talk between both symbionts and manipulation of plant defence responses (reviewed in Garcia, Delaux, Cope, & Ané, 2015; Martin, Uroz, & Barker, 2017; Mello & Balestrini, 2018), PCWDEs, and proteins responsible for the transport of nutrients (reviewed in Casieri et al., 2013; Garcia et al., 2016). Interestingly, orthologues of a large set of these up-regulated genes are also found in saprotrophic species, indicating their origin in saprotrophic fungi and their conservation by ECM fungi (Kohler et al., 2015). However, their precise role in the ECM symbiosis remains to be determined.

2.2 Diversity of the ectomycorrhizal symbiosis

Based on morphological, molecular and isotopic studies estimates, around 20,000 fungal species can form ectomycorrhizas (Tedersoo et al., 2012).

Although many of them are partly saprotrophic, they often rely on photosynthetically fixed carbon from their associated host trees for their sexual reproduction through the production of carpophores (Fig. 1). It has been estimated that around 6000 woody plant species from 30 families can interact with ECM fungi, including many economically important crop trees such as pines, beeches, oaks, eucalypts, dipterocarps or poplars (Heijden et al., 2015). Strict host specificity is rare in ECM fungi (Churchland & Grayston, 2014), and many ECM fungi can simultaneously colonise dozens of plants (Horton & van der Heijden, 2008). Interestingly, recent evidence also illustrated that ECM host specificity patterns depend on the mode of colonisation (Lofgren, Nguyen, & Kennedy, 2018), and that tree genetics influences the composition of ECM communities (Gehring, Sthultz, Flores-Rentería, Whipple, & Whitham, 2017). A low specificity can provide plants with a competitive advantage in the uptake of nutrients in comparison to non-host species, or plant species with a high specificity. This means that trees in forests are interconnected through the mycelium of many ECM fungi forming a living belowground network called the "Wood-Wide Web" (Newman, 1988; Simard et al., 1997).

3. Structure of ectomycorrhizas

3.1 Root colonisation by ectomycorrhizal fungi

Although ECM fungi penetrate into the roots of their host, their hyphae never enter the host cells, and the mycelium develops intercellularly in the root cortex (Fig. 1). Ectomycorrhizal roots can be distinguished into three structural components: extra-radical hyphae exploring the soil, hyphal mantle or sheath that encloses the root, and intra-radical hyphae, or the Hartig net, that forms a specialised interface with the host root (Fig. 2; Bonfante & Genre, 2010; Peterson & Massicotte, 2004). Interestingly, Hacquard et al. (2013) reported distinct transcriptional profiles between these three types of hyphae, indicating that all have different functions during the symbiosis.

Development of a functioning ectomycorrhiza follows a series of distinct steps that have extensively been reviewed (Balestrini & Kottke, 2016; Smith & Read, 2008). In brief, the symbiosis starts with an exchange of diffusible signals between host roots and the fungus, followed by the attachment of the hyphae to root surface, and the development of the fungal mantle. Finally, hyphae invade the root cortex and establish the Hartig net where nutrients are exchanged between both partners (Fig. 2). Fungal development within the host root is accompanied by changes in root architecture and

Fig. 1 Fruiting bodies of three model ECM fungi and ectomycorrhizas formed with their host tree. (A) *Laccaria bicolor* fruiting bodies under *Pseudotsuga menziesii* at Toll Bridge Park, Mount Hood, OR, USA, and (B) ectomycorrhizas after inoculation of *Populus trichocarpa* (Oak Ridge National Laboratory, Oak Ridge, TN, USA). (C) *Hebeloma cylindrosporum* fruiting bodies under *Pinus* sp. in the Landes forest, France, and (D) ectomycorrhiza in association with *Pinus pinaster* (Biochemistry and Plant Molecular Physiology Laboratory in Montpellier, France). (E) Fruiting body of the invasive European *Amanita muscaria* under *Quercus humboldtii* in Colombia, and (F) ectomycorrhizas on the same host species. *Images courtesy of Jay Chen and Jesse Labbé (A, B—Oak Ridge National Laboratory, USA), Claude Plassard and Jacques Guinberteau (C—French National Institute for Agricultural Research, INRA, France), Natalia Vargas Estupiñan and Anne Pringle (E, F—Universidad de Los Andes, Colombia, and University of Wisconsin-Madison, USA, respectively).*

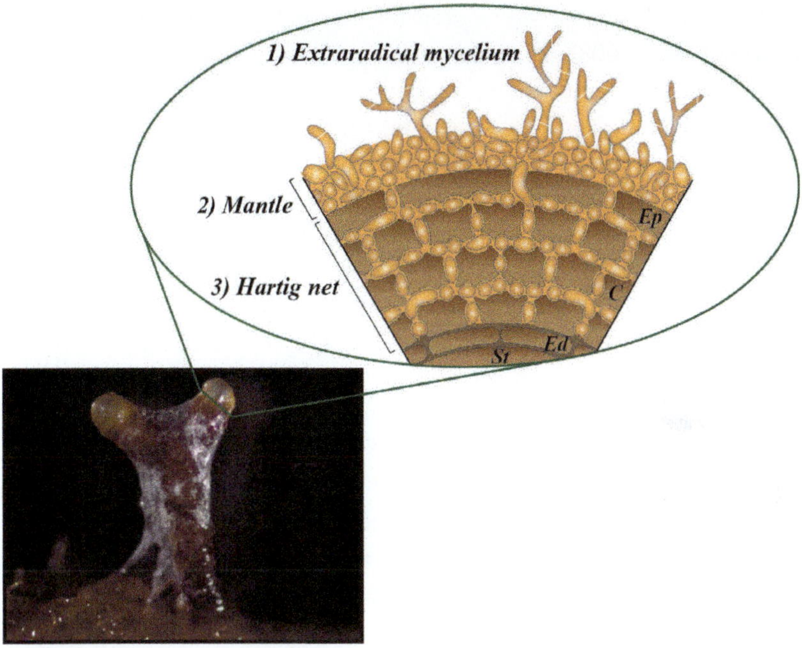

Fig. 2 Structure of a plant root colonised by an ectomycorrhizal fungus. Ectomycorrhizas can be divided into three structural components: (1) extra-radical mycelium with hyphae that explore the soil for nutrients and water, (2) hyphal mantle or sheath that encloses the root, and can represent a significant apoplastic barrier, and (3) the Hartig net in the root cortex, the site of nutrient and water exchange between both symbiotic partners. Abbreviations: *C*, cortex; *Ed*, endodermis; *Ep*, epidermis; *St*, stele.

morphology of the host by a reorganisation of the plant cytoskeleton (Timonen, Finlay, Olsson, & Söderström, 1996), and an increased development of lateral roots, and an inhibition of root hair formation (Ditengou, Béguiristain, & Lapeyrie, 2000).

3.2 The soil-fungus interface

The extra-radical mycelium of ECM fungi can represent up to 32% of total microbial biomass and 700–900 kg ha^{-1} in forest soils (Wallander, Nilsson, Hagerberg, & Bååth, 2001). It can be composed of individual hyphae with similar structure, or differentiated into rhizomorphs, complex aggregates of parallel hyphae (Agerer, 2001) that are involved in the long-distance transport of water and nutrients from hyphal tips to host roots. The structure of rhizomorphs can range from loose arrangements of relatively undifferentiated hyphae to aggregates of highly differentiated hyphae (Smith & Read, 2008).

Another important feature of ectomycorrhizas is the fungal mantle encompassing the roots. Its structural composition ranges from relatively thin, loosely arranged assemblages of hyphae, to very thick, multi-layered and pseudoparenchymatous mantles. The surface of the mantle can be compact and smooth, or rough with numerous emerging hyphae, hyphal strands, or rhizomorphs (Agerer, 2001). The mantle isolates colonised roots from the soil, protects them against desiccation and pathogens, and can play a role in nutrient acquisition (Hacquard et al., 2013). Hyphal mantles also represent a significant apoplastic barrier for the entry of nutrients into the root cortex and can put nutrient uptake under the control of the fungal symplast (Behrmann & Heyser, 1992; Bücking, Kuhn, Schröder, & Heyser, 2002). Although poorly understood, the fungal mantle plays a key role in storage and partitioning of carbohydrates from the host plant, and nutrients acquired from the soil by extra-radical hyphae (Bücking, Liepold, & Ambilwade, 2012; Vesk, Ashford, Markovina, & Allaway, 2000).

3.3 The symbiotic plant-fungus interface

The intra-radical network of hyphae, or Hartig net, surrounds epidermal root cells in angiosperms, or both epidermal and cortical roots cells in gymnosperms up to the endodermis (Peterson & Massicotte, 2004; Smith & Read, 2008). Hyphae from the Hartig net display a complex and dense labyrinthine architecture, increasing the contact area between both partners and form an efficient interface for the bidirectional transport of nutrients (Finlay, 2008; Peterson & Massicotte, 2004). In the Hartig net, plant and fungal cells share a common apoplast in which solutes can freely move from one partner to another. The plasma membranes of both partners in the Hartig net show high activities of ATPases (Lei & Dexheimer, 1988) and many specific or up-regulated transport systems (Garcia et al., 2016), facilitating active or passive transport of nutrients and water across this mycorrhizal interface.

4. Importance of ectomycorrhizal fungi for forest ecosystems

ECM networks play a key role in the biochemical cycles of forest ecosystems and actively participate in carbon sequestration and mineral cycling (Clemmensen et al., 2013). For example, ECM fungi can contribute with up to 80% to the total plant nitrogen (N) and phosphorus (P) uptake, increasing plant productivity (Casarin, Plassard, Hinsinger, & Arvieu, 2004; Chalot & Brun, 1998; Chalot et al., 2002; Perez-Moreno & Read, 2000; Torres Aquino & Plassard, 2004; Van Tichelen & Colpaert, 2000). ECM fungi can also reduce nutrient losses by leaching and denitrification (Midgley & Phillips, 2014). Besides their role in nutrient cycling, the occurrence of ECM associations can drastically improve seedling establishment (Nara, 2006), litter decomposition (Lindahl et al., 2006), soil formation, soil aggregation (Rillig & Mummey, 2006) and provides tolerance to numerous biotic and abiotic stresses (reviewed in Finlay, 2008). However, these effects are often variable depending on the fungal and plant species and environmental conditions (Heijden et al., 2015).

Plant availability of nutrients in boreal and temperate forest ecosystems is usually low and many nutrients are in organic form and integrated in humus rich soil organic matter (SOM) and litter in different stages of decomposition (Read & Perez-Moreno, 2003). SOM is composed of relatively small biomolecules originating from the decomposition of plant and microbial biomass such as lignocellulosic components (Berg & McClaugherty, 2014; Simpson et al., 2007). ECM fungi were long seen primarily as mutualists that absorb nutrients released during litter decomposition by saprotrophic fungi (Read & Perez-Moreno, 2003). However, recent analyses of fungal genomes revealed that ECM fungi still have the ability, even if limited, to decompose complex SOM (Kohler et al., 2015). For example, *H. cylindrosporum* and *Cortinarius glaucopus* still harbour genes coding for class II peroxidases involved in lignin modification and degradation (Kohler et al., 2015; Martin et al., 2017). Field and microcosm studies have shown that ECM systems can access nutrients sequestered in cellulose and polyphenols (Bending & Read, 1995; Read & Perez-Moreno, 2003). ECM fungi are also able to intensively exploit litter to reduce and uptake N and P (Perez-Moreno & Read, 2000), indicating that they are still able to degrade SOM.

Studies on saprotrophic fungi revealed that white-rot and brown-rot fungi use distinct mechanisms to decompose lignocellulose (Hatakka & Hammel, 2010). White-rot fungi rely on enzymatic systems for the degradation of lignocellulose, whereas brown-rot fungi employ a non-enzymatic mechanism due to their lack of oxidative enzymes and PCWDEs (Eastwood et al., 2011; Floudas et al., 2012). This non-enzymatic mechanism, called the Fenton reaction ($H_2O_2 + Fe^{2+} + H^+ \rightarrow H_2O + Fe^{3+} + OH$), allows the release of sequestered carbon without the need for complete lignin degradation (Arantes, Jellison, & Goodell, 2012). Interestingly, the ECM fungus

Paxillus involutus can also assimilate organic N from the decomposition of lignocellulosic material in SOM through a Fenton-based oxidation mechanism (Op De Beeck, Troein, Peterson, Persson, & Tunlid, 2018; Rineau et al., 2012). However, ECM fungi are not able to acquire significant amounts of carbon from SOM, and primarily rely on the carbon provided by their host (Lindahl & Tunlid, 2015). Rineau et al. (2013) demonstrated that adding glucose to organic material stimulated the expression of transcripts for enzymes involved in the Fenton reaction. This indicates that ECM fungi use carbon resources derived from the host to mobilise N and P trapped in recalcitrant organic matter complexes (Tunlid, Floudas, Koide, & Rineau, 2016). This is further supported by both the overall reduction of genes coding for PCWDEs in ECM species (Kohler et al., 2015) and field studies showing that less than 2% of carbon in the fungal biomass originated from litter decomposition (Treseder, Torn, & Masiello, 2006). However, some studies demonstrated that ECM fungi express genes encoding for carbohydrate-active enzymes (CAZymes) upon growth on SOM (Shah et al., 2015). In conclusion, the importance of SOM decomposition for the acquisition of carbon is still under debate for ECM fungi, and further observations are necessary to determine the contribution of carbon derived from SOM decomposition to the nutrition of ECM fungi.

5. The hydro-mineral nutrition of ectomycorrhizal trees

5.1 The symbiotic acquisition of nitrogen

N is one of the most important growth-limiting factors in boreal and temperate forests, and ECM fungi play a critical role in its acquisition, assimilation and transport to their host plants (Franklin, Näsholm, Högberg, & Högberg, 2014). N is available in the soil in an organic form such as peptides or amino acids from decomposing plant or animal matter, or in an inorganic form through N fixation, precipitation, or surface runoff. Proteins for the uptake of inorganic and organic N sources have been characterised in both extra-radical and intra-radical hyphae of ectomycorrhizas, highlighting the importance of ECM for tree N nutrition (Fig. 3).

5.1.1 Inorganic nitrogen uptake by fungal hyphae

5.1.1.1 Ammonium transporters

Ammonium (NH_4^+) is the major source of N for ECM fungi even though the soil concentration of NH_4^+ in forest soils is generally lower than that of nitrate (NO_3^-). NH_4^+ is often preferred by ECM fungi due to its lower

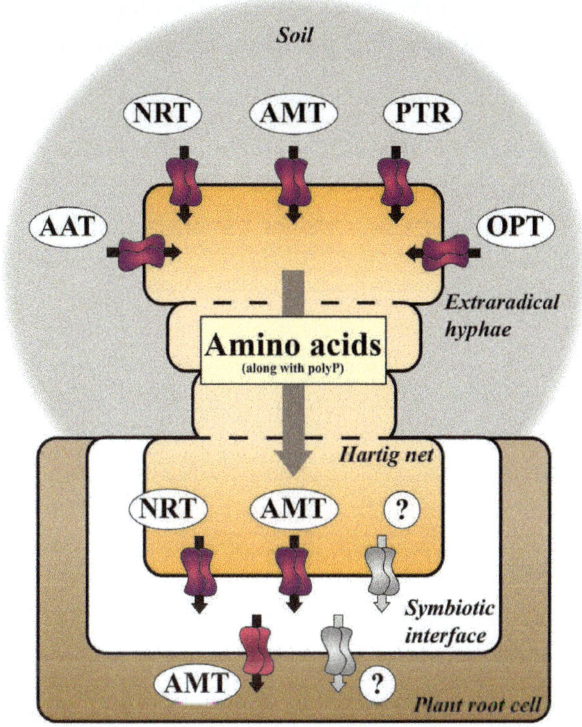

Fig. 3 Schematic representation of the molecular players governing the transport of nitrogen in ectomycorrhizas. Transport proteins are localised in the extra-radical hyphae for the uptake of ammonium (AMT), nitrate (NRT), amino acids (AAT), peptides (PTR) and oligopeptides (OPT). After uptake, nitrogen is translocated from extra-radical hyphae to the Hartig net in vacuoles presumably associated with polyphosphates (polyP). Fungal nitrate (NRT) and ammonium (AMT) transporters are localised in the fungal membrane and release nitrogen into the symbiotic interface. Other forms of nitrogen could also be transferred into the mycorrhizal interface through currently unidentified proteins. Nitrogen is taken up by plant cortical cells from the symbiotic interface via ammonium transporters (AMT). Other forms of nitrogen could also be released to the interface and taken up, but the transporters are currently unknown.

assimilation costs compared to the reduction of NO_3^- to NH_4^+ (Courty, Smith, Koegel, Redecker, & Wipf, 2015). Hyphal NH_4^+ acquisition can account for up to 45% of the total plant N uptake under N deficiency (Javelle, André, Marini, & Chalot, 2003). ECM fungi express high- and low-affinity transporters of the AMT family that take up NH_4^+ under low or high N supply conditions, respectively (Fig. 3; Garcia et al., 2016).

A bioinformatical analysis suggests the existence of eight putative AMTs, three high-affinity and five low-affinity transporters, in the transcriptome of *Laccaria bicolor* (Lucic et al., 2008). Laser capture microdissection and microarray analysis demonstrated three putative NH_4^+ uptake transporters in ectomycorrhizas of *Tuber melanosporum*, and one of these transporters is specifically upregulated in the fungal mantle, but down-regulated in the Hartig net (Hacquard et al., 2013). High-affinity AMTs have been described for several ECM fungi, including *H. cylindrosporum* (*HcAMT1* and *HcAMT2*: Javelle, André, et al., 2003; Javelle, Morel, et al., 2003; Javelle et al., 2001), *Amanita muscaria* (*AmAMT2*: Willmann, Weiß, & Nehls, 2007), and *Tuber borchii* (*TbAMT1*: Montanini, Moretto, Soragni, Percudani, & Ottonello, 2002).

In general, the expression levels of these transporters are regulated by the external N supply conditions. The high-affinity NH_4^+ transporter of *A. muscaria AmAMT2*, for example, is highly expressed under low N supply conditions, but down-regulated when easily metabolisable N sources, such as NH_4^+ or amino acids, are readily available (Willmann et al., 2007). Transcript levels of *AmAMT2* are also high when NO_3^- is supplied, indicating that *A. muscaria* cannot utilise NO_3^- as a N source. The fact that *AmAMT2* is highly expressed in the extra-radical mycelium but down-regulated in both mantle and Hartig net of ECM poplar roots indicates that AmAMT2 is involved in NH_4^+ acquisition from the soil but not in its transfer towards the host (Willmann et al., 2007). The expression levels of *AMT1* and *AMT2* of *Hebeloma crustuliniforme* are not only controlled by the external availability of N sources but also by the internal levels of glutamine (Javelle, Morel, et al., 2003). Glutamine serves as an essential amino N donor for the synthesis of many essential metabolites, such as nucleic acids, amino sugars, histidine, tyrosine, asparagine, and various co-factors. The intracellular concentrations of glutamine have also been shown in *Arabidopsis* and *Saccharomyces cerevisiae* to repress the expression of AMTs (Rawat, Silim, Kronzucker, Siddiqi, & Glass, 1999; Slaughter, McKernan, & Saita, 1990). In contrast, the low-affinity AMT of *H. cylindrosporum HcAMT3* is not affected by the internal glutamine levels, what could ensure a basal level of NH_4^+ uptake under different N supply conditions (Javelle, Morel, et al., 2003).

Interestingly, some aquaporins from *L. bicolor* (LbAQPs) that are normally involved in water transport (Section 5.5) can also facilitate the membrane transport of other small molecules, such as urea and NH_4^+/NH_3^- (Dietz, von Bülow, Beitz, & Nehls, 2011; Selle et al., 2005). Candidate aquaporins that could play a role for N uptake were also found in *T. melanosporum* (Hacquard et al., 2013).

5.1.1.2 Nitrate transporters

Only a few NO_3^- transporters have so far been identified in ECM fungi (Fig. 3). Their low representation in fungal genomes is likely due to the fungal preference for NH_4^+. NH_4^+ is generally energetically favoured by fungi, since NO_3^- first needs to be reduced to nitrite and eventually to NH_4^+ by the activity of NO_3^- and nitrite reductases, respectively. Fungal uptake of secondary N sources, such as NO_3^-, is closely regulated by the fungal high affinity NO_3^- assimilation cluster (fhant-AC), which consists of a high-affinity transporter (*Nrt*), a reductase (*nr*), and a ferredoxin-independent nitrite reductase (NAD(P)-H-*nir*) (Jargeat, Gay, Debaud, & Marmeisse, 2000; Jargeat et al., 2003; Kemppainen & Pardo, 2013). All genes within this cluster share the same regulation pattern that is generally characterised by a down-regulation of their transcript levels by NH_4^+. NO_3^- availability on the other hand leads to an up-regulation of the fhant-AC gene cluster but does not neutralise its strong suppression by NH_4^+ (Kemppainen, Alvarez Crespo, & Pardo, 2010).

HcNRT2 of *H. cylindrosporum* was the first high-affinity NO_3^- transporter that was characterised in ECM fungi. Consistent with the general characteristics of the fhant-AC, *HcNRT2* is not induced by NO_3^- but is suppressed by an external supply of NH_4^+ (Jargeat et al., 2003). Interestingly, fungal strains with only a minimal detectable transcription of *LbNrt* after silencing were affected in their colonisation of poplar and showed only a loose hyphal growth on the root surface, and no Hartig net development (Kemppainen & Pardo, 2013). This illustrates that despite the fact that ECM fungi generally prefer NH_4^+ over NO_3^-, expression of this transporter and an uncompromised N metabolism of the fungal partner is necessary for the stability of the ECM symbiosis.

TbNRT2 is a high-affinity NO_3^- transporter of *T. borchii* with a transport activity also for nitrite. *TbNRT2* is up-regulated by NO_3^- and nitrite but also by N starvation through a NO_3^- independent de-repression mechanism. In contrast, the adjacent NO_3^- reductase in the same gene cluster is only expressed at low levels under N-starvation, and requires NO_3^- for induction (Montanini et al., 2006). This dual expression pattern of *TbNRT* could provide *T. borchii* with an advantage for N uptake under fluctuating N supply conditions. Though the NO_3^- independent gene expression pattern would lead to higher costs for NO_3^- assimilation when NH_4^+ is available, it would also ensure a certain level of N uptake when both inorganic N sources are limited.

Many ECM fungi have the ability to grow on media with NO_3^- as the sole N source (Nygren et al., 2008), but the contribution of NO_3^- to tree nutrition is still under debate. Since ECM fungi generally favour NH_4^+ as N source, it was generally believed that NO_3^- plays only a minor role for the N nutrition of trees. However, NO_3^- has a higher mobility in soils than other N sources, and a recent study suggests that NO_3^- may play a more prominent role for N nutrition than previously been expected. In the presence of mass flow, NO_3^- can represent the dominating form of mobile N even in nutrient rich soils, and plant transpiration can be an important driver for NO_3^- fluxes and N acquisition of boreal forests (Oyewole, Inselsbacher, Näsholm, & Jämtgårda, 2017).

5.1.2 Organic nitrogen uptake by fungal hyphae

5.1.2.1 Amino acid transporters

In addition to NH_4^+ and NO_3^-, other forms of N sources can be taken up and assimilated by fungal hyphae and transferred to host plants through the fungal mantle and Hartig net, including amino acids, peptides and other organic N sources (Fig. 3). Amino acids can dominate soil N fluxes in both nutrient rich and poor forest ecosystems and can thereby form an important N source for ECM fungi (Oyewole et al., 2017). Recently, it has been shown that N deposition and the shift from inorganic to organic N sources in the soil is a determining factor for the diversity of ECM communities (van der Linde et al., 2018). Fungi that use organic N, such as *Cortinarius*, *Piloderma* and *Tricholoma* species, show a low abundance at sites with high N deposition, while species that rely mainly on inorganic N sources, such as *Laccaria* and *Elaphomyces* species, become dominant species in these ECM communities.

AmAPP1 was the first amino acid transporter that was identified in the ECM fungus *A. muscaria* (Nehls, Kleber, Wiese, & Hampp, 1999). *AmAPP1* encodes a general amino acid transporter with a preference and high affinity for basic and aromatic amino acids and a lower affinity for acidic or neutral amino acids. Furthermore, its expression is enhanced by the absence of inorganic N sources but repressed when inorganic N sources are available (Nehls et al., 1999). *HcGAP1* from *H. cylindrosporum* codes for a general amino acid permease, like *AmAAP1*, and is down-regulated under high N supply but up-regulated under N limitation (Wipf, Benjdia, Tegeder, & Frommer, 2002). Most fungal amino acid transporters belong to the amino-acid-polyamine-organocation (APC) superfamily that mediates the transfer of a broad spectrum of amino acids (Casieri et al., 2013). Bioinformatical analysis revealed an expansion of the APC superfamily

(29 gene models of 4 different families) in *L. bicolor*, compared to basidiomycetes with a saprobic or parasitic lifestyle. This could indicate a higher capability of ECM fungi to utilise organic N sources, including amino acids, and could be related to their dual lifestyle (symbiotic vs. saprobic) (Lucic et al., 2008).

5.1.2.2 Peptide transporters

H. cylindrosporum can take up di- and tripeptides and utilise them as sole N source (Benjdia et al., 2006). After uptake, peptides can be rapidly converted by peptidases and utilised as source for N, amino acids or carbon. Several fungal peptide and oligopeptide transporters (PTRs and OPTs, respectively) have been identified in genomes of ECM fungi, and two di- and tripeptide transporters of *H. cylindrosporum*, *HcPTR2A* and *HcPTR2B*, have so far been functionally characterised (Benjdia et al., 2006). *HcPTR2A* is involved in high-affinity uptake of peptides under N limiting conditions. *HcPTR2A* is strongly expressed under N limitation, or when only NO_3^- is available as N source, but down-regulated when other organic sources become available. *HcPTR2A* and *HcGAP1* (Wipf et al., 2002) show similar expression patterns and are both controlled by a system that senses external N availabilities, but are also regulated by a feedback mechanism that is dependent on internal amino acid pools. In contrast, *HcPTR2B* is constitutively expressed, and ensures a basal level of peptide uptake activity under various N supply conditions (Benjdia et al., 2006). In the same fungus, the oligopeptide transporter *HcOPT1* is also constitutively expressed, but this transporter is not fully characterised yet (Lambilliotte et al., 2004; Müller et al., 2007).

In the genome of *L. bicolor*, two genes coding for putative peptide transporters and eight for oligopeptide transporters were also found (Lucic et al., 2008). Similar to *HcPRT2b*, both peptide transporters are constitutively expressed, indicating that they could facilitate a basal uptake of peptides under various N supply conditions. Of the nine putative oligopeptide transporters (*LbOPTs*), four of them are constitutively expressed in all tissues, two are specifically upregulated in fungal fruitbodies, and two in fruitbodies and ECM mycelium. This suggests that all fungal hyphae might have the capability for peptide uptake either for fungal nutrition or for tissue redistribution (Lucic et al., 2008).

5.1.2.3 Urea transporters

Urea is another important nutrient in forest soils and one of the most abundant nitrogenous compounds found in ECM fungi (Smith & Read, 2008).

Urea can represent up to 73% of the total soluble N compounds in extra-radical mycelium of the *P. involutus–Betula pendula* symbiosis under acidic soil conditions or when N is limited (Morel et al., 2005). Urea is first metabolised to $NH_4{}^+$ by urease activity and then converted into glutamine by glutamine synthetase (Morel et al., 2008). Microarrays showed that the urease coding gene was up-regulated in *L. bicolor* (Martin et al., 2008). *PiDUR3* encodes a urea transporter of *P. involutus*, and has been characterised as a permease, transmembrane transport protein of the sodium: solute symporter (SSS) family. This transporter family catalyses the uptake of a wide range of solutes, such as sugars, amino acids, nucleosides, urea or anions via Na^+ symport. Urea uptake activity by this transporter is highly pH-dependent, and external protons stimulate urea transport (Morel et al., 2008). High glutamine levels in the cell lead to lower transcript levels and a catabolic repression of *PiDUR3* (Morel et al., 2008). The repression of *PiDUR3* by high intracellular glutamine concentrations reduces the import of urea when $NH_4{}^+$, the favoured N source, is available (Morel et al., 2008). In *T. melanosporum*, three putative transporters were identified that could participate in the transport of $NH_4{}^+$, urea, and water (Hacquard et al., 2013). In *L. bicolor*, aquaporins were identified that are also permeable for urea (Dietz et al., 2011; Hacquard et al., 2013; Selle et al., 2005).

5.1.3 Nitrogen metabolism in ectomycorrhizal fungi

In plants and fungi, absorbed $NH_4{}^+$ is first assimilated into the key N donors glutamate and glutamine via the glutamine synthetase/glutamate synthesis (GS/GOGAT) pathway or the glutamate dehydrogenase (GDH) pathway (Chalot, Blaudez, & Brun, 2006). In the GS/GOGAT pathway, glutamate is aminated into glutamine by glutamine synthetase (GS), and the amino group of glutamine is transferred to 2-oxoglutarate to yield two molecules of glutamate by glutamate synthase. In the GDH pathway, glutamate dehydrogenase (NADP-GDH) catalyses the reversible amination of 2-oxoglutarate to glutamate. Most organisms possess both pathways to produce glutamate.

In *H. cylindrosporum* GS and GDH activity can be found, and the expression of both genes is up-regulated by low $NH_4{}^+$ availabilities in the soil, but GDH is down-regulated under high $NH_4{}^+$ availabilities. However, not all ECM fungi have a functioning NADP-GDH pathway. GDH-positive fungi such as *H. cylindrosporum* have both GS and GDH activities, while GDH-negative fungi such as *P. involutus*, *A. muscaria*, and *Suillus bovinus* have only GS activities (Morel, Buée, Chalot, & Brun, 2006). The authors suggest that the evolutionary loss of GDH in some fungi is an adaptation to their environment. GDH-negative fungi are common in humus rich soils or soil layers, in which organic N sources such as amino acids are the dominant N sources. Direct uptake of amino acids including glutamate from the environment would make GDH activities dispensable (Morel et al., 2006). Genes involved in the GS/GOGAT pathway, the urea cycle, and the provision of carbon skeletons for NH_4^+ assimilation via β-oxidation and the glyoxylate cycle have been shown to be highly expressed in rhizomorphs and the extra-radical mycelium of *P. involutus* (Wright, Johansson, Le Quéré, Söderström, & Tunlid, 2005).

Another unique feature in *L. bicolor* is the capability to synthesise and utilise allantoin via urea. This capability is represented by uric acid oxidase, allantoinase, and allantoicase proteins encoded and expressed in the genome (Larsen et al., 2011). Allantoate can be produced by purine and pyrimidine catabolism and transported by an allantoate (*DAL5*) transporter or degraded into urea by the enzymatic activity of allantoicase (Morel et al., 2005). Allantoin can be used by some fungi, bacteria, and plants as carbon and N source, and it has been suggested that allantoin or a derivative of the allantoin pathway may contribute to the fungal capability to provide N to the host plant (Larsen et al., 2011). In several ECM fungi, allantoate/allantoin transporters have been identified (Larsen et al., 2011; Rineau et al., 2013).

5.1.4 Nitrogen transport across the mycorrhizal interface to the host

NH_4^+, amino acids and peptides have been proposed as candidates for N transport across the mycorrhizal interface to the host plant in ECM symbiosis. Amino acids were long seen as the main N source that is transferred across the symbiotic interface. This assumption was mainly based on the capability of ECM fungi to take up amino acids from the soil. However, a transport of amino acids across the mycorrhizal interface after uptake of inorganic N sources and assimilation in the extra-radical mycelium would also result in a flux of carbon skeletons back to the plant host. So far, no clear candidate for fungal amino acids efflux carriers has been identified in the genome of ECM fungi that could be involved in amino acid release from the Hartig net hyphae (Nehls & Plassard, 2018). However, in both *L. bicolor* and *H. cylindrosporum* genomes and mycorrhizal root tips, homologues of yeast AQR1 (Acids Quinidine Resistance 1) were identified (Casieri et al., 2013). AQR1 serves in yeast as an amino acid excretion transporter, that is particularly expressed when amino acids accumulate intracellularly, and is involved in the secretion of homoserine, threonine, and other amino acids from the cell (Velasco, Tenreiro, Calderon, & André, 2004). Whether this transporter could facilitate a secretion of amino acids from the Hartig net hyphae into the mycorrhizal interface needs to be elucidated.

There is increasing evidence, however, that the transport of NH_4^+/NH_3 across the interface may play a more significant role for N transport than previously has been suggested. Amino acids could be assimilated in the extra-radical mycelium, transferred to the Hartig net hyphae, presumably via polyphosphates (polyPs) (Bücking & Heyser, 1999), and broken down to release NH_4^+ into the symbiotic interface.

It has been suggested that N compounds are released into the mycorrhizal apoplast either via passive diffusion, active transport, or exocytosis (Chalot et al., 2006). Potential candidates for active transport proteins for the release NH_4^+ into the mycorrhizal interface could be ATO proteins (for ammonia transport outward) or aquaporins. ATO proteins have been identified in *A. muscaria* (Selle et al., 2005) and *L. bicolor* (Lucic et al., 2008). Fungal aquaporins generally facilitate water transport across membranes (see below), but some are also permeable for NH_4^+. Fungal aquaporins have been identified in *L. bicolor* and *T. melanosporum*, and have been shown to play a key role for ectomycorrhiza development and functioning (Dietz et al., 2011; Hacquard et al., 2013). In addition, voltage-dependent cation channels could be involved in the release of inorganic NH_4^+ to the apoplast (Chalot et al., 2006).

The proposed NH_4^+ transport is consistent with (1) the observed decrease in the amino acid pools (Blaudez, Chalot, Dizengremel, & Botton, 1998) and a strong repression of the GS expression in hyphae of the Hartig net (Wright et al., 2005); (2) the high expression of genes for ammonia-permeable aquaporins in basidiomycota (*L. bicolor*, Dietz et al., 2011) and ascomycota (*Cenococcum geophilum*, Peter et al., 2016); and (3) the up-regulation of plant NH_4^+ transporters.

There is also evidence that an allantoin transporter gene or allantoin permease is expressed in *L. bicolor* and *T. melanosporum* (Hacquard et al., 2013; Martin et al., 2008) which could be involved in the export of allantoate into the symbiotic interface. Additional investigation is needed to determine the contribution of different N sources to N transport in the ECM symbiosis and the precise function and localisation of these proteins in the different symbiotic tissues.

5.2 The symbiotic acquisition of phosphorus

P limits the productivity of plants in many terrestrial ecosystems, particularly in forests (Batjes, 1997). Orthophosphate ions (HPO_4^{2-}; $H_2PO_4^-$; P_i) are the main forms of P taken up by roots from the soil solution. The concentration of free P_i in the soil is generally very low, and often does not exceed 1–10 μM (Hinsinger, 2001). This is due to its strong affinity for combining with cations and clays, resulting in the formation of insoluble P complexes unavailable for plants (Bieleski, 1973; Hinsinger, 2001). Organic P (P_o) can represent the major form of P in soils, particularly in tropical and temperate forests, and can contribute with up to 30–65%, and sometimes to even more than 90% to the total soil P (Condron & Tiessen, 2005; Tibbett, 2002). Most of the P_o is in the form of phosphate esters (C–O–P bond), such as phosphate monoesters (e.g., sugar phosphates) and phosphate diesters (nucleic acids, phospholipids; Turner, 2008), but access to these sources requires specific hydraulic capacities. To cope with this very low availability, trees acquire a significant part of their P from the symbiosis with ECM fungi. ECM associations are therefore determinant factors for the functioning of forest ecosystems and facilitate P movements and cycling.

5.2.1 Inorganic phosphate acquisition by fungal hyphae

Active absorption of P results in the generation of depletion zones around plant roots. Mycelium of ECM fungi allows trees to overcome this issue by providing roots with access to a larger volume of soil, beyond depletion zones (Cairney, 2011; Smith & Read, 2008). Van Tichelen and Colpaert (2000) demonstrated that the P_i uptake capacity of pine roots was strongly increased in plants with ECM associations, compared to non-mycorrhizal plants. Accumulation of P in ECM plants is correlated to the soil exploration capacity of external hyphae, indicating that P can be directly transferred from the fungus to the roots (Torres Aquino & Plassard, 2004).

At the molecular level, extra-radical hyphae acquire P_i by expressing plasma membrane transporters (Fig. 4; Becquer, Trap, Irshad, Ali, & Claude, 2014; Casieri et al., 2013; Garcia et al., 2016). Several genes putatively encoding P_i transporters have been identified in ECM fungi (Casieri et al., 2013; Kohler et al., 2015) based on the homology with *PHO84* and *PHO89* P transporters from yeast (Bun-Ya, Nishimura, Harashima, & Oshima, 1991; Martinez & Persson, 1998). Most sequenced ECM fungi harbour three to five putative H^+/P_i symporters belonging to the *Pht1* subfamily, suggesting that the ability of ECM fungal cells to take up P_i is dependent on the external pH. So far, only one Na^+/P_i symporter from the *Pht2* family has been identified in the ECM fungus, *T. melanosporum*, suggesting that at least in this fungal species an additional pathway for the acquisition of P might exist (Casieri et al., 2013). Among all P_i transporters

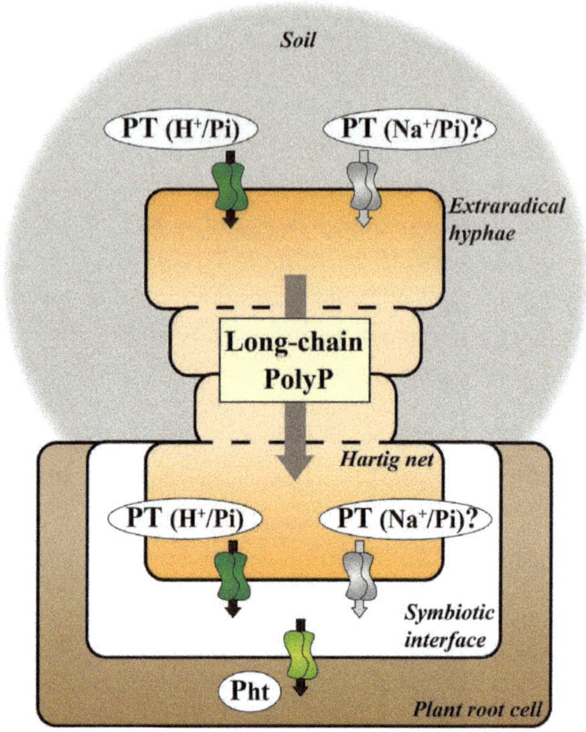

Fig. 4 Schematic representation of the molecular processes that control the transport of inorganic phosphate (P_i) in ectomycorrhizas. P_i is acquired from the soil by extra-radical hyphae through H^+/P_i and possibly Na^+/P_i PT transporters. P_i is condensed into short or long-chain polyphosphates (polyPs) in the fungal vacuoles. PolyPs are translocated from the extra-radical hyphae to the Hartig net via motile tubular vacuoles. Fungal H^+/P_i and potentially Na^+/P_i PT transporters are localised at the fungal membrane at the mycorrhizal interface for the release of P_i. P_i is taken up from the symbiotic interface by plant cortical cells through transporters belonging to the Pht family.

identified to date in ECM fungi, only two proteins from *H. cylindrosporum* (HcPT1.1 and HcPT2), one from *Boletus edulis* (BePT), one from *Rhizopogon luteolus* (RlPT) and one from *Leucocortinarius bulbiger* (LbPT) have been characterised by heterologous expression in yeast (Tatry et al., 2009; Wang, Li, Wu, & Zhao, 2014; Zheng et al., 2016). Of the two P_i transporters initially found in the extra-radical mycelium of *H. cylindrosporum*, HcPT1.1 mediates P_i uptake under P limiting conditions and HcPT2 under high P availabilities (Becquer, Garcia, Amenc et al., 2018; Garcia et al., 2013; Tatry et al., 2009). Another H^+/P_i transporter, HcPT1.2, has also been recently described in *H. cylindrosporum* as participating in P_i uptake from the external medium under high and low P_i conditions (Becquer, Garcia, & Plassard, 2018).

5.2.2 Mineral and organic P mobilisation from the rhizosphere

To release P_i from insoluble minerals, ECM fungi secrete low-molecular-weight organic acids (mainly oxalate) and protons (Courty et al., 2010; Rineau & Garbaye, 2010). This capability of ECM fungi and the resulting P accumulation in host tissues were demonstrated by using apatite (Casarin et al., 2004; Smits, Bonneville, Haward, & Leake, 2008; Wallander, Wickman, & Jacks, 1997) or rock phosphate (Liu, Loganathan, Hedley, & Grace, 2008) as a sole source of P. There is a strong positive correlation between the available P_i levels, P accumulation in colonised plants, and the oxalate concentrations near ECM root tips (Wallander, 2000). Given that the ability of ECM fungi to release organic acids depends on the fungal species and the host species involved, a high variability in the improvement of P nutrition by ECM fungi can be observed (Arvieu, Leprince, & Plassard, 2003; Casarin et al., 2004; Courty et al., 2010; Lapeyrie, Ranger, & Vairelles, 1991).

ECM fungi are also able to produce phosphatases to remove P_i groups from P_o compounds (Plassard & Dell, 2010). ECM fungi can deplete P from birch, pine or beech litters (Perez-Moreno & Read, 2000), from pollen (Read & Perez-Moreno, 2003), and even from nematode necromass (Perez-Moreno & Read, 2002). Dependent on the external source of P that was examined in microcosm experiments, a reduction of 30–97% of the initial litter P content was measured, and 7–33% of this released P was transferred to the host plant through ECM fungi (Quiquampoix & Mousain, 2005). It was suggested that phosphatases released by ECM fungi more likely play a role for the release of P_i from P_o compounds contained in fresh organic matter rather than P that is associated with mineral soil particles (Ali, Louche, Legname, Duchemin, & Plassard, 2009). Phosphatase activities have been shown to be enhanced at low P_i levels in both axenic cultures (Tibbett, Sanders, & Cairney, 1998) and in soils (Louche et al., 2010). This enhancement was also observed when the host plant was in demand of P (Alvarez et al., 2012; van Aarle & Plassard, 2010), suggesting that molecular signals coming from the plant could trigger the expression and excretion of phosphatases by the fungal partner. Lastly, Rennenberg and Herschbach (2013) proposed that ECM fungi could also be able to absorb P_o. However, the molecular players controlling P_o acquisition, movement, and delivery to the host, including transporters and regulators, are still unknown.

5.2.3 Synthesis and regulation of fungal polyphosphates

After uptake, P_i is rapidly incorporated into the active P pool, including phosphorylated primary metabolites, structural molecules and nucleic acids, or condensed into polyP chains stored in vacuoles (Ashford, Ryde, & Barrow, 1994). PolyP chains are linear polymers of three (short-chained, $\leq$20 units) to hundreds of P_i residues (long-chained $\geq$21 units) connected by energy-rich phospho-anhydride bonds (Wood & Clark, 1988). Mobile short-chained polyP can be detected in fungal tissues using nuclear magnetic resonance (NMR) spectroscopy (Pfeffer, Bago, & Shachar-Hill, 2001). Metachromatic granules in ECM fungi have also been observed to consist of mobile and immobile short- and long-chained polyP by either chemical treatments (Ashford, Peterson, Dwarte, & Chilvers, 1986; Ashford, Vesk, Orlovich, Markovina, & Allaway, 1999) or energy dispersive X-ray micro-analysis (Bücking & Heyser, 2000).

The different steps involved in P uptake and transfer are the (i) preferential accumulation and transport of newly absorbed P_i as long-chained polyP by P-starved mycelia, (ii) conversion of long-chained polyP pool into shorter chain lengths in the Hartig net, (iii) hydrolysis of short-chained polyP to P_i to replenish the metabolically active P_i pool in fungal hyphae, (iv) specific release of fungal P_i into the symbiotic interface, and (v) uptake by cortical cells (Torres-Aquino et al., 2017). It has been assumed that short-chained or long-chained polyP could be transferred from the extra-radical mycelium to Hartig net hyphae through a motile tubular vacuole system (Fig. 4) (Bücking & Heyser, 1999; Cairney, 2011). Mobilisation of long-chained polyP is probably under the control of a specific signal released by host plant cells, although its characterisation remains to be determined (Torres-Aquino et al., 2017).

The carbohydrate supply of the host plant serves as an important trigger for fungal P transport. Long-chained polyPs accumulated in the Hartig net hyphae and were not remobilised when photosynthetic activity of the host and carbohydrate transfer to the fungal partner were reduced (Bücking & Heyser, 2003). It has been suggested that there is a bidirectional transport of carbohydrates and of P_i across the same interface (Bücking & Heyser, 2001), and that the fungal degradation of polyPs in the Hartig net plays an important role for the control of the P_i flux across the mycorrhizal interface. Transcripts of an endopolyphosphatase and exopolyphosphatase and their respective activities that could be involved in the breakdown of long-chained polyPs were demonstrated in arbuscular mycorrhizal fungi (Ezawa, Smith, & Smith, 2002; Tisserant et al., 2011), but never in ECM fungi. However, orthologues of these enzymes can be found in the genome of *H. cylindrosporum*, suggesting that comparable mechanisms might exist in ECM symbiosis (Plassard, pers. comm.).

5.2.4 Phosphate delivery to colonised roots

Although it is known for more than half of a century that ECM fungi transfer P to host trees (Melin & Nilsson, 1950), the molecular mechanisms that control the allocation from fungi to root cells remain largely unknown (Becquer et al., 2014; Nehls & Plassard, 2018). Usually, P_i is transported across membranes by H^+/P_i or Na^+/P_i symporters depending on proton-motive force and electrochemical gradients for P_i (Preuss, Huang, & Tyerman, 2011). Since passive transport proteins that could act as channels have not yet been identified, the release of P_i from P_i-rich fungal cells towards the P_i-poor mycorrhizal interface is puzzling. However, a recent *in silico* analysis in arbuscular mycorrhizal interactions predicted that proton-coupled transporters might not only be suited for the uptake of nutrients but also for their release towards the host plant (Schott et al., 2016). Consequently, P_i transporters may transport their substrate along coupled electrochemical gradients without rectification preferences (Ai et al., 2009; Cubero et al., 2009; Fristedt, Weinander, Martinsson, & Persson, 1999; Hürlimann, Pinson, Stadler-Waibel, Zeeman, & Freimoser, 2009; Preuss, Huang, Gilliham, & Tyerman, 2010). For example, a homologous high affinity phosphate transporter (Pho84) has been identified in yeast that ensures a bidirectional transport of P_i along localised pH and P_i gradients at each side of the membrane (Fristedt et al., 1999). Similarly, Schott et al. (2016) proposed a model that high P_i concentrations in the fungal cytosol might be a factor that stimulates P_i fluxes from fungus to the plant through transporters. A P_i gradient which stimulates P_i efflux may be the result of a facilitated polyP degradation within the Hartig net (Torres-Aquino et al., 2017) triggered by the host plant supply with carbohydrates (Bücking, 2004). Anion channels could also mediate P_i efflux. In the genome of *H. cylindrosporum*, two anion channels of ClC family are present but have not yet functionally described.

Supporting the view of a bidirectional P_i transport, two H^+/P_i transporters from the ECM fungus *H. cylindrosporum*, HcPT1.1 and HcPT2, were localised in both the extra-radical mycelium and within the Hartig net (Becquer, Garcia, Amenc et al., 2018; Garcia et al., 2013). Although HcPT1.1 seems to be primarily involved in P_i acquisition under low P_i conditions (Tatry et al., 2009), the key role of HcPT2 in both P_i uptake from the soil and release to the symbiotic apoplast has recently been revealed (Becquer, Garcia, Amenc et al., 2018). The efflux of P_i from fungus to plant through HcPT2 was dependent on an unknown signal originating from the host (Becquer, Garcia, Amenc et al., 2018). The artificial down-regulation of *HcPT2* inhibits ectomycorrhiza formation, indicating that a less cooperative fungus can be sanctioned by the host and that P_i itself can be a key regulator of ectomycorrhiza formation (Garcia et al., 2015). However, we cannot exclude that a transport of P_i could also be mediated by other mechanisms, such as vesicular trafficking or the activity of other transport proteins (Becquer, Garcia, Amenc et al., 2018; Becquer, Garcia, & Plassard, 2018).

Plant cortical cells surrounded by the Hartig net need to specifically express and/or regulate transporters to take up P_i that has been delivered by the fungus to the symbiotic interface. To date, no plant transporter ensuring this role has been functionally characterised, but some candidates have been identified in poplar (Loth-Pereda et al., 2011) and pine (Plassard, pers. comm.). Although *PtPT9* and *PtPT12* were up-regulated in poplar roots colonised by the ECM fungus *L. bicolor* and the arbuscular mycorrhizal fungus *R. irregularis* (Loth-Pereda et al., 2011), further studies are necessary to assess their importance in these symbioses.

5.3 The symbiotic acquisition of potassium

Potassium (K^+) is one of the most important macronutrients for plant growth. This cation is essential for various physiological and cellular functions and for resistance against environmental stressors. In plant roots, K^+ is taken up by specialised high or low affinity transport systems (Anschütz, Becker, & Shabala, 2014; Nieves-Cordones, Alemán, Martínez, & Rubio, 2014; Shabala & Pottosin, 2014). Since K^+ is a major constituent of many rock-forming minerals, its availability in natural ecosystems is often drastically limited (Zörb, Senbayram, & Peiter, 2014). Due to their physiological and biochemical properties, fungal hyphae can better solubilise K^+ from rocks than plant roots. ECM fungi can solubilise K^+ through the secretion of protons and organic acids, and thereby significantly contribute to host plant K^+ nutrition (Rygiewicz & Bledsoe, 1984). For example, amount of exchangeable K^+ through weathering found in soils colonised with *Piloderma* spp. was higher than in non-ECM soils (Arocena, Glowa, Massicotte, & Lavkulich, 1999). The ECM fungus *Suillus variegatus* also increases the concentration of organic acids in soils when biotite is used as K^+ source. However, no significant effect on plant K^+ nutrition was observed in seedlings that were colonised with this fungus compared to non-mycorrhizal control plants (Wallander & Wickman, 1999). *P. involutus* is also able to weather muscovite, to release K^+ and to enhance plant nutrition (Van Schöll, Smits, & Hoffland, 2006). However, not all

ECM fungal species show this capability, indicating that the effect on K^+ nutrition can vary depending on the fungus involved in this interaction and on environmental conditions. For example, it has been reported that *P. involutus* contributes 5–6% of total K^+ of its host *Picea abies* (Jentschke, Brandes, Kuhn, Schröder, & Godbold, 2001). Another fungus, *Pisolithus albus*, enhanced K^+ acquisition of *Acacia spirorbis* and *Eucalyptus globulus* by 21–38% when grown in a New Caledonian topsoil that was contaminated with toxic metals (Jourand et al., 2014).

Under K^+ limiting conditions, *P. pinaster* seedlings colonised by *H. cylindrosporum* showed an up to 35% higher K^+ content (Garcia et al., 2014). Insights into the molecular players that are involved in K^+ transport in the ECM symbiosis have become possible by the increasing number of sequencing programs for ECM fungi, as cited above. Consequently, four families of putative K^+ transport systems have recently been identified in ECM fungi (Garcia & Zimmermann, 2014). The current model suggests that transporters are responsible for K^+ absorption from the soil by extra-radical or mantle hyphae, while ion channels contribute to the release of K^+ from the Hartig net into the mycorrhizal interface (Fig. 5).

5.3.1 Potassium uptake from the soil through fungal transporters

So far, two conserved families of K^+ transporters have been identified in fungi, Trk (Trk/Ktr/HKT; Transporter of K^+) and HAK (KT/KUP/HAK; High-Affinity K^+ uptake) (Benito, Garciadeblás, Fraile-Escanciano, & Rodríguez-Navarro, 2011). Trk transporters were first identified in *S. cerevisiae* (Gaber, Styles, & Fink, 1988; Ko & Gaber, 1991) and were later also described for bacteria, plants, and fungi as K^+/H^+ or Na^+/K^+ symporters, or Na^+ uniporters (Corratgé-Faillie et al., 2010; Rodrı´guez-Navarro, 2000). This functional diversity indicates that Trk proteins may also be involved in other cellular functions than K^+ transport, including the acquisition of Na^+ and its compartmentalization in the vacuole for detoxification.

In all genomes of ECM fungi that have so far been sequenced, at least one member of the Trk family has been identified (Garcia & Zimmermann, 2014). In *H. cylindrosporum*, two K^+ transporters HcTrk1 and HcTrk2 were found but only one of them (HcTrk1) has so far been functionally characterised and studied in ECM roots (Corratgé et al., 2007; Garcia et al., 2014). Due to its localisation at uptake active sites of ECM roots of *P. pinaster*, it has been suggested that HcTrk1 plays a major role in K^+ acquisition from soils. Colonisation by *H. cylindrosporum* increased the K^+ contents in the shoots of pine seedlings by about 35% (Garcia et al., 2014).

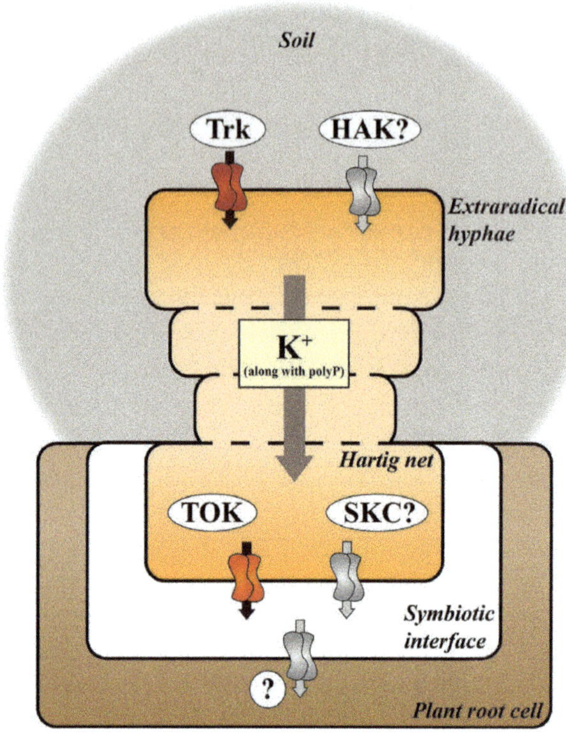

Fig. 5 Schematic representation of the molecular mechanisms that control the transport of potassium (K^+) in ectomycorrhizas. K^+ is acquired from the soil by Trk and possibly HAK transporters localised in extra-radical hyphae. K^+ is translocated from the extra-radical hyphae to the Hartig net in vacuoles in association with polyPs. TOK and potentially SKC channels are localised at the symbiotic interface and release K^+ into the mycorrhizal interface. K^+ is transported from the symbiotic interface to plant cortical cells through unidentified transporters.

However, this effect was not observed in seedlings that were colonised by transgenic fungal strains overexpressing *HcTrk1* (Garcia et al., 2014). It is likely that the overexpression of *HcTrk1* triggered an ectopic accumulation of the corresponding proteins in the Hartig net that resulted in a competition with plant cells for K^+ acquisition from the symbiotic interface (Garcia et al., 2014). Similarly, plants that were colonised with these overexpressing strains displayed reduced P concentrations, indicating that K^+ homeostasis is a crucial component for fungal P allocation to the host. K^+ serves as one of the major counter-ions for the formation and stabilisation of polyPs in ECM fungi (Bücking & Heyser, 1999).

HAK transporters are well characterised in fungi and plants and are involved in K^+ uptake under deficient conditions (Cabrera, Álvarez, Martín, Siverio, & Ramos, 2012; Rivetta, Allen, Slayman, & Slayman, 2013). One to four HAK members were found in the genome of ECM fungi (Garcia & Zimmermann, 2014), but their role for K^+ transport in the symbiosis is still unknown. Based on their function in other organisms and their level of expression in ECM tissues, we can hypothesise that they might play a significant role for the K^+ nutrition of ECM plants under K^+ limitation (Garcia and Zimmermann, unpublished).

Besides Trk and HAK transporters, members of the ACU P-type ATPase family (alkali cation uptake ATPase) with K^+ and Na^+ transport activities were found in most fungal genomes (Benito et al., 2011; Benito, Garciadeblás, Schreier, & Rodríguez-Navarro, 2004). However, their roles in both free-living mycelia or symbiotic conditions are still unknown.

5.3.2 Potassium movements within the fungus

Although the long-distance transport of K^+ from uptake to release sites of ectomycorrhizas is only poorly understood, it seems closely related to the movements of polyP through fungal hyphae (Benito & González-Guerrero, 2014). K^+ has been detected by elemental analyses in polyP granules of fungal hyphae and has been identified as one of the major counter-ions (Bücking & Heyser, 1999). The correlation between P_i and K^+ concentrations in hyphae of *H. cylindrosporum* supports the view of a joined transport within the fungus (Garcia et al., 2014). This is currently the only mechanism hypothesised so far since no diffusion process along concentration gradients was observed in fungi (Fig. 5).

5.3.3 Potassium transfer from the Hartig net to colonised roots—The missing step

Although the symbiotic transport of K^+ from the Hartig net to root cortical cells remains largely unknown, promising candidates have been identified in EST libraries (Lambilliotte et al., 2004) and fungal genomes (Kohler et al., 2015). The two most promising candidate families that could be involved in the release of K^+ from fungal hyphae into the mycorrhizal apoplast are Shaker-like K^+ channels (SKC) and tandem-pore outward-rectifying K^+ (TOK) channels (Garcia & Zimmermann, 2014). SKCs belong to the voltage-gated ion channel (VIC) superfamily, and only members from animal and plant species have so far been characterised (Papazian, Schwarz, Tempel, Jan, & Jan, 1987; Sentenac et al., 1992). These channels have also been found in fungi, including ECM fungi, from basal species (neocallimastigomycota, chytridiomycota, mucoromycota, zoopagomycota) to basidiomycota, but not in ascomycota (Garcia & Zimmermann, 2014). It has been hypothesised that SKC proteins might have a role in K^+ release across the fungal plasma membrane towards the apoplast, but no report describes their involvement in ECM symbiosis, and no ECM fungal member has so far been functionally characterised.

Members of the fungal-specific TOK channel family are also promising candidates for K^+ efflux in ectomycorrhizas (Garcia & Zimmermann, 2014; Guerrero-Galán, Delteil, et al., 2018, Guerrero-Galán, Garcia, Houdinet, & Zimmermann, 2018). These channels harbour in each polypeptide two pore domains in tandem that are assembled as dimers. They are described as outward rectifiers in yeast (Ketchum, Joiner, Sellers, Kaczmarek, & Goldstein, 1995), and as ion channels, rectification drives the current preferentially in one direction. Interestingly, small inward currents can also be observed dependent on the K^+ concentration and membrane potential. Inward currents can explain the uptake of K^+, the growth and observed complementation of TOK yeast mutants (Bertl et al., 2003). Two TOK channel subfamilies have been found in ascomycota and basidiomycota, including some ECM species (Garcia & Zimmermann, 2014).

Three TOK channels HcTOK1, HcTOK2.1 and HcTOK2.2 have been identified in *H. cylindrosporum* based on their sequence similarities and secondary structures. Given the nature of already studied TOK channels in yeast (Ketchum et al., 1995) and *Neurospora crassa* (Roberts, 2003), their homologues in *H. cylindrosporum* could be interesting candidates for K^+ transfer from the fungus to plant apoplast. Recent functional studies have demonstrated outward K^+ currents in *Xenopus laevis* oocytes that express HcTOK1 and HcTOK2.1, and growth complementation by HcTOK1 and HcTOK2.2 in yeast mutants (Guerrero-Galán, Delteil, et al., 2018). An enhanced expression of *HcTOK2.2* in ECM roots and its localisation in the Hartig net region implies that this channel could be involved in the K^+ transfer towards host roots. Consistently, pine seedlings that were colonised with transgenic fungal strains overexpressing *HcTOK2.2* displayed a larger accumulation of K^+ in their shoots than control plants (Fig. 5; Guerrero-Galán, Delteil, et al., 2018). It has also recently been demonstrated that the plasma membrane channel *HcTOK1* plays a key role in fungal K^+ homeostasis, and ultimately in the symbiotic transfer of K^+ to *P. pinaster* (Guerrero-Galán, Garcia, et al., 2018).

K^+ released at the symbiotic interface needs to be taken up by transport systems that are localised at the plasma membrane of plant cortical cells. Dozens of K^+ transporters from various model plant species have been characterised, but only very few were identified in the roots of trees and so far no transporter has been identified that is involved in the uptake of K^+ from the mycorrhizal interface. A SKOR-type outward Shaker-like K^+ channel was highly up-regulated in the roots of *Pinus sylvestris* that were colonised by the ECM fungi *C. geophilum*, *Suillus granulatus* and *Rhizopogon roseolus* (Peter et al., 2016). Typically, these channels mediate K^+ loading to the xylem, suggesting that they might have a similar role in colonised roots. More research is needed to clearly identify those plant transport systems in the plasma membrane of cortical cells responsible for the uptake of K^+ from the symbiotic interface.

5.4 The symbiotic acquisition of other ions

ECM fungi are also able to provide their host with other essential ions and can protect the roots against high metal concentrations or against salt stress (Colpaert, Wevers, Krznaric, & Adriaensen, 2011). This protective function from the fungal partner can partly be explained by the presence of a fungal sheath surrounding the roots (Bücking et al., 2002; Jentschke & Godbold, 2000), but also by the contribution of highly selective ion transport systems.

5.4.1 Other macronutrients

An improved nutrition by ECM fungi with other macronutrients such as magnesium, calcium, and sulphur has also been reported (Casieri et al., 2013). However, in comparison to N, P and K^+, the molecular mechanisms that control these transport processes are largely unknown. Organic acid secretion by ECM fungi can also mobilise other essential nutrients by mineral weathering and can for example release calcium from apatite (Blum et al., 2002; Landeweert, Hoffland, Finlay, Kuyper, & van Breemen, 2001). Formation of calcium oxalate crystals in colonised eucalypt roots indicates a higher calcium uptake of ECM plants (Pylro et al., 2013). A symbiotic transport of magnesium, an essential element for forest productivity that is particularly limiting in acidified forest soils, has also been observed in *P. abies* seedlings colonised by *P. involutus* (Jentschke & Godbold, 2000). *P. involutus* led also to an improved calcium and magnesium nutrition in *Populus deltoides* seedlings under aluminium stress (Khosla, Kaur, & Sudhakara Reddy, 2009). In nickel-rich soils from New Caledonia, calcium acquisition was significantly enhanced in *A. spirorbis* and *E. globulus* plants by the ECM fungus *P. albus*, while magnesium uptake was significantly reduced (Jourand et al., 2014). A decrease in sulphur starvation of ECM plants has also been observed, and sulphate and methionine permeases have been found in fungal genomes (Casieri et al., 2013; Garcia et al., 2016).

5.4.2 Micronutrients

ECM fungi can also transfer micronutrients, including iron, zinc (Zn^{2+}), copper, and manganese from the soil to its host root. At the same time, ECM fungi can also protect their host against toxic concentrations of these metals in their environment (Colpaert & van Assche, 1987; Colpaert et al., 2011; Galli, Schüepp, & Brunold, 1994). This requires a tight regulation of cytosolic concentrations of these micronutrients in fungal hyphae through specialised transport systems. Recent studies demonstrated the improvement of plant micronutrient nutrition by ECM fungi and identified several of the involved fungal transporters. For example, Zn^{2+} is acquired from the soil through transporters belonging to the ZIP family (Zrt/Irt-like proteins). SlZRT1 of *Suillus luteus* is expressed at the plasma membrane and facilitates the cellular uptake of Zn from the soil (Fig. 6; Coninx et al., 2017). When Zn^{2+} concentrations in the environment are too high, *S. bovinus* is able to limit its transfer to the host plant *P. sylvestris* and protect the plant against its toxic effects (Adriaensen, Vangronsveld, & Colpaert, 2006; Bücking & Heyser, 1994). Involved in the protection against toxic metal concentrations could be the accumulation and detoxification of Zn^{2+} in fungal vacuoles mediated by vacuolar transporters such as SlZnT1, a Cation Diffusion Facilitator (CDF) that has been identified in *S. luteus* (Ruytinx et al., 2017). High concentrations of Zn^{2+} in the fungal vacuole could be detoxified by binding to long-chained polyPs (Bücking & Heyser, 1999). Interestingly, *H. cylindrosporum* stores excess Zn^{2+} in ER-derived vesicles, probably through the activity of HcZnT1 (Blaudez & Chalot, 2011). Similar transporters have been identified in *Russula atropurpurea* that could mediate both vacuolar Zn^{2+} storage (RaCDF1) and bidirectional metal transport (RaCDF2) at the plasma membrane (Sácký, Leonhardt, & Kotrba, 2016).

5.5 The symbiotic acquisition of water

It is known for almost a century that ECM fungi can help trees in the acquisition of water, particularly during the recovering after a drought period (Cromer, 1935; Goss, 1960; Reid, 1978; Zevora, 1955). Transport of water from ECM fungi to their host plant has been demonstrated for the first time using tritiated water in associations between *S. bovinus* and

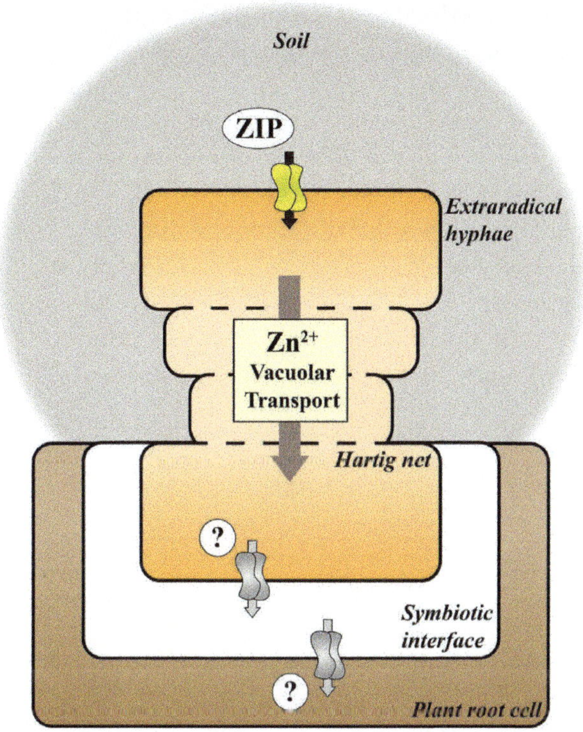

Fig. 6 Schematic representation of the molecular mechanisms that control the transport of zinc (Zn^{2+}) in ectomycorrhizal roots. Extra-radical hyphae take up Zn^{2+} from the soil by Zrt/Irt-like proteins (ZIP) transporters. Zn^{2+} is transported into vacuoles through cation diffusion facilitators (see main text) and then translocated from the extra-radical mycelium to the Hartig net. Unidentified fungal and plant proteins are responsible for the release of Zn^{2+} at the symbiotic interface and its uptake by cortical cells.

P. sylvestris (Duddridge, Malibari, & Read, 1980). Several tree species show an improved performance under water stress by the ECM symbiosis (Garbaye, 2000; Lehto & Zwiazek, 2011). Symplastic and apoplastic pathways for water transport in ectomycorrhizas have been discussed in relation to the hydrophilic or hydrophobic properties of fungal cell walls (Behrmann & Heyser, 1992; Bücking, Hans, & Heyser, 2007; Bücking et al., 2002; Lehto & Zwiazek, 2011).

5.5.1 Water uptake by fungal hyphae

In addition to the larger volume of soil that is explored by fungal hyphae, specific water channels, called aquaporins, have been attributed to the higher water uptake of mycorrhizal plants. Aquaporins of the major intrinsic protein (MIP) family are found in all organisms and mediate water transport across membranes (Maurel, Verdoucq, Luu, & Santoni, 2008; Nehls & Dietz, 2014). They are divided into five subfamilies and among them plasma membrane intrinsic proteins (PIPs) and tonoplast intrinsic proteins (TIPs) that contribute the most to water transport.

In the genome of several ECM fungi, including *L. bicolor*, different aquaporin families were discovered (Fig. 7; Dietz et al., 2011), and it becomes increasingly clear that these aquaporins can play an essential role during ECM root development. For example, knockout mutants of *L. bicolor*, with a reduced expression of the aquaporin *LbAQP1*, showed

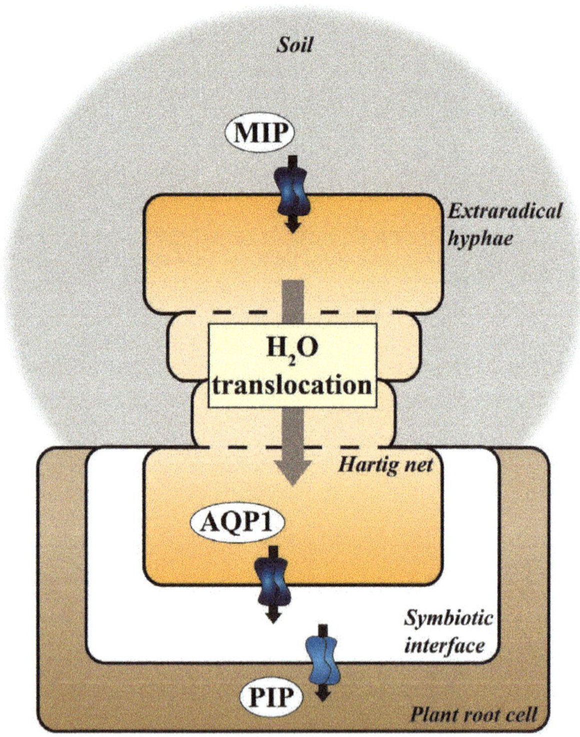

Fig. 7 Schematic representation of molecular mechanisms that control the transport of water (H_2O) in ectomycorrhizal roots. Extra-radical hyphae take up water from the soil through major intrinsic protein (MIP) aquaporins. Water is translocated from the extra-radical hyphae to the Hartig net where aquaporins (AQP1) are responsible for its release into the symbiotic interface. H_2O is transported from the symbiotic interface to plant cortical cells through plant aquaporins belonging to the plasma membrane intrinsic protein (PIP) family.

no Hartig net development in *Populus tremuloides* seedlings and a reduced expression of the mycorrhizal effector protein MiSSP7 (Navarro-Ródenas, Xu, Kemppainen, Pardo, & Zwiazek, 2015). Overexpression of JQ585595, another aquaporin of *L. bicolor*, led to higher hydraulic conductivities in roots of mycorrhizal *Picea glauca* plants, and revealed the importance of fungal aquaporins for the transport of water from the fungus to the roots (Xu et al., 2015). Recently, transcriptional profiling of *C. geophilum* demonstrated that one aquaglyceroporin and one classical aquaporin were highly up-regulated in ECM *P. sylvestris* roots (Peter et al., 2016). Under drought conditions, these two aquaporins were down-regulated in both ectomycorrhizal root and free-living mycelia, whereas another classical aquaporin was highly up-regulated (Peter et al., 2016). These observations support the hypothesis that fungal aquaporins are required for symbiotic transport of water and that their expression is regulated by the water demand of the host plant.

5.5.2 Water transfer towards colonised roots

The use of tracer techniques with isotopes and dyes allowed to follow the water exchange between the roots of woody plants and ECM fungi (Plamboeck et al., 2007). Increase of root hydraulic conductance in ECM roots of *Ulmus americana* and *Pinus banksiana* has been attributed to a decrease of the apoplastic water flow resistance and to an increase of aquaporin-mediated water transport (Lee, Calvo-Polanco, Chung, & Zwiazek, 2010; Muhsin & Zwiazek, 2002). Interestingly, the increase in water transport capacity of ECM poplar roots was correlated with the up-regulation of aquaporin-encoding PIP genes (Marjanović et al., 2005). A link between fungal water transport and expression of root aquaporins was established in *P. glauca* roots colonised by transgenic *L. bicolor* strains with an over-expression of JQ585595 (Xu et al., 2015). This overexpression resulted in the up-regulation of a plant root aquaporin, suggesting a positive correlation between fungal and plant water transport. However, there are also reports in which the root colonisation with *P. involutus* led to a downregulation of two plant aquaporins (*BpPIP1* and *BpPIP2*) of birch (Le Quéré, Wright, Söderström, Tunlid, & Johansson, 2005). Although there are results suggesting that a tight regulation of aquaporins from both partners is needed for water fluxes in ECM symbiosis, the water transport regulation may still depend on the plant and fungal species involved, and the water demand from the host (Fig. 7).

6. Future challenges and concluding remarks

Identifying the molecular mechanisms controlling the transport of water and nutrients from the soil to tree roots through ECM fungi is crucial to understand resource dynamics in forest ecosystems. Several proteins of the "ECM transportome" have already been described in both plant and fungal partners, but many others particularly at the symbiotic interface are still not identified and their regulation unknown.

Although ECM associations are the dominant mycorrhizal form in temperate and boreal forests, other forms of mycorrhizal symbioses co-exist in these ecosystems. For example, the arbuscular mycorrhizal symbiosis is one of the most important associations in terms of its economic impact and its ecology due to the high number of plant species that form this symbiosis, including many agronomically important crops (Lee, Eo, Ka, & Eom, 2013; Wang & Qiu, 2006). Interestingly, some woody plants like poplar can simultaneously be colonised by ECM and arbuscular mycorrhizal fungi. Some transporters are up-regulated in roots colonised by both root symbionts, whereas others are only specifically expressed during one symbiosis (Loth-Pereda et al., 2011). However, the actual involvement of these transporters in tree nutrition remains unclear. Although far less abundant than the two other associations mentioned above, ectendomycorrhizal mycorrhizas can also be found in shrub species and tree roots that are colonised by ECM fungi. This symbiosis is defined by the presence of a fungal mantle, and a Hartig net, but in contrast to ECM fungi, in the ectendomycorrhizal symbiosis an intracellular penetration of the root cortex and a different interface is involved in the hydro-mineral nutrition of woody plants (Smith & Read, 2008). However, no molecular data of this symbiosis form are yet available.

A remaining challenge in the study of mycorrhizal interactions is that our current understanding of many transport processes is only based on the symbiosis of one host plant with one fungal species. However, plants are simultaneously colonised by multiple fungi, and tree nutrition is therefore determined by communities of fungi, that may more or less overlap in their functionality of the host, and also compete for carbon supply from the host. The same is true for the fungal partner that can simultaneously colonise multiple hosts from one or more species and can choose to which host certain nutrients are delivered. Developing projects that investigate tripartite interactions with more than one symbiosis form will allow the identification of shared and specific allocation pathways for nutrients and water in very different associations.

Another impediment for the identification of molecular processes involved in ECM associations is that currently only a handful of transformable fungal and plant species, and only a limited number of tools for species transformation, are available. The development of more transformable fungal and plant models, particularly in the group of gymnosperms, will allow the assessment of both molecular diversity and common features driving the formation and functioning of ECM symbiosis. Although the use of reverse genetics has been successful in some species and provided with some insights into the effects of an over-expression or RNA-silencing of candidate genes, the lack of knock-out mutants limits the description of new and specific phenotypes. Interestingly, the increasing application of CRISPR/Cas9-based genome editing in many organisms, including trees (Fan et al., 2015) or filamentous fungi (Shi et al., 2017), opens a path towards its use in ECM species. Subsequent efforts in the development of original molecular tools, and additional fungal and plant models are necessary to unravel the development, functioning and evolution of ECM symbiosis.

Acknowledgements

We wish to thank Dr. Claude Plassard for her valuable input to the manuscript. We also thank her and Jacques Guinberteau for sharing a photo of *H. cylindrosporum* fruiting bodies. We wish to thank Dr. Anne Pringle and Dr. Natalia Vargas Estupiñan, and Dr. Jessy Labbé and Dr. Jay Chen for sharing photos of fruiting bodies and ectomycorrhizal roots of *A. muscaria* and *L. bicolor*, respectively. A.B. was supported by an INRA CJS grant "Contrat Jeune Scientifique". J.L.E. and H.B. acknowledge funding from the USDA (2017-67014-26530), and the Agricultural Experiment Station at South Dakota State University. C.G.-G. and G.H. were supported by fellowships from the French Ministry of Higher Education and Research. K.G. acknowledges support of the North Carolina Agriculture Research Service (NCARS).

References

Adriaensen, K., Vangronsveld, J., & Colpaert, J. V. (2006). Zinc-tolerant *Suillus bovinus* improves growth of Zn-exposed *Pinus sylvestris* seedlings. *Mycorrhiza*, *16*(8), 553–558. https://doi.org/10.1007/s00572-006-0072-7.

Agerer, R. (2001). Exploration types of ectomycorrhizae. *Mycorrhiza*, *11*(2), 107–114. https://doi.org/10.1007/s005720100108.

Ai, P., Sun, S., Zhao, J., Fan, X., Xin, W., Guo, Q., et al. (2009). Two rice phosphate transporters, OsPht1;2 and OsPht1;6, have different functions and kinetic properties in uptake and translocation. *The Plant Journal*, *57*(5), 798–809. https://doi.org/10.1111/j.1365-313X.2008.03726.x.

Ali, M. A., Louche, J., Legname, E., Duchemin, M., & Plassard, C. (2009). *Pinus pinaster* seedlings and their fungal symbionts show high plasticity in phosphorus acquisition in acidic soils. *Tree Physiology*, *29*(12), 1587–1597. https://doi.org/10.1093/treephys/tpp088.

Alvarez, M., Huygens, D., Díaz, L. M., Villanueva, C. A., Heyser, W., & Boeckx, P. (2012). The spatial distribution of acid phosphatase activity in ectomycorrhizal tissues depends on soil fertility and morphotype, and relates to host plant phosphorus uptake. *Plant, Cell & Environment*, *35*(1), 126–135. https://doi.org/10.1111/j.1365-3040.2011.02422.x.

Anschütz, U., Becker, D., & Shabala, S. (2014). Going beyond nutrition: Regulation of potassium homoeostasis as a common denominator of plant adaptive responses to environment. *Journal of Plant Physiology, 171*(9), 670–687. https://doi.org/10.1016/j.jplph.2014.01.009.

Arantes, V., Jellison, J., & Goodell, B. (2012). Peculiarities of brown-rot fungi and biochemical Fenton reaction with regard to their potential as a model for bioprocessing biomass. *Applied Microbiology and Biotechnology, 94*(2), 323–338. https://doi.org/10.1007/s00253-012-3954-y.

Arocena, J. M., Glowa, K. R., Massicotte, H. B., & Lavkulich, L. (1999). Chemical and mineral composition of ectomycorrhizosphere soils of subalpine fir (Abieslasiocarpa (Hook.) Nutt.) in the Ae horizon of a Luvisol. *Canadian Journal of Soil Science, 79*(1), 25–35. https://doi.org/10.4141/S98-037.

Arvieu, J.-C., Leprince, F., & Plassard, C. (2003). Release of oxalate and protons by ectomycorrhizal fungi in response to P-deficiency and calcium carbonate in nutrient solution. *Annals of Forest Science, 60*(8), 815–821. https://doi.org/10.1051/forest:2003076.

Ashford, A. E., Peterson, R. L., Dwarte, D., & Chilvers, G. A. (1986). Polyphosphate granules in eucalypt mycorrhizas: Determination by energy dispersive x-ray microanalysis. *Canadian Journal of Botany, 64*(3), 677–687. https://doi.org/10.1139/b86-086.

Ashford, A. E., Ryde, S., & Barrow, K. D. (1994). Demonstration of a short chain polyphosphate in *Pisolithus tinctorius* and the implications for phosphorus transport. *New Phytologist, 126*(2), 239–247. https://doi.org/10.1111/j.1469-8137.1994.tb03942.x.

Ashford, A. E., Vesk, P. A., Orlovich, D. A., Markovina, A.-L., & Allaway, W. G. (1999). Dispersed polyphosphate in fungal vacuoles in *Eucalyptus pilularis/Pisolithus tinctorius* ectomycorrhizas. *Fungal Genetics and Biology, 28*(1), 21–33. https://doi.org/10.1006/fgbi.1999.1140.

Balestrini, R., & Kottke, I. (2016). Structure and development of ectomycorrhizal roots. In F. Martin (Ed.), *Molecular mycorrhizal symbiosis* (pp. 47–61). Chichester, UK: John Wiley & Sons. https://doi.org/10.1002/9781118951446.ch4.

Batjes, N. H. (1997). A world dataset of derived soil properties by FAO–UNESCO soil unit for global modelling. *Soil Use and Management, 13*(1), 9–16. https://doi.org/10.1111/j.1475-2743.1997.tb00550.x.

Becquer, A., Garcia, K., Amenc, L., Rivard, C., Doré, J., Trives-Segura, C., et al. (2018). The *Hebeloma cylindrosporum* HcPT2 Pi transporter plays a key role in the ectomycorrhizal symbiosis. *New Phytologist, 220*(4), 1185–1199. https://doi.org/10.1111/nph.15281.

Becquer, A., Garcia, K., & Plassard, C. (2018). HcPT1.2 participates in Pi acquisition in *Hebeloma cylindrosporum* external hyphae of ectomycorrhizas under high and low phosphate conditions. *Plant Signaling & Behavior, 13*(10), e1525997. https://doi.org/10.1080/15592324.2018.1525997.

Becquer, A., Trap, J., Irshad, U., Ali, M. A., & Claude, P. (2014). From soil to plant, the journey of P through trophic relationships and ectomycorrhizal association. *Frontiers in Plant Science, 5*, 548. https://doi.org/10.3389/fpls.2014.00548.

Behrmann, P., & Heyser, W. (1992). Apoplastic transport through the fungal sheath of *Pinus sylvestris/Suillus bovinus* ectomycorrhizae. *Botanica Acta, 105*(6), 427–434. https://doi.org/10.1111/j.1438-8677.1992.tb00324.x.

Bending, G. D., & Read, D. J. (1995). The structure and function of the vegetative mycelium of ectomycorrhizal plants. *New Phytologist, 130*(3), 411–417. https://doi.org/10.1111/j.1469-8137.1995.tb01835.x.

Benito, B., Garciadeblás, B., Fraile-Escanciano, A., & Rodríguez-Navarro, A. (2011). Potassium and sodium uptake systems in fungi. The transporter diversity of *Magnaporthe oryzae*. *Fungal Genetics and Biology, 48*(8), 812–822. https://doi.org/10.1016/j.fgb.2011.03.002.

Benito, B., Garciadeblás, B., Schreier, P., & Rodríguez-Navarro, A. (2004). Novel P-type ATPases mediate high-affinity potassium or sodium uptake in fungi. *Eukaryotic Cell, 3*(2), 359–368. https://doi.org/10.1128/EC.3.2.359-368.2004.

Benito, B., & González-Guerrero, M. (2014). Unravelling potassium nutrition in ectomycorrhizal associations. *New Phytologist, 201*(3), 707–709. https://doi.org/10.1111/nph.12659.

Benjdia, M., Rikirsch, E., Müller, T., Morel, M., Corratgé, C., Zimmermann, S., et al. (2006). Peptide uptake in the ectomycorrhizal fungus *Hebeloma cylindrosporum*: Characterization of two di- and tripeptide transporters (HcPTR2A and B). *New Phytologist, 170*(2), 401–410. https://doi.org/10.1111/j.1469-8137.2006.01672.x.

Berg, B., & McClaugherty, C. (2014). *Plant litter: Decomposition, humus formation, carbon sequestration* (3rd ed.). Heidelberg, Germany: Springer. https://doi.org/10.1007/978-3-642-38821-7.

Bertl, A., Ramos, J., Ludwig, J., Lichtenberg-Fraté, H., Reid, J., Bihler, H., et al. (2003). Characterization of potassium transport in wild-type and isogenic yeast strains carrying all combinations of *trk1*, *trk2* and *tok1* null mutations. *Molecular Microbiology, 47*(3), 767–780. http://www.ncbi.nlm.nih.gov/pubmed/12535075.

Bieleski, R. L. (1973). Phosphate pools, phosphate transport, and phosphate availability. *Annual Review of Plant Physiology, 24*(1), 225–252. https://doi.org/10.1146/annurev.pp.24.060173.001301.

Blaudez, D., & Chalot, M. (2011). Characterization of the ER-located zinc transporter ZnT1 and identification of a vesicular zinc storage compartment in *Hebeloma cylindrosporum*. *Fungal Genetics and Biology, 48*(5), 496–503. https://doi.org/10.1016/j.fgb.2010.11.007.

Blaudez, D., Chalot, M., Dizengremel, P., & Botton, B. (1998). Structure and function of the ectomycorrhizal association between *Paxillus involutus* and *Betula pendula*. II. Metabolic changes during mycorrhiza formation. *New Phytologist, 138*(3), 543–552. http://www.jstor.org/stable/2588350.

Blum, J. D., Klaue, A., Nezat, C. A., Driscoll, C. T., Johnson, C. E., Siccama, T. G., et al. (2002). Mycorrhizal weathering of apatite as an important calcium source in base-poor forest ecosystems. *Nature, 417*, 729–731. https://doi.org/10.1038/nature00793.

Bonfante, P., & Genre, A. (2010). Mechanisms underlying beneficial plant-fungus interactions in mycorrhizal symbiosis. *Nature Communications, 1*, 48. https://doi.org/10.1038/ncomms1046.

Branco, S., Gladieux, P., Ellison, C. E., Kuo, A., LaButti, K., Lipzen, A., et al. (2015). Genetic isolation between two recently diverged populations of a symbiotic fungus. *Molecular Ecology, 24*(11), 2747–2758. https://doi.org/10.1111/mec.13132.

Brundrett, M. C. (2002). Coevolution of roots and mycorrhizas of land plants. *New Phytologist, 154*(2), 275–304. https://doi.org/10.1046/j.1469-8137.2002.00397.x.

Bücking, H. (2004). Phosphate absorption and efflux of three ectomycorrhizal fungi as affected by external phosphate, cation and carbohydrate concentrations. *Mycological Research, 108*(6), 599–609. https://doi.org/10.1017/S0953756204009992.

Bücking, H., Hans, R., & Heyser, W. (2007). The apoplast of ectomycorrhizal roots—Site of nutrient uptake and nutrient exchange between the symbiotic partners. In B. Sattelmacher & W. J. Horst (Eds.), *The apoplast of higher plants: Compartment of storage, transport and reactions: The significance of the apoplast for the mineral nutrition of higher plants*. Dordrecht: Springer Netherlands. https://doi.org/10.1007/978-1-4020-5843-1_7.

Bücking, H., & Heyser, W. (1994). The effect of ectomycorrhizal fungi on Zn uptake and distribution in seedlings of *Pinus sylvestris* L. *Plant and Soil, 167*(2), 203–212. https://doi.org/10.1007/BF00007946.

Bücking, H., & Heyser, W. (1999). Elemental composition and function of polyphosphates in ectomycorrhizal fungi—An X-ray microanalytical study. *Mycological Research, 103*(1), 31–39. https://doi.org/10.1017/S0953756298006935.

Bücking, H., & Heyser, W. (2000). Subcellular compartmentation of elements in non-mycorrhizal and mycorrhizal roots of *Pinus sylvestris*: An X-ray microanalytical study. I. The distribution of phosphate. *New Phytologist, 145*(2), 311–320. http://www.jstor.org/stable/2588956.

Bücking, H., & Heyser, W. (2001). Microautoradiographic localization of phosphate and carbohydrates in mycorrhizal roots of *Populus tremula* × *Populus alba* and the implications for transfer processes in ectomycorrhizal associations. *Tree Physiology*, *21*(2–3), 101–107. https://doi.org/10.1093/treephys/21.2-3.101.
Bücking, H., & Heyser, W. (2003). Uptake and transfer of nutrients in ectomycorrhizal associations: Interactions between photosynthesis and phosphate nutrition. *Mycorrhiza*, *13*(2), 59–68. https://doi.org/10.1007/s00572-002-0196-3.
Bücking, H., Kuhn, A. J., Schröder, W. H., & Heyser, W. (2002). The fungal sheath of ectomycorrhizal pine roots: An apoplastic barrier for the entry of calcium, magnesium, and potassium into the root cortex? *Journal of Experimental Botany*, *53*(374), 1659–1669. https://doi.org/10.1093/jxb/erf011.
Bücking, H., Liepold, E., & Ambilwade, P. (2012). The role of the mycorrhizal symbiosis in nutrient uptake of plants and the regulatory mechanisms underlying these transport processes. In N. K. Dhal (Ed.), *Plant science*. IntechOpen. https://doi.org/10.5772/52570.
Bun-Ya, M., Nishimura, M., Harashima, S., & Oshima, Y. (1991). The *PHO84* gene of *Saccharomyces cerevisiae* encodes an inorganic phosphate transporter. *Molecular and Cellular Biology*, *11*(6), 3229–3238. http://www.ncbi.nlm.nih.gov/pmc/articles/PMC360175/.
Cabrera, E., Álvarez, M. C., Martín, Y., Siverio, J. M., & Ramos, J. (2012). K(+) uptake systems in the yeast *Hansenula polymorpha*. Transcriptional and post-translational mechanisms involved in high-affinity K(+) transporter regulation. *Fungal Genetics and Biology*, *49*(9), 755–763. https://doi.org/10.1016/j.fgb.2012.06.008.
Cairney, J. W. G. (2011). Ectomycorrhizal fungi: The symbiotic route to the root for phosphorus in forest soils. *Plant and Soil*, *344*(1), 51–71. https://doi.org/10.1007/s11104-011-0731-0.
Casarin, V., Plassard, C., Hinsinger, P., & Arvieu, J. (2004). Quantification of ectomycorrhizal fungal effects on the bioavailability and mobilization of soil P in the rhizosphere of *Pinus pinaster*. *New Phytologist*, *163*(1), 177–185. https://doi.org/10.1111/j.1469-8137.2004.01093.x.
Casieri, L., Lahmidi, N. A., Doidy, J., Veneault-Fourrey, C., Migeon, A., Bonneau, L., et al. (2013). Biotrophic transportome in mutualistic plant–fungal interactions. *Mycorrhiza*, *23*, 597–625. https://doi.org/10.1007/s00572-013-0496-9.
Chalot, M., Blaudez, D., & Brun, A. (2006). Ammonia: A candidate for nitrogen transfer at the mycorrhizal interface. *Trends in Plant Science*, *11*(6), 263–266. https://doi.org/10.1016/j.tplants.2006.04.005.
Chalot, M., & Brun, A. (1998). Physiology of organic nitrogen acquisition by ectomycorrhizal fungi and ectomycorrhizas. *FEMS Microbiology Reviews*, *22*(1), 21–44. https://doi.org/10.1111/j.1574-6976.1998.tb00359.x.
Chalot, M., Javelle, A., Blaudez, D., Lambilliote, R., Cooke, R., Sentenac, H., et al. (2002). An update on nutrient transport processes in ectomycorrhizas. *Plant and Soil*, *244*(1), 165–175. https://doi.org/10.1023/A:1020240709543.
Churchland, C., & Grayston, S. J. (2014). Specificity of plant-microbe interactions in the tree mycorrhizosphere biome and consequences for soil C cycling. *Frontiers in Microbiology*, *5*, 261. https://doi.org/10.3389/fmicb.2014.00261.
Clemmensen, K. E., Bahr, A., Ovaskainen, O., Dahlberg, A., Ekblad, A., Wallander, H., et al. (2013). Roots and associated fungi drive long-term carbon sequestration in boreal forest. *Science (New York, NY)*, *339*(6127), 1615–1618. https://doi.org/10.1126/science.1231923.
Colpaert, J. V., & van Assche, J. A. (1987). Heavy metal tolerance in some ectomycorrhizal fungi. *Functional Ecology*, *1*(4), 415–421. https://doi.org/10.2307/2389799.
Colpaert, J. V., Wevers, J. H. L., Krznaric, E., & Adriaensen, K. (2011). How metal-tolerant ecotypes of ectomycorrhizal fungi protect plants from heavy metal pollution. *Annals of Forest Science*, *68*(1), 17–24. https://doi.org/10.1007/s13595-010-0003-9.

Condron, L. M., & Tiessen, H. (2005). Organic phosphorus in terrestrial ecosystems. In B. L. Turner, E. Frossard, & D. S. Baldwin (Eds.), *Organic phosphorus in the environment* (pp. 295–307). Wallingford CABI Publishing. https://doi.org/10.1079/9780851998220.0000.

Coninx, L., Thoonen, A., Slenders, E., Morin, E., Arnauts, N., Op De Beeck, M., et al. (2017). The *SlZRT1* gene encodes a plasma membrane-located ZIP (Zrt-, Irt-like protein) transporter in the ectomycorrhizal fungus *Suillus luteus*. *Frontiers in Microbiology*, *8*, 2320. https://doi.org/10.3389/fmicb.2017.02320.

Corratgé, C., Zimmermann, S., Lambilliotte, R., Plassard, C., Marmeisse, R., Thibaud, J.-B., et al. (2007). Molecular and functional characterization of a Na(+)-K(+) transporter from the Trk family in the ectomycorrhizal fungus *Hebeloma cylindrosporum*. *The Journal of Biological Chemistry*, *282*(36), 26057–26066. https://doi.org/10.1074/jbc.M611613200.

Corratgé-Faillie, C., Jabnoune, M., Zimmermann, S., Véry, A.-A., Fizames, C., & Sentenac, H. (2010). Potassium and sodium transport in non-animal cells: The Trk/Ktr/HKT transporter family. *Cellular and Molecular Life Sciences*, *67*(15), 2511–2532. https://doi.org/10.1007/s00018-010-0317-7.

Courty, P.-E., Buée, M., Diedhiou, A. G., Frey-Klett, P., Le Tacon, F., Rineau, F., et al. (2010). The role of ectomycorrhizal communities in forest ecosystem processes: New perspectives and emerging concepts. *Soil Biology and Biochemistry*, *42*(5), 679–698. https://doi.org/10.1016/j.soilbio.2009.12.006.

Courty, P.-E., Doidy, J., Garcia, K., Wipf, D., & Zimmermann, S. D. (2016). The transportome of mycorrhizal systems. In F. Martin (Ed.), *Molecular mycorrhizal symbiosis*. (pp. 239–256). https://doi.org/10.1002/9781118951446.ch14.

Courty, P.-E., Smith, P., Koegel, S., Redecker, D., & Wipf, D. (2015). Inorganic nitrogen uptake and transport in beneficial plant root-microbe interactions. *Critical Reviews in Plant Sciences*, *34*(1–3), 4–16. https://doi.org/10.1080/07352689.2014.897897

Cromer, D. A. N. (1935). The significance of the mycorrhiza of *Pinus radiata*. In *Commonwealth Forestry Bureau, Canberra.*Australia Bull. 16 (15 pp).

Cubero, B., Nakagawa, Y., Jiang, X.-Y., Miura, K.-J., Li, F., Raghothama, K. G., et al. (2009). The phosphate transporter PHT4;6 is a determinant of salt tolerance that is localized to the Golgi apparatus of Arabidopsis. *Molecular Plant*, *2*(3), 535–552. https://doi.org/10.1093/mp/ssp013.

Dietz, S., von Bülow, J., Beitz, E., & Nehls, U. (2011). The aquaporin gene family of the ectomycorrhizal fungus *Laccaria bicolor*: Lessons for symbiotic functions. *New Phytologist*, *190*(4), 927–940. https://doi.org/10.1111/j.1469-8137.2011.03651.x.

Ditengou, F. A., Béguiristain, T., & Lapeyrie, F. (2000). Root hair elongation is inhibited by hypaphorine, the indole alkaloid from the ectomycorrhizal fungus *Pisolithus tinctorius*, and restored by indole-3-acetic acid. *Planta*, *211*(5), 722–728. https://doi.org/10.1007/s004250000342.

Doré, J., Kohler, A., Dubost, A., Hundley, H., Singan, V., Peng, Y., et al. (2017). The ectomycorrhizal basidiomycete *Hebeloma cylindrosporum* undergoes early waves of transcriptional reprogramming prior to symbiotic structures differentiation. *Environmental Microbiology*, *19*(3), 1338–1354. https://doi.org/10.1111/1462-2920.13670.

Doré, J., Perraud, M., Dieryckx, C., Kohler, A., Morin, E., Henrissat, B., et al. (2015). Comparative genomics, proteomics and transcriptomics give new insight into the exoproteome of the basidiomycete *Hebeloma cylindrosporum* and its involvement in ectomycorrhizal symbiosis. *New Phytologist*, *208*(4), 1169–1187. https://doi.org/10.1111/nph.13546.

Duddridge, J. A., Malibari, A., & Read, D. J. (1980). Structure and function of mycorrhizal rhizomorphs with special reference to their role in water transport. *Nature*, *287*, 834–836. https://doi.org/10.1038/287834a0.

Eastwood, D. C., Floudas, D., Binder, M., Majcherczyk, A., Schneider, P., Aerts, A., et al. (2011). The plant cell wall–decomposing machinery underlies the functional diversity of forest fungi. *Science*, *333*(6043), 762–765. http://science.sciencemag.org/content/333/6043/762.abstract.

Ezawa, T., Smith, S. E., & Smith, F. A. (2002). P metabolism and transport in AM fungi. *Plant and Soil*, *244*(1), 221–230. https://doi.org/10.1023/A:1020258325010.

Fan, D., Liu, T., Li, C., Jiao, B., Li, S., Hou, Y., et al. (2015). Efficient CRISPR/Cas9-mediated targeted mutagenesis in *Populus* in the first generation. *Scientific Reports*, *5*, 12217. https://doi.org/10.1038/srep12217.

Finlay, R. D. (2008). Ecological aspects of mycorrhizal symbiosis: With special emphasis on the functional diversity of interactions involving the extraradical mycelium. *Journal of Experimental Botany*, *59*(5), 1115–1126. https://doi.org/10.1093/jxb/ern059.

Floudas, D., Binder, M., Riley, R., Barry, K., Blanchette, R. A., Henrissat, B., et al. (2012). The Paleozoic origin of enzymatic lignin decomposition reconstructed from 31 fungal genomes. *Science*, *336*(6089), 1715–1719. https://doi.org/10.1126/science.1221748.

Franklin, O., Näsholm, T., Högberg, P., & Högberg, M. N. (2014). Forests trapped in nitrogen limitation—An ecological market perspective on ectomycorrhizal symbiosis. *New Phytologist*, *203*(2), 657–666. https://doi.org/10.1111/nph.12840.

Fristedt, U., Weinander, R., Martinsson, H.-S., & Persson, B. L. (1999). Characterization of purified and unidirectionally reconstituted Pho84 phosphate permease of *Saccharomyces cerevisiae*. *FEBS Letters*, *458*(1), 1–5. https://doi.org/10.1016/S0014-5793(99)01108-4.

Gaber, R. F., Styles, C. A., & Fink, G. R. (1988). TRK1 encodes a plasma membrane protein required for high-affinity potassium transport in *Saccharomyces cerevisiae*. *Molecular and Cellular Biology*, *8*(7), 2848–2859. http://www.ncbi.nlm.nih.gov/pmc/articles/PMC363504/.

Galli, U., Schüepp, H., & Brunold, C. (1994). Heavy metal binding by mycorrhizal fungi. *Physiologia Plantarum*, *92*(2), 364–368. https://doi.org/10.1111/j.1399-3054.1994.tb05349.x.

Garbaye, J. (2000). The role of ectomycorrhizal symbiosis in the resistance of forests to water stress. *Outlook on Agriculture*, *29*(1), 63–69. https://doi.org/10.5367/000000000101293068.

Garcia, K., & Ané, J.-M. (2016). Comparative analysis of secretomes from ectomycorrhizal fungi with an emphasis on small-secreted proteins. *Frontiers in Microbiology*, *7*, 1734. https://doi.org/10.3389/fmicb.2016.01734.

Garcia, K., Delaux, P.-M., Cope, K. R., & Ané, J.-M. (2015). Molecular signals required for the establishment and maintenance of ectomycorrhizal symbioses. *New Phytologist*, *208*(1), 79–87. https://doi.org/10.1111/nph.13423.

Garcia, K., Delteil, A., Conéjéro, G., Becquer, A., Plassard, C., Sentenac, H., et al. (2014). Potassium nutrition of ectomycorrhizal *Pinus pinaster*: Overexpression of the *Hebeloma cylindrosporum* HcTrk1 transporter affects the translocation of both K(+) and phosphorus in the host plant. *New Phytologist*, *201*(3), 951–960. https://doi.org/10.1111/nph.12603.

Garcia, K., Doidy, J., Zimmermann, S. D., Wipf, D., & Courty, P.-E. (2016). Take a trip through the plant and fungal transportome of mycorrhiza. *Trends in Plant Science*, *21*(11), 937–950. https://doi.org/10.1016/j.tplants.2016.07.010.

Garcia, K., Haider, M. Z., Delteil, A., Corratgé-Faillie, C., Conéjero, G., Tatry, M.-V., et al. (2013). Promoter-dependent expression of the fungal transporter HcPT1.1 under Pi shortage and its spatial localization in ectomycorrhiza. *Fungal Genetics and Biology*, *58–59*, 53–61. https://doi.org/10.1016/j.fgb.2013.06.007.

Garcia, K., & Zimmermann, S. D. (2014). The role of mycorrhizal associations in plant potassium nutrition. *Frontiers in Plant Science*, *5*, 1–9. https://doi.org/10.3389/fpls.2014.00337.

Gehring, C. A., Sthultz, C. M., Flores-Rentería, L., Whipple, A. V., & Whitham, T. G. (2017). Tree genetics defines fungal partner communities that may confer drought tolerance. *Proceedings of the National Academy of Sciences of the United States of America, 114*(42), 11169–11174. http://www.pnas.org/content/114/42/11169.abstract.

Goss, R. W. (1960). Mycorrhizae of ponderosa pine in Nebraska grassland soils. *Research Bulletin. Nebraska Agricultural Experiment Station, 192*, 47.

Guerrero-Galán, C., Delteil, A., Garcia, K., Houdinet, G., Conéjéro, G., Gaillard, I., et al. (2018). Plant potassium nutrition in ectomycorrhizal symbiosis: Properties and roles of the three fungal TOK potassium channels in *Hebeloma cylindrosporum*. *Environmental Microbiology, 20*(5), 1873–1887. https://doi.org/10.1111/1462-2920.14122.

Guerrero-Galán, C., Garcia, K., Houdinet, G., & Zimmermann, S. D. (2018). HcTOK1 participates in the maintenance of K+ homeostasis in the ectomycorrhizal fungus *Hebeloma cylindrosporum*, which is essential for the symbiotic K+ nutrition of *Pinus pinaster*. *Plant Signaling & Behavior, 13*(6), e1480845. https://doi.org/10.1111/pce.13359.

Hacquard, S., Tisserant, E., Brun, A., Legué, V., Martin, F., & Kohler, A. (2013). Laser microdissection and microarray analysis of *Tuber melanosporum* ectomycorrhizas reveal functional heterogeneity between mantle and Hartig net compartments. *Environmental Microbiology, 15*(6), 1853–1869. https://doi.org/10.1111/1462-2920.12080.

Hatakka, A., & Hammel, K. E. (2010). Fungal biodegradation of lignocelluloses. In *Industrial applications. Mycota* (2nd ed., pp. 319–340). Berlin, Heidelberg: Springer. ISBN: 9783642114571 (rel): 3642114571 (rel).

Heijden, M. G. A., Martin, F. M., Selosse, M., & Sanders, I. R. (2015). Mycorrhizal ecology and evolution: The past, the present, and the future. *New Phytologist, 205*(4), 1406–1423. https://doi.org/10.1111/nph.13288.

Hibbett, D. S., Gilbert, L. B., & Donoghue, M. J. (2000). Evolutionary instability of ectomycorrhizal symbioses in basidiomycetes. *Nature, 407*(6803), 506–508. https://doi.org/10.1038/35035065.

Hinsinger, P. (2001). Bioavailability of soil inorganic P in the rhizosphere as affected by root-induced chemical changes: A review. *Plant and Soil, 237*(2), 173–195. https://doi.org/10.1023/A:1013351617532.

Horton, T. R., & Van der Heijden, M. G. A. (2008). The role of symbioses in seedling establishment and survival. In M. A. Leck, R. L. Simpson, & V. T. Parker (Eds.), *Seedling ecology and evolution* (pp. 189–214). Cambridge: Cambridge University Press. https://doi.org/10.1017/CBO9780511815133.011.

Hürlimann, H. C., Pinson, B., Stadler-Waibel, M., Zeeman, S. C., & Freimoser, F. M. (2009). The SPX domain of the yeast low-affinity phosphate transporter Pho90 regulates transport activity. *EMBO Reports, 10*(9), 1003–1008. http://embor.embopress.org/content/10/9/1003.abstract.

James, T. Y., Kauff, F., Schoch, C. L., Matheny, P. B., Hofstetter, V., Cox, C. J., et al. (2006). Reconstructing the early evolution of fungi using a six-gene phylogeny. *Nature, 443*, 818. https://doi.org/10.1038/nature05110.

Jargeat, P., Gay, G., Debaud, J.-C., & Marmeisse, R. (2000). Transcription of a nitrate reductase gene isolated from the symbiotic basidiomycete fungus *Hebeloma cylindrosporum* does not require induction by nitrate. *Molecular and General Genetics, 263*(6), 948–956. https://doi.org/10.1007/PL00008695.

Jargeat, P., Rekangalt, D., Verner, M.-C., Gay, G., Debaud, J.-C., Marmeisse, R., et al. (2003). Characterisation and expression analysis of a nitrate transporter and nitrite reductase genes, two members of a gene cluster for nitrate assimilation from the symbiotic basidiomycete *Hebeloma cylindrosporum*. *Current Genetics, 43*(3), 199–205. https://doi.org/10.1007/s00294-003-0387-2.

Javelle, A., André, B., Marini, A.-M., & Chalot, M. (2003). High-affinity ammonium transporters and nitrogen sensing in mycorrhizas. *Trends in Microbiology*, *11*(2), 53–55. https://doi.org/10.1016/S0966-842X(02)00012-4.

Javelle, A., Morel, M., Rodríguez-Pastrana, B. R., Botton, B., André, B., Marini, A. M., et al. (2003). Molecular characterization, function and regulation of ammonium transporters (Amt) and ammonium-metabolizing enzymes (GS, NADP-GDH) in the ectomycorrhizal fungus *Hebeloma cylindrosporum*. *Molecular Microbiology*, *47*(2), 411–430. https://doi.org/10.1046/j.1365-2958.2003.03303.x.

Javelle, A., Rodrı´guez-Pastrana, B.-R., Jacob, C., Botton, B., Brun, A., André, B., et al. (2001). Molecular characterization of two ammonium transporters from the ectomycorrhizal fungus *Hebeloma cylindrosporum*. *FEBS Letters*, *505*(3), 393–398. https://doi.org/10.1016/S0014-5793(01)02802-2.

Jentschke, G., Brandes, B., Kuhn, A. J., Schröder, W. H., & Godbold, D. L. (2001). Interdependence of phosphorus, nitrogen, potassium and magnesium translocation by the ectomycorrhizal fungus *Paxillus involutus*. *New Phytologist*, *149*(2), 327–337. https://doi.org/10.1046/j.1469-8137.2001.00014.x.

Jentschke, G., & Godbold, D. L. (2000). Metal toxicity and ectomycorrhizas. *Physiologia Plantarum*, *109*(2), 107–116. https://doi.org/10.1034/j.1399-3054.2000.100201.x.

Jourand, P., Hannibal, L., Majorel, C., Mengant, S., Ducousso, M., & Lebrun, M. (2014). Ectomycorrhizal *Pisolithus albus* inoculation of *Acacia spirorbis* and *Eucalyptus globulus* grown in ultramafic topsoil enhances plant growth and mineral nutrition while limits metal uptake. *Journal of Plant Physiology*, *171*(2), 164–172. https://doi.org/10.1016/j.jplph.2013.10.011.

Kemppainen, M. J., Alvarez Crespo, M. C., & Pardo, A. G. (2010). fHANT-AC genes of the ectomycorrhizal fungus *Laccaria bicolor* are not repressed by L-glutamine allowing simultaneous utilization of nitrate and organic nitrogen sources. *Environmental Microbiology Reports*, *2*(4), 541–553. https://doi.org/10.1111/j.1758-2229.2009.00111.x.

Kemppainen, M. J., & Pardo, A. G. (2013). LbNrt RNA silencing in the mycorrhizal symbiont *Laccaria bicolor* reveals a nitrate-independent regulatory role for a eukaryotic NRT2-type nitrate transporter. *Environmental Microbiology Reports*, *5*(3), 353–366. https://doi.org/10.1111/1758-2229.12029.

Ketchum, K. A., Joiner, W. J., Sellers, A. J., Kaczmarek, L. K., & Goldstein, S. A. N. (1995). A new family of outwardly rectifying potassium channel proteins with two pore domains in tandem. *Nature*, *376*, 690. https://doi.org/10.1038/376690a0.

Khosla, B., Kaur, H., & Sudhakara Reddy, M. (2009). Influence of ectomycorrhizal colonization on the growth and mineral nutrition of *Populus deltoides* under aluminum toxicity. *Journal of Plant Interactions*, *4*(2), 93–99. https://doi.org/10.1080/17429140802382953.

Ko, C. H., & Gaber, R. F. (1991). *TRK1* and *TRK2* encode structurally related K+ transporters in *Saccharomyces cerevisiae*. *Molecular and Cellular Biology*, *11*(8), 4266–4273. http://www.ncbi.nlm.nih.gov/pmc/articles/PMC361257/.

Kohler, A., Kuo, A., Nagy, L. G., Morin, E., Barry, K. W., Buscot, F., et al. (2015). Convergent losses of decay mechanisms and rapid turnover of symbiosis genes in mycorrhizal mutualists. *Nature Genetics*, *47*, 410–415. https://doi.org/10.1038/ng.3223.

Lambilliotte, R., Cooke, R., Samson, D., Fizames, C., Gaymard, F., Plassard, C., et al. (2004). Large-scale identification of genes in the fungus *Hebeloma cylindrosporum* paves the way to molecular analyses of ectomycorrhizal symbiosis. *New Phytologist*, *164*(3), 505–513. https://doi.org/10.1111/j.1469-8137.2004.01185.x.

Landeweert, R., Hoffland, E., Finlay, R. D., Kuyper, T. W., & van Breemen, N. (2001). Linking plants to rocks: Ectomycorrhizal fungi mobilize nutrients from minerals. *Trends in Ecology & Evolution*, *16*(5), 248–254. https://doi.org/10.1016/S0169-5347(01)02122-X.

Lapeyrie, F., Ranger, J., & Vairelles, D. (1991). Phosphate-solubilizing activity of ectomycorrhizal fungi in vitro. *Canadian Journal of Botany*, *69*(2), 342–346. https://doi.org/10.1139/b91-046.

Larsen, P. E., Sreedasyam, A., Trivedi, G., Podila, G. K., Cseke, L. J., & Collart, F. R. (2011). Using next generation transcriptome sequencing to predict an ectomycorrhizal metabolome. *BMC Systems Biology*, *5*(1), 1–14. https://doi.org/10.1186/1752-0509-5-70.

Lee, S. H., Calvo-Polanco, M., Chung, G. C., & Zwiazek, J. J. (2010). Role of aquaporins in root water transport of ectomycorrhizal jack pine (*Pinus banksiana*) seedlings exposed to NaCl and fluoride. *Plant, Cell & Environment*, *33*(5), 769–780. https://doi.org/10.1111/j.1365-3040.2009.02103.x.

Lee, E.-H., Eo, J.-K., Ka, K.-H., & Eom, A.-H. (2013). Diversity of arbuscular mycorrhizal fungi and their roles in ecosystems. *Mycobiology*, *41*(3), 121–125. https://doi.org/10.5941/MYCO.2013.41.3.121.

Lehto, T., & Zwiazek, J. J. (2011). Ectomycorrhizas and water relations of trees: A review. *Mycorrhiza*, *21*(2), 71–90. https://doi.org/10.1007/s00572-010-0348-9.

Lei, J., & Dexheimer, J. (1988). Ultrastructural localization of ATPase activity in the *Pinus sylvestris*/*Laccaria laccata* ectomycorrhizal association. *New Phytologist*, *108*(3), 329–334. https://doi.org/10.1111/j.1469-8137.1988.tb04170.x.

Le Quéré, A., Wright, D. P., Söderström, B., Tunlid, A., & Johansson, T. (2005). Global patterns of gene regulation associated with the development of ectomycorrhiza between birch (*Betula pendula* Roth.) and *Paxillus involutus* (Batsch) Fr. *Molecular Plant-Microbe Interactions*, *18*(7), 659–673. https://doi.org/10.1094/MPMI-18-0659.

Liang, Y., Chen, H., Tang, M., & Shen, S. (2007). Proteome analysis of an ectomycorrhizal fungus *Boletus edulis* under salt shock. *Mycological Research*, *111*(8), 939–946. https://doi.org/10.1016/j.mycres.2007.06.005.

Lindahl, B. D., Ihrmark, K., Boberg, J., Trumbore, S. E., Högberg, P., Stenlid, J., et al. (2006). Spatial separation of litter decomposition and mycorrhizal nitrogen uptake in a boreal forest. *New Phytologist*, *173*(3), 611–620. https://doi.org/10.1111/j.1469-8137.2006.01936.x.

Lindahl, B. D., & Tunlid, A. (2015). Ectomycorrhizal fungi—Potential organic matter decomposers, yet not saprotrophs. *New Phytologist*, *205*(4), 1443–1447. https://doi.org/10.1111/nph.13201.

Liu, Q., Loganathan, P., Hedley, M. J., & Grace, L. J. (2008). Effect of mycorrhizal inoculation on rhizosphere properties, phosphorus uptake and growth of pine seedlings treated with and without a phosphate rock fertilizer. *Journal of Plant Nutrition*, *31*(1), 137–156. https://doi.org/10.1080/01904160701742063.

Lofgren, L., Nguyen, N. H., & Kennedy, P. G. (2018). Ectomycorrhizal host specificity in a changing world: Can legacy effects explain anomalous current associations? *New Phytologist*, *220*(4), 1273–1284. https://doi.org/10.1111/nph.15008. https://doi.org/10.1111/nph.15008.

Loth-Pereda, V., Orsini, E., Courty, P.-E., Lota, F., Kohler, A., Diss, L., et al. (2011). Structure and expression profile of the phosphate *Pht1* transporter gene family in mycorrhizal *Populus trichocarpa*. *Plant Physiology*, *156*(4), 2141–2154. https://doi.org/10.1104/pp.111.180646.

Louche, J., Ali, M. A., Cloutier-Hurteau, B., Sauvage, F.-X., Quiquampoix, H., & Plassard, C. (2010). Efficiency of acid phosphatases secreted from the ectomycorrhizal fungus *Hebeloma cylindrosporum* to hydrolyse organic phosphorus in podzols. *FEMS Microbiology Ecology*, *73*(2), 323–335. http://femsec.oxfordjournals.org/content/73/2/323.abstract.

Lucic, E., Fourrey, C., Kohler, A., Martin, F., Chalot, M., & Brun-Jacob, A. (2008). A gene repertoire for nitrogen transporters in *Laccaria bicolor*. *New Phytologist*, *180*(2), 343–364. https://doi.org/10.1111/j.1469-8137.2008.02580.x.

Marjanović, Ž., Uehlein, N., Kaldenhoff, R., Zwiazek, J., Weiß, M., Hampp, R., et al. (2005). Aquaporins in poplar: What a difference a symbiont makes! *Planta*, *222*(2), 258–268. https://doi.org/10.1007/s00425-005-1539-z.

Marmeisse, R., & Girlanda, M. (2016). Mycorrhizal fungi and the soil carbon and nutrient cycling. In I. S. Druzhinina & C. P. Kubicek (Eds.), *Environmental and microbial relationships* (pp. 189–203). Cham: Springer International Publishing. https://doi.org/10.1007/978-3-319-29532-9_10.

Martin, F. M., Aerts, A., Ahren, D., Brun, A., Danchin, E. G. J., Duchaussoy, F., et al. (2008). The genome of *Laccaria bicolor* provides insights into mycorrhizal symbiosis. *Nature, 452*(7183), 88–92. https://doi.org/10.1038/nature06556.

Martin, F., Duplessis, S., Ditengou, F., Lagrange, H., Voiblet, C., & Lapeyrie, F. (2001). Developmental cross talking in the ectomycorrhizal symbiosis: Signals and communication genes. *New Phytologist, 151*(1), 145–154. https://doi.org/10.1046/j.1469-8137.2001.00169.x.

Martin, F., Kohler, A., Murat, C., Balestrini, R., Coutinho, P. M., Jaillon, O., et al. (2010). Perigord black truffle genome uncovers evolutionary origins and mechanisms of symbiosis. *Nature, 464*(7291), 1033–1038. https://doi.org/10.1038/nature08867.

Martin, F. M., Uroz, S., & Barker, D. G. (2017). Ancestral alliances: Plant mutualistic symbioses with fungi and bacteria. *Science, 356*(6340), eaad4501. http://science.sciencemag.org/content/356/6340/eaad4501.abstract.

Martinez, P., & Persson, B. L. (1998). Identification, cloning and characterization of a derepressible Na+-coupled phosphate transporter in *Saccharomyces cerevisiae. Molecular and General Genetics, 258*(6), 628–638. https://doi.org/10.1007/s004380050776.

Martino, E., Morin, E., Grelet, G., Kuo, A., Kohler, A., Daghino, S., et al. (2018). Comparative genomics and transcriptomics depict ericoid mycorrhizal fungi as versatile saprotrophs and plant mutualists. *New Phytologist, 217*(3), 1213–1229. https://doi.org/10.1111/nph.14974.

Maurel, C., Verdoucq, L., Luu, D.-T., & Santoni, V. (2008). Plant aquaporins: Membrane channels with multiple integrated functions. *Annual Review of Plant Biology, 59*(1), 595–624. https://doi.org/10.1146/annurev.arplant.59.032607.092734.

Melin, E., & Nilsson, H. (1950). Transfer of radioactive phosphorus to pine seedlings by means of mycorrhizal hyphae. *Physiologia Plantarum, 3*(1), 88–92. https://doi.org/10.1111/j.1399-3054.1950.tb07495.x.

Mello, A., & Balestrini, R. (2018). Recent insights on biological and ecological aspects of ectomycorrhizal fungi and their interactions. *Frontiers in Microbiology, 9*, 216. https://doi.org/10.3389/fmicb.2018.00216.

Midgley, M. G., & Phillips, R. P. (2014). Mycorrhizal associations of dominant trees influence nitrate leaching responses to N deposition. *Biogeochemistry, 117*(2), 241–253. https://doi.org/10.1007/s10533-013-9931-4.

Montanini, B., Moretto, N., Soragni, E., Percudani, R., & Ottonello, S. (2002). A high-affinity ammonium transporter from the mycorrhizal ascomycete *Tuber borchii. Fungal Genetics and Biology, 36*(1), 22–34. https://doi.org/10.1016/S1087-1845(02)00001-4.

Montanini, B., Viscomi, A. R., Bolchi, A., Martin, Y., Siverio, J. M., Balestrini, R., et al. (2006). Functional properties and differential mode of regulation of the nitrate transporter from a plant symbiotic ascomycete. *Biochemical Journal, 394*(1), 125–134. https://doi.org/10.1042/BJ20051199.

Morel, M., Buée, M., Chalot, M., & Brun, A. (2006). NADP-dependent glutamate dehydrogenase: A dispensable function in ectomycorrhizal fungi. *New Phytologist, 169*(1), 179–190. https://doi.org/10.1111/j.1469-8137.2005.01556.x.

Morel, M., Jacob, C., Fitz, M., Wipf, D., Chalot, M., & Brun, A. (2008). Characterization and regulation of PiDur3, a permease involved in the acquisition of urea by the ectomycorrhizal fungus *Paxillus involutus. Fungal Genetics and Biology, 45*(6), 912–921. https://doi.org/10.1016/j.fgb.2008.01.002.

Morel, M., Jacob, C., Kohler, A., Johansson, T., Martin, F., Chalot, M., et al. (2005). Identification of genes differentially expressed in extraradical mycelium and ectomycorrhizal roots during *Paxillus involutus-Betula pendula* ectomycorrhizal symbiosis. *Applied and Environmental Microbiology, 71*(1), 382–391. https://doi.org/10.1128/AEM.71.1.382-391.2005.

Moyersoen, B. (2006). Pakaraimaea dipterocarpacea is ectomycorrhizal, indicating an ancient Gondwanaland origin for the ectomycorrhizal habit in dipterocarpaceae. *New Phytologist, 172*(4), 753–762. https://doi.org/10.1111/j.1469-8137.2006.01860.x.

Muhsin, T. M., & Zwiazek, J. J. (2002). Ectomycorrhizas increase apoplastic water transport and root hydraulic conductivity in *Ulmus americana* seedlings. *New Phytologist, 153*(1), 153–158. https://doi.org/10.1046/j.0028-646X.2001.00297.x.

Mujic, A. B., Kuo, A., Tritt, A., Lipzen, A., Chen, C., Johnson, J., et al. (2017). Comparative genomics of the ectomycorrhizal sister species *Rhizopogon vinicolor* and *Rhizopogon vesiculosus* (basidiomycota: boletales) reveals a divergence of the mating type B locus. *G3: Genes|Genomes|Genetics*, 7(6), 1775–1789. https://doi.org/10.1534/g3.117.039396.

Müller, T., Avolio, M., Olivi, M., Benjdia, M., Rikirsch, E., Kasaras, A., et al. (2007). Nitrogen transport in the ectomycorrhiza association: The *Hebeloma cylindrosporum-Pinus pinaster* model. *Phytochemistry, 68*(1), 41–51. https://doi.org/10.1016/j.phytochem.2006.09.021.

Nara, K. (2006). Ectomycorrhizal networks and seedling establishment during early primary succession. *New Phytologist, 169*(1), 169–178. https://doi.org/10.1111/j.1469-8137.2005.01545.x.

Navarro-Ródenas, A., Xu, H., Kemppainen, M., Pardo, A. G., & Zwiazek, J. J. (2015). *Laccaria bicolor* aquaporin LbAQP1 is required for Hartig net development in trembling aspen (*Populus tremuloides*). *Plant, Cell & Environment, 38*(11), 2475–2486. https://doi.org/10.1111/pce.12552.

Nehls, U., & Dietz, S. (2014). Fungal aquaporins: Cellular functions and ecophysiological perspectives. *Applied Microbiology and Biotechnology, 98*(21), 8835–8851. https://doi.org/10.1007/s00253-014-6049-0.

Nehls, U., Kleber, R., Wiese, J., & Hampp, R. (1999). Isolation and characterization of a general amino acid permease from the ectomycorrhizal fungus *Amanita muscaria*. *New Phytologist, 144*(2), 343–349. https://doi.org/10.1046/j.1469-8137.1999.00513.x.

Nehls, U., & Plassard, C. (2018). Nitrogen and phosphate metabolism in ectomycorrhizas. *New Phytologist, 220*(4), 1047–1058. https://doi.org/10.1111/nph.15257.

Newman, E. I. (1988). Mycorrhizal links between plants: Their functioning and ecological significance. In M. Begon, A. H. Fitter, E. D. Ford, & E. R. Macfadyen (Eds.), *Advances in ecological research: Vol. 18* (pp. 243–270). Academic Press. https://doi.org/10.1016/S0065-2504(08)60182-8.

Nieves-Cordones, M., Alemán, F., Martínez, V., & Rubio, F. (2014). K+ uptake in plant roots. The systems involved, their regulation and parallels in other organisms. *Journal of Plant Physiology, 171*(9), 688–695. https://doi.org/10.1016/j.jplph.2013.09.021.

Nygren, C. M., Eberhardt, U., Karlsson, M., Parrent, J. L., Lindahl, B. D., & Taylor, A. F. (2008). Growth on nitrate and occurrence of nitrate reductase-encoding genes in a phylogenetically diverse range of ectomycorrhizal fungi. *New Phytologist, 180*(4), 875–889. https://doi:10.1111/j.1469-8137.2008.02618.x.

Op De Beeck, M., Troein, C., Peterson, C., Persson, P., & Tunlid, A. (2018). Fenton reaction facilitates organic nitrogen acquisition by an ectomycorrhizal fungus. *New Phytologist, 218*(1), 335–343. https://doi.org/10.1111/nph.14971.

Oyewole, O. A., Inselsbacher, E., Näsholm, T., & Jämtgårda, S. (2017). Incorporating mass flow strongly promotes N flux rates in boreal forest soils. *Soil Biology and Biochemistry, 114*, 263–269. https://doi.org/10.1016/j.soilbio.2017.07.021.

Papazian, D. M., Schwarz, T. L., Tempel, B. L., Jan, Y. N., & Jan, L. Y. (1987). Cloning of genomic and complementary DNA from Shaker, a putative potassium channel gene from Drosophila. *Science*, *237*(4816), 749–753. http://science.sciencemag.org/content/237/4816/749.abstract.

Pellegrin, C., Morin, E., Martin, F. M., & Veneault-Fourrey, C. (2015). Comparative analysis of secretomes from ectomycorrhizal fungi with an emphasis on small-secreted proteins. *Frontiers in Microbiology*, *6*, 1278. https://doi.org/10.3389/fmicb.2015.01278.

Perez-Moreno, J., & Read, J. D. (2000). Mobilization and transfer of nutrients from litter to tree seedlings via the vegetative mycelium of ectomycorrhizal plants. *New Phytologist*, *145*(2), 301–309. https://doi.org/10.1046/j.1469-8137.2000.00569.x.

Perez-Moreno, J., & Read, D. J. (2002). Nutrient transfer from soil nematodes to plants: A direct pathway provided by the mycorrhizal mycelial network. *Plant, Cell & Environment*, *24*(11), 1219–1226. https://doi.org/10.1046/j.1365-3040.2001.00769.x.

Peter, M., Kohler, A., Ohm, R. A., Kuo, A., Krützmann, J., Morin, E., et al. (2016). Ectomycorrhizal ecology is imprinted in the genome of the dominant symbiotic fungus *Cenococcum geophilum*. *Nature Communications*, 7, 12662. https://doi.org/10.1038/ncomms12662.

Peterson, R. L., & Massicotte, H. B. (2004). Exploring structural definitions of mycorrhizas, with emphasis on nutrient-exchange interfaces. *Canadian Journal of Botany*, *82*(8), 1074–1088. https://doi.org/10.1139/b04-071.

Pfeffer, P. E., Bago, B., & Shachar-Hill, Y. (2001). Exploring mycorrhizal function with NMR spectroscopy. *New Phytologist*, *150*(3), 543–553. https://doi.org/10.1046/j.1469-8137.2001.00139.x.

Plamboeck, A. H., Dawson, T. E., Egerton-Warburton, L. M., North, M., Bruns, T. D., & Querejeta, J. I. (2007). Water transfer via ectomycorrhizal fungal hyphae to conifer seedlings. *Mycorrhiza*, *17*(5), 439–447. https://doi.org/10.1007/s00572-007-0119-4.

Plassard, C., & Dell, B. (2010). Phosphorus nutrition of mycorrhizal trees. *Tree Physiology*, *30*(9), 1129–1139. https://doi.org/10.1093/treephys/tpq063.

Preuss, C. P., Huang, C. Y., Gilliham, M., & Tyerman, S. D. (2010). Channel-like characteristics of the low-affinity barley phosphate transporter PHT1;6 when expressed in *Xenopus* oocytes. *Plant Physiology*, *152*(3), 1431–1441. http://www.plantphysiol.org/content/152/3/1431.abstract.

Preuss, C. P., Huang, C. Y., & Tyerman, S. D. (2011). Proton-coupled high-affinity phosphate transport revealed from heterologous characterization in *Xenopus* of barley-root plasma membrane transporter, HvPHT1;1. *Plant, Cell & Environment*, *34*(4), 681–689. https://doi.org/10.1111/j.1365-3040.2010.02272.x.

Pylro, V. S., de Freitas, A. L. M., Otoni, W. C., da Silva, I. R., Borges, A. C., & Costa, M. D. (2013). Calcium oxalate crystals in eucalypt ectomycorrhizae: Morphochemical characterization. *PLoS One*, *8*(7), e67685 https://doi.org/10.1371/journal.pone.0067685.

Quiquampoix, H., & Mousain, D. (2005). Enzymatic hydrolysis of organic phosphorus. In B. L. Turner, E. Frossard, & D. S. Baldwin (Eds.), *Organic phosphorus in the environment* (pp. 89–112). Wallingford, UK: CAB International. https://doi.org/10.1079/9780851998220.0089.

Quirk, J., Andrews, M. Y., Leake, J. R., Banwart, S. A., & Beerling, D. J. (2014). Ectomycorrhizal fungi and past high CO(2) atmospheres enhance mineral weathering through increased below-ground carbon-energy fluxes. *Biology Letters*, *10*(7), 20140375. https://doi.org/10.1098/rsbl.2014.0375.

Rawat, S. R., Silim, S. N., Kronzucker, H. J., Siddiqi, M. Y., & Glass, A. D. (1999). *AtAMT1* gene expression and NH_4^+ uptake in roots of *Arabidopsis thaliana*: Evidence for regulation by root glutamine levels. *The Plant Journal*, *19*(2), 143–152. https://doi.org/10.1046/j.1365-313X.1999.00505.x.

Read, D. J., & Perez-Moreno, J. (2003). Mycorrhizas and nutrient cycling in ecosystems—A journey towards relevance. *New Phytologist*, *157*(3), 475–492. https://doi.org/10.1046/j.1469-8137.2003.00704.x.

Reid, C. P. (1978). Mycorrhizae and water stress. In A. Riedacker & J. Gagnaire-Michard (Eds.), *Root physiology and symbiosis, IUFRO Symposium Proceedings, Nancy, France* (pp. 392–408).

Rennenberg, H., & Herschbach, C. (2013). Phosphorus nutrition of woody plants: Many questions—Few answers. *Plant Biology*, *15*(5), 785–788. https://doi.org/10.1111/plb.12078.

Rillig, M. C., & Mummey, D. L. (2006). Mycorrhizas and soil structure. *New Phytologist*, *171*(1), 41–53. https://doi.org/10.1111/j.1469-8137.2006.01750.x.

Rineau, F., & Garbaye, J. (2010). Effects of liming on potential oxalate secretion and iron chelation of beech ectomycorrhizal root tips. *Microbial Ecology*, *60*(2), 331–339. https://doi.org/10.1007/s00248-010-9697-3.

Rineau, F., Roth, D., Shah, F., Smits, M., Johansson, T., Canbäck, B., et al. (2012). The ectomycorrhizal fungus *Paxillus involutus* converts organic matter in plant litter using a trimmed brown-rot mechanism involving Fenton chemistry. *Environmental Microbiology*, *14*(6), 1477–1487. https://doi.org/10.1111/j.1462-2920.2012.02736.x.

Rineau, F., Shah, F., Smits, M. M., Persson, P., Johansson, T., Carleer, R., et al. (2013). Carbon availability triggers the decomposition of plant litter and assimilation of nitrogen by an ectomycorrhizal fungus. *ISME Journal*, 7(10), 2010–2022. https://doi.org/10.1038/ismej.2013.91.

Rivetta, A., Allen, K. E., Slayman, C. W., & Slayman, C. L. (2013). Coordination of K+ transporters in neurospora: TRK1 is scarce and constitutive, while HAK1 is abundant and highly regulated. *Eukaryotic Cell*, *12*(5), 684–696. https://doi.org/10.1128/EC.00017-13.

Roberts, S. K. (2003). TOK homologue in *Neurospora crassa*: First cloning and functional characterization of an ion channel in a filamentous fungus. *Eukaryotic Cell*, *2*(1), 181–190. https://doi.org/10.1128/EC.2.1.181-190.2003.

Rodrı´guez-Navarro, A. (2000). Potassium transport in fungi and plants. *Biochimica et Biophysica Acta (BBA)—Reviews on Biomembranes*, *1469*(1), 1–30. https://doi.org/10.1016/S0304-4157(99)00013-1.

Ruytinx, J., Coninx, L., Nguyen, H., Smisdom, N., Morin, E., Kohler, A., et al. (2017). Identification, evolution and functional characterization of two Zn CDF-family transporters of the ectomycorrhizal fungus *Suillus luteus*. *Environmental Microbiology Reports*, *9*(4), 419–427. https://doi.org/10.1111/1758-2229.12551.

Rygiewicz, P. T., & Bledsoe, C. S. (1984). Mycorrhizal effects on potassium fluxes by northwest coniferous seedlings. *Plant Physiology*, *76*(4), 918–923. http://www.ncbi.nlm.nih.gov/pmc/articles/PMC1064406/.

Sácký, J., Leonhardt, T., & Kotrba, P. (2016). Functional analysis of two genes coding for distinct cation diffusion facilitators of the ectomycorrhizal Zn-accumulating fungus *Russula atropurpurea*. *Biometals*, *29*(2), 349–363. https://doi.org/10.1007/s10534-016-9920-x.

Schott, S., Valdebenito, B., Bustos, D., Gomez-Porras, J. L., Sharma, T., & Dreyer, I. (2016). Cooperation through competition—Dynamics and microeconomics of a minimal nutrient trade system in arbuscular mycorrhizal symbiosis. *Frontiers in Plant Science*, 7, 912. https://doi.org/10.3389/fpls.2016.00912.

Selle, A., Willmann, M., Grunze, N., Geßler, A., Weiß, M., & Nehls, U. (2005). The high-affinity poplar ammonium importer PttAMT1.2 and its role in ectomycorrhizal symbiosis. *New Phytologist*, *168*(3), 697–706. https://doi.org/10.1111/j.1469-8137.2005.01535.x.

Sentenac, H., Bonneaud, N., Minet, M., Lacroute, F., Salmon, J. M., Gaymard, F., et al. (1992). Cloning and expression in yeast of a plant potassium ion transport system. *Science*, *256*(5057), 663–665. http://science.sciencemag.org/content/256/5057/663.abstract.

Shabala, S., & Pottosin, I. (2014). Regulation of potassium transport in plants under hostile conditions: Implications for abiotic and biotic stress tolerance. *Physiologia Plantarum, 151*(3), 257–279. https://doi.org/10.1111/ppl.12165.

Shah, F., Nicolás, C., Bentzer, J., Ellström, M., Smits, M., Rineau, F., et al. (2015). Ectomycorrhizal fungi decompose soil organic matter using oxidative mechanisms adapted from saprotrophic ancestors. *New Phytologist, 209*(4), 1705–1719. https://doi.org/10.1111/nph.13722.

Shi, T.-Q., Liu, G.-N., Ji, R.-Y., Shi, K., Song, P., Ren, L.-J., et al. (2017). CRISPR/Cas9-based genome editing of the filamentous fungi: The state of the art. *Applied Microbiology and Biotechnology, 101*(20), 7435–7443. https://doi.org/10.1007/s00253-017-8497-9.

Simard, S. W., Perry, D. A., Jones, M. D., Myrold, D. D., Durall, D. M., & Molina, R. (1997). Net transfer of carbon between ectomycorrhizal tree species in the field. *Nature, 388*, 579. https://doi.org/10.1038/41557.

Simpson, A. J., Song, G., Smith, E., Lam, B., Novotny, E. H., & Hayes, H. B. (2007). Unraveling the structural components of soil humin by use of solution-state nuclear magnetic resonance spectroscopy. *Environmental Science & Technology, 41*(3), 876–883. https://doi.org/10.1021/es061576c.

Slaughter, J. C., McKernan, G., & Saita, M. (1990). Intracellular asparagine pool as a factor in control of ammonium uptake by *Saccharomyces cerevisiae*. *Mycological Research, 94*(8), 1009–1012. https://doi.org/10.1016/S0953-7562(09)81324-5.

Smith, S. E., & Read, D. (2008). *Mycorrhizal symbiosis* (3rd ed.). London: Academic Press. https://doi.org/10.1016/B978-012370526-6.50017-9.

Smits, M. M., Bonneville, S., Haward, S., & Leake, J. R. (2008). Ectomycorrhizal weathering, a matter of scale? *Mineralogical Magazine, 72*(1), 131–134. https://doi.org/10.1180/minmag.2008.072.1.131.

Strullu-Derrien, C., Selosse, M.-A., Kenrick, P., & Martin, F. M. (2018). The origin and evolution of mycorrhizal symbioses: From palaeomycology to phylogenomics. *New Phytologist, 220(4)*, 1012–1030. https://doi.org/10.1111/nph.15076.

Tatry, M.-V., El Kassis, E., Lambilliotte, R., Corratgé, C., van Aarle, I., Amenc, L. K., et al. (2009). Two differentially regulated phosphate transporters from the symbiotic fungus *Hebeloma cylindrosporum* and phosphorus acquisition by ectomycorrhizal *Pinus pinaster*. *The Plant Journal, 57*(6), 1092–1102. https://doi.org/10.1111/j.1365-313X.2008.03749.x.

Tedersoo, L., Bahram, M., Toots, M., Diedhiou, A. G., Henkel, T. W., Kjoller, R., et al. (2012). Towards global patterns in the diversity and community structure of ectomycorrhizal fungi. *Molecular Ecology, 21*(17), 4160–4170. https://doi.org/10.1111/j.1365-294X.2012.05602.x.

Tedersoo, L., May, T. W., & Smith, M. E. (2010). Ectomycorrhizal lifestyle in fungi: Global diversity, distribution, and evolution of phylogenetic lineages. *Mycorrhiza, 20*(4), 217–263. https://doi.org/10.1007/s00572-009-0274-x.

Tedersoo, L., & Smith, M. E. (2017). Ectomycorrhizal fungal lineages: Detection of four new groups and notes on consistent recognition of ectomycorrhizal taxa in high-throughput sequencing studies. In L. Tedersoo (Ed.), *Biogeography of mycorrhizal symbiosis* (pp. 125–142). Cham: Springer International Publishing. https://doi.org/10.1007/978-3-319-56363-3_6.

Tibbett, M. (2002). Considerations on the use of the p-nitrophenyl phosphomonoesterase assay in the study of the phosphorus nutrition of soil born fungi. *Microbiological Research, 157*(3), 221–231. https://doi.org/10.1078/0944-5013-00154.

Tibbett, M., Sanders, F. E., & Cairney, J. W. G. (1998). The effect of temperature and inorganic phosphorus supply on growth and acid phosphatase production in arctic and temperate strains of ectomycorrhizal *Hebeloma* spp. in axenic culture. *Mycological Research, 102*(2), 129–135. https://doi.org/10.1017/S0953756297004681.

Timonen, S., Finlay, R. D., Olsson, S., & Söderström, B. (1996). Dynamics of phosphorus translocation in intact ectomycorrhizal systems: Non-destructive monitoring using a β-scanner. *FEMS Microbiology Ecology, 19*(3), 171–180. https://doi.org/10.1111/j.1574-6941.1996.tb00210.x.

Tisserant, E., Kohler, A., Dozolme-Seddas, P., Balestrini, R., Benabdellah, K., Colard, A., et al. (2011). The transcriptome of the arbuscular mycorrhizal fungus *Glomus intraradices* (DAOM 197198) reveals functional tradeoffs in an obligate symbiont. *New Phytologist, 193*(3), 755–769. https://doi.org/10.1111/j.1469-8137.2011.03948.x.

Torres Aquino, M., & Plassard, C. (2004). Dynamics of ectomycorrhizal mycelial growth and P transfer to the host plant in response to low and high soil P availability. *FEMS Microbiology Ecology, 48*(2), 149–156. https://doi.org/10.1016/j.femsec.2004.01.008.

Torres-Aquino, M., Becquer, A., Le Guernevé, C., Louche, J., Amenc, L. K., Staunton, S., et al. (2017). The host plant *Pinus pinaster* exerts specific effects on phosphate efflux and polyphosphate metabolism of the ectomycorrhizal fungus *Hebeloma cylindrosporum*: A radiotracer, cytological staining and 31P NMR spectroscopy study. *Plant, Cell & Environment, 40*(2), 190–202. https://doi.org/10.1111/pce.12847.

Treseder, K. K., Torn, M. S., & Masiello, C. A. (2006). An ecosystem-scale radiocarbon tracer to test use of litter carbon by ectomycorrhizal fungi. *Soil Biology and Biochemistry, 38*(5), 1077–1082. https://doi.org/10.1016/j.soilbio.2005.09.006.

Tunlid, A., Floudas, D., Koide, R., & Rineau, F. (2016). Soil organic matter decomposition mechanisms in ectomycorrhizal fungi. In *Molecular mycorrhizal symbiosis* (pp. 257–275): Wiley-Blackwell. https://doi.org/10.1002/9781118951446.ch15.

Turner, B. L. (2008). Resource partitioning for soil phosphorus: A hypothesis. *Journal of Ecology, 96*(4), 698–702. https://doi.org/10.1111/j.1365-2745.2008.01384.x.

van Aarle, I. M., & Plassard, C. (2010). Spatial distribution of phosphatase activity associated with ectomycorrhizal plants is related to soil type. *Soil Biology and Biochemistry, 42*(2), 324–330. https://doi.org/10.1016/j.soilbio.2009.11.011.

van der Linde, S., Suz, L. M., Orme, C. D. L., Cox, F., Andreae, H., Asi, E., et al. (2018). Environment and host as large-scale controls of ectomycorrhizal fungi. *Nature, 558*, 243–248. https://doi.org/10.1038/s41586-018-0189-9.

Van Schöll, L., Smits, M. M., & Hoffland, E. (2006). Ectomycorrhizal weathering of the soil minerals muscovite and hornblende. *New Phytologist, 171*(4), 805–814. https://doi.org/10.1111/j.1469-8137.2006.01790.x.

Van Tichelen, K. K., & Colpaert, V. J. (2000). Kinetics of phosphate absorption by mycorrhizal and non-mycorrhizal Scots pine seedlings. *Physiologia Plantarum, 110*(1), 96–103. https://doi.org/10.1034/j.1399-3054.2000.110113.x.

Velasco, I., Tenreiro, S., Calderon, I. L., & André, B. (2004). Saccharomyces cerevisiae Aqr1 Is an internal-membrane transporter involved in excretion of amino acids. *Eukaryotic Cell, 3*(6), 1492–1503. https://doi.org/10.1128/EC.3.6.1492-1503.2004.

Vesk, P. A., Ashford, A. E., Markovina, A.-L., & Allaway, W. G. (2000). Apoplasmic barriers and their significance in the exodermis and sheath of *Eucalyptus pilularis-Pisolithus tinctorius* ectomycorrhizas. *New Phytologist, 145*(2), 333–346. http://www.jstor.org/stable/2588958.

Wallander, H. (2000). Uptake of P from apatite by *Pinus sylvestris* seedlings colonised by different ectomycorrhizal fungi. *Plant and Soil, 218*(1), 249–256. https://doi.org/10.1023/A:1014936217105.

Wallander, H., Nilsson, L. O., Hagerberg, D., & Bååth, E. (2001). Estimation of the biomass and seasonal growth of external mycelium of ectomycorrhizal fungi in the field. *New Phytologist, 151*(3), 753–760. https://doi.org/10.1046/j.0028-646x.2001.00199.x.

Wallander, H., & Wickman, T. (1999). Biotite and microcline as potassium sources in ectomycorrhizal and non-mycorrhizal *Pinus sylvestris* seedlings. *Mycorrhiza*, *9*(1), 25–32. https://doi.org/10.1007/s005720050259.

Wallander, H., Wickman, T., & Jacks, G. (1997). Apatite as a P source in mycorrhizal and non-mycorrhizal *Pinus sylvestris* seedlings. *Plant and Soil*, *196*(1), 123–131. https://doi.org/10.1023/A:1004230525164.

Wang, J., Li, T., Wu, X., & Zhao, Z. (2014). Molecular cloning and functional analysis of a H (+)-dependent phosphate transporter gene from the ectomycorrhizal fungus *Boletus edulis* in southwest China. *Fungal Biology*, *118*(5–6), 453–461. https://doi.org/10.1016/j.funbio.2014.03.003.

Wang, B., & Qiu, Y.-L. (2006). Phylogenetic distribution and evolution of mycorrhizas in land plants. *Mycorrhiza*, *16*(5), 299–363. https://doi.org/10.1007/s00572-005-0033-6.

Willmann, A., Weiß, M., & Nehls, U. (2007). Ectomycorrhiza-mediated repression of the high-affinity ammonium importer gene *AmAMT2* in *Amanita muscaria*. *Current Genetics*, *51*(2), 71–78. https://doi.org/10.1007/s00294-006-0106-x.

Wipf, D., Benjdia, M., Tegeder, M., & Frommer, W. B. (2002). Characterization of a general amino acid permease from *Hebeloma cylindrosporum*. *FEBS Letters*, *528*(1–3), 119–124. https://doi.org/10.1016/S0014-5793(02)03271-4.

Wolfe, B. E., Tulloss, R. E., & Pringle, A. (2012). The irreversible loss of a decomposition pathway marks the single origin of an ectomycorrhizal symbiosis. *PLoS One*, 7(7), 1–9. https://doi.org/10.1371/journal.pone.0039597.

Wood, H. G., & Clark, J. E. (1988). Biological aspects of inorganic polyphosphates. *Annual Review of Biochemistry*, *57*(1), 235–260. https://doi.org/10.1146/annurev.bi.57.070188.001315.

Wright, D. P., Johansson, T., Le Quéré, A., Söderström, B., & Tunlid, A. (2005). Spatial patterns of gene expression in the extramatrical mycelium and mycorrhizal root tips formed by the ectomycorrhizal fungus *Paxillus involutus* in association with birch (*Betula pendula*) seedlings in soil microcosms. *New Phytologist*, *167*(2), 579–596. https://doi.org/10.1111/j.1469-8137.2005.01441.x.

Xu, H., Kemppainen, M., El Kayal, W., Lee, S. H., Pardo, A. G., Cooke, J. E. K., et al. (2015). Overexpression of *Laccaria bicolor* aquaporin JQ585595 alters root water transport properties in ectomycorrhizal white spruce (*Picea glauca*) seedlings. *New Phytologist*, *205*(2), 757–770. https://doi.org/10.1111/nph.13098.

Zevora, M. Y. (1955). Mykorrhiza formation in forest trees of the Ukrainian SSR. In A. Imshenetskii (Ed.), *Mycotrophy in plants*. Washington, DC: USDA and NSF [translated from Russian by Israel Programme for Scientific Translocations, 1967].

Zheng, R., Wang, J., Liu, M., Duan, G., Gao, X., Bai, S., et al. (2016). Molecular cloning and functional analysis of two phosphate transporter genes from *Rhizopogon luteolus* and *Leucocortinarius bulbiger*, two ectomycorrhizal fungi of *Pinus tabulaeformis*. *Mycorrhiza*, *26*(7), 633–644. https://doi.org/10.1007/s00572-016-0702-7.

Zörb, C., Senbayram, M., & Peiter, E. (2014). Potassium in agriculture—Status and perspectives. *Journal of Plant Physiology*, *171*(9), 656–669. https://doi.org/10.1016/j.jplph.2013.08.008.

CHAPTER FOUR

Nitrogen storage and cycling in trees

Gen Li, Gary D. Coleman*
Department of Plant Science and Landscape Architecture, University of Maryland, College Park, MD, United States
*Corresponding author: e-mail address: gcoleman@umd.edu

Contents

Abstract

Seasonal nitrogen cycling is an important trait and adaptive strategy for trees to conserve and reuse nitrogen. This review examines the current progress towards understanding the molecular and physiological basis of seasonal nitrogen cycling in trees. Since most of what is known about nitrogen cycling in trees comes from research in *Populus*, this review relies heavily on research in this genus. We also highlight major knowledge gaps in understanding the regulatory pathways and transport mechanism that govern seasonal N cycling.

1. Introduction

Trees have existed on earth for over 370 million years. Estimates place the number of trees on earth at approximately 3.04 trillion; 1.30 trillion in tropical and subtropical forests, 0.74 trillion in boreal regions and 0.66

Advances in Botanical Research, Volume 89
ISSN 0065-2296
https://doi.org/10.1016/bs.abr.2018.11.004

trillion in temperate regions (Crowther et al., 2015). Trees are also long-lived organism and the ring-width chronologies of a bristlecone pine tree (*Pinus longaeva*) in California indicates that this tree has lived for over 5000 years (Ferguson, 1968; Salzer, Hughes, Bunn, & Kipfmueller, 2009). One of physiological mechanisms that contributes to the longevity of trees is internal seasonal nutrient cycling including the internal recycling of nitrogen (N) (Eckstein, Karlsson, & Weih, 1999; Fife & Nambiar, 1982; Nambiar & Fife, 1991; Rennenberg & Schmidt, 2010).

Nitrogen (N) is an essential nutrient for plant growth, development and reproduction as it is a basic constitute of amino acids, the building blocks of proteins. N is also found in other important compounds such as amides, nucleic acids, nucleotides, alkaloids, vitamins, chlorophyll and phytohormones. Plant productivity is impacted by N deficiency which can impact photosynthesis by reducing chlorophyll production and enzyme activity and can further affect sugar metabolism and/or carbohydrate partitioning between source and sink tissues as well as root morphology (Hermans, Hammond, White, & Verbruggen, 2006).

N is one of the major limiting nutrients for worldwide agricultural production. Since the global population is predicted to approach 9 billion by 2050, expanding food demands will require an annual increase in food production of nearly 44 million metric tons (Godfray et al., 2010). Accordingly, N fertilizer application is predicted to increase by threefold over the next 40 years unless nitrogen use efficiency (NUE) is greatly improved (Xu, Fan, & Miller, 2012). Like herbaceous annual plants trees must acquire N from the soil but because trees are perennials a significant portion of N demand is met by seasonal internal cycling. Because the adaptive strategy of seasonal N cycling occurs annually, this recursive process increasingly contributes to the overall N budget as trees age (Millard & Grelet, 2010; Millard, Wendler, Grassi, Grelet, & Tagliavini, 2006). Thus, internal N cycling is a major factor that contributes to the N economy of trees and provides a competitive advantage in N limited environments (Chapin, Schulze, & Mooney, 1990; Millard & Grelet, 2010; Rennenberg & Schmidt, 2010; Vitousek, 1982). Consequently, seasonal N cycling is important to the ecology of N use in trees and makes a major contribution to NUE (Aerts, 1990; Birk & Vitousek, 1986; Boerner, 1984; Chapin & Kedrowski, 1983; Kost & Boerner, 1985; Ryan & Bormann, 1982; Vitousek, 1982).

In temperate trees seasonal N cycling involves a number of processes including: (1) uptake and assimilation of N from the soil, (2) transport and partitioning associated with N sink demands during growth, (3) autumn redistribution of N from leaves during senescence and storage in perennial tissues and, (4) remobilization and transport of stored nitrogen during renewed growth in the spring. Thus, different spatial and temporal scales of N transport and N fluxes are important in controlling N cycling and partitioning in trees (Fig. 1). A hallmark of seasonal N cycling in temperate trees are the changes in N source-sink relations that occur between leaves and storage tissues during this process (Fig. 1). During spring growth leaves and expanding shoots are N sinks while storage tissues are N sources but in autumn the source-sink status of these tissues change with leaves serving as N sources and storage tissues as N sinks. In this review we focus on our current understanding of the physiological and molecular aspects of seasonal N storage and cycling in trees. Although we touch on aspects that regulate N transport and storage, this is not the focus of this review since this topic has been recently reviewed (Babst & Coleman, 2018). We also focus primarily on temperate tree species since these are the species that the majority of studies are focused. Further, since the majority of research related to seasonal N cycling in trees has been conducted in *Populus* (poplars) much of what we present relies on research in this genus.

2. Physiology of N storage and cycling

2.1 Fruit trees

Since the 1920s, the importance of nitrogen reserves to spring growth has been recognized in fruit trees (Roberts, 1921; Thomas, 1927). In deciduous fruit trees, N cycling involves translocation of leaf N to woody tissues which serves as an N reservoir in autumn and winter. These N reserves are then remobilized and used during growth in the following spring. Apple (*Malus domestica*) is one of the earliest and most extensively studied fruit tree species with respect to seasonal N cycling (Roberts, 1921; Thomas, 1927; Titus & Kang, 1982). N released from senescing apple leaves is translocated to spurs and twigs, bark, and roots (Mason & Whitfield, 1960; Murneek & Logan, 1932) resulting in increased N content of these tissues in autumn. Although N content increased in these various tissues during autumn, a greater proportion of N was stored in bark compared to other perennial tissues (i.e. wood and roots) (Mason & Whitfield, 1960; O'Kennedy, Hennerty, & Titus, 1975). During apple leaf senescence leaf protein contents declined by about one-half while at the same time the total protein content in bark increased by 240% (Kang & Titus, 1980a). The majority of the decline in leaf protein content during senescence was attributed to an 80% decline in the leaf content of ribulose 1,5-bisphosphate carboxylase/oxygenase

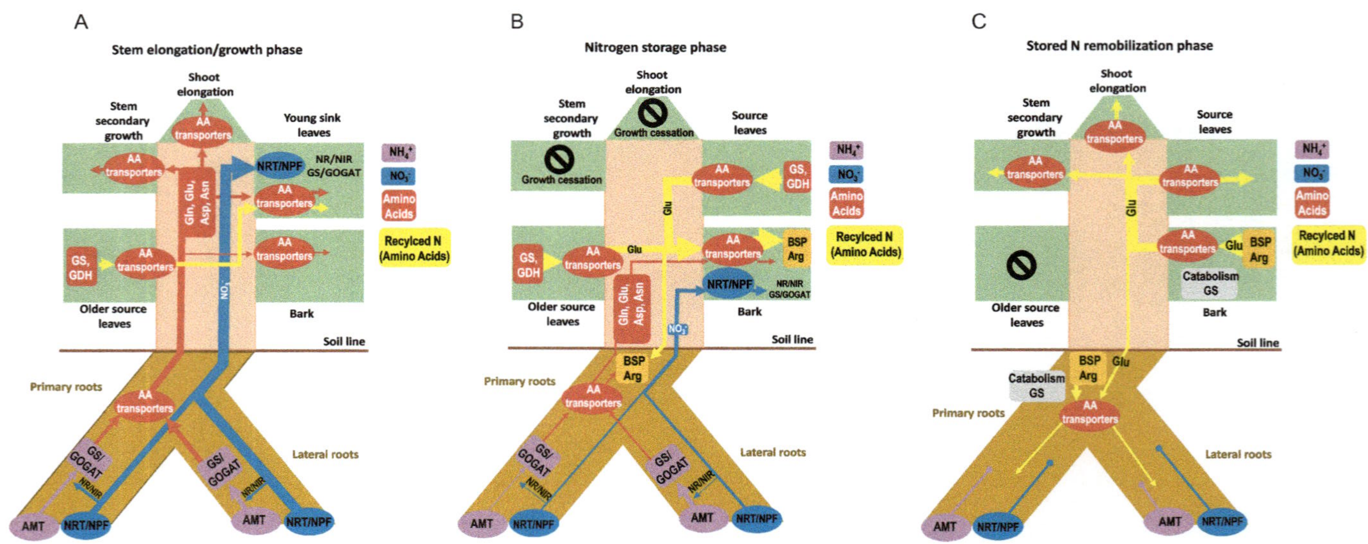

Fig. 1 Schematic routes and fluxes of N during stages of seasonal N cycling in trees. (A) N transport routes and fluxes during growth following N remobilization. During this stage N is acquired from the soil and transported to N sinks associated with growth. (B) N transport routes and fluxes during autumn N remobilization and storage. N from senescing leaves (source) is transported to storage tissues (sinks) in the stem and root. (C) Transport routes and fluxes during N remobilization in the spring. N from storage tissues (source) is remobilized to expanding buds and shoots (sinks). Possible factors involved (transporters and enzymes) and the chemical nature of N is indicated in the figures. Abbreviations: AA, amino acid; Gln, glutamine; Glu, glutamate; Asp, aspartate; Asn, asparagine; NRT, nitrate transporter; NPF, NITRATE TRANSPORTER1/PEPTIDE TRANSPORTER family; NR, nitrate reductase; NiR, nitrite reductase; GS, glutamine synthetase; GDH, glutamate dehydrogenase; GOGAT, glutamine-2-oxoglutarate aminotransferase; AMT, ammonium transporter; BSP, bark storage protein; Arg, arginine.

(RUBISCO) (Millard & Thomson, 1989). Therefore, besides its role in photosynthesis and carbon fixation, RUBISCO also serves as a major leaf storage pool of N which is a N source during autumn N remobilization from source (leaf) to sink (bark).

Some controversy still seems to exist as to the chemical nature of N storage in apple (Cheng, Ma, & Ranwala, 2004; Titus & Kang, 1982). Early studies in apple (Oland, 1959), peach (Muñoz, Guerri, Legaz, & Primo-Millo, 1993; Taylor, 1967; Taylor & May, 1967; Taylor & van den Ende, 1969, 1970) and pear (Taylor, van den Ende, & Canterford, 1975) indicated that N was stored as a soluble form mainly as free amino acids. In contrast, Kang and Titus (1980a, 1980b) reported that nearly 90% of N stored in apple bark during that late stages of leaf senescence was in protein. In addition, protein was also found to be the major form of N storage and N reservoir in citrus trees (Kato, Yamagata, Tsukahara, & Shikoku, 1984). The form of N storage in apple may also be related to N status or availability since proteins were observed to be a main form of N storage and independent of N supply while the proportion of stored free amino acids increased as external N supply increased (Cheng et al., 2004). Storing free amino acids may be more metabolically efficient compared to protein and storage of high N containing amino acids (i.e. arginine) when N supplies are abundant suggests flexibility of the N storage pool relative to N supply (Cheng et al., 2004). To the best of our knowledge, how N fluxes between amino acid and protein storage pools are regulated in apple is not known but raises some intriguing questions. Since proteins are stored in protein storage vacuoles (Herman & Larkins, 1999; Müntz, 1998, 2007) it might be possible that under abundant N conditions that protein storage sites or systems become "saturated" which could feedback and alter N fluxes to amino acid storage. Since amino acid storage would need also need to be in the vacuole in order to maintain cellular homeostasis, the regulation of a shift from protein to amino acid storage in relation to N supply could be an important regulatory step in N storage and partitioning.

In fruit trees the timing of fertilization impacts fruit production with late summer and autumn N fertilization commonly used to improve productivity. Fertilization in late summer and autumn appears to increase N storage over winter resulting in increased productivity when N is remobilized during flowering and regrowth the following spring (Bravo et al., 2017; Quartieri, Millard, & Tagliavini, 2002). Further, autumn foliar N fertilization results in greater N storage in apple bark compared to soil N fertilization, supporting that autumn N redistribution from senescing leaves contributes to a greater proportion of N storage than N uptake from soil (O'Kennedy et al., 1975). This has important implications in the management of N nutrition and fruit tree productivity and it is possible that crop management strategies or genetic alterations that enhance seasonal N storage could be effective in enhancing fruit tree NUE.

2.2 Forest trees

Similar to fruit trees, leaf N levels decline throughout the fall in temperate forest trees and is accompanied by increased bark N content (Côté & Dawson, 1986; Côté, Vogel, & Dawson, 1989). For many deciduous temperate tree species the autumn increase in bark N levels is closely associated with the accumulation of specific proteins termed bark storage proteins (BSP and BSPs). Over the past 30 years significant progress has been made in identification of storage proteins in many temperate tree species for both angiosperms and gymnosperms. BSPs are characterized by increased abundance during autumn and their disappearance in spring when growth resumes. For example, specific BSPs whose abundance fluctuates seasonally have be reported for *Acer saccharum*, *Salix × smithiana*, *Populus deltoides* (Wetzel, Demmers, & Greenwood, 1989), *Larix decidua*, *Pinus sylvestris*, *Pinus strobus* (Wetzel & Greenwood, 1989), *Taxodium distichum* (Harms & Sauter, 1991), *Alnus glutinosa*, *Betula papyrifera*, *Quercus rubra*, *Fagus sylvatica*, *Fraxinus americana*, *Tilia americana* and *Fagus glutinosa* (Wetzel & Greenwood, 1991a). The wide diversity of tree species from which seasonally fluctuating bark proteins have been observed indicates that seasonal storage of N as protein is extensively used as an adaptive mechanism for the reuse of assimilated N in trees. However, in temperate tree species the location or tissue that serves as an N storage site appears to be associated with leaf habit (Millard & Grelet, 2010). While stems and roots are the major N storage tissues in deciduous trees, evergreen trees appear to store N in leaves and needles that are retained over the winter (for review see Millard & Grelet, 2010). Although most of the research associated with seasonal N storage has been performed in temperate trees species, evidence indicates that tropical and subtropical forest trees also store N and internal cycling is related to phenological changes that occur from the dry season to the rainy season and vice versa (Gérant, Pluchon, Mareschal, Koutika, & Epron, 2017; Schmidt & Stewart, 1998).

3. Biochemical and molecular aspects of N storage and cycling

3.1 Bark storage proteins (BSPs)

Bark storage proteins (BSPs) are a class of vegetative storage proteins (VSPs) and are a central component of seasonal N storage and cycling in a wide range of temperate tree species (Millard & Grelet, 2010). For temperate deciduous trees BSPs accumulate in bark and root tissues during autumn and winter and accumulated BSP then disappears from storage tissues once growth resumes in the spring. Presumably, the disappearance of BSP in the spring involves protein catabolism and amino biosynthesis although little is known about these metabolic processes during spring N remobilization. Because of the importance of BSPs to seasonal N cycling in temperate trees, we present a review of BSPs and aspects that regulate their accumulation and catabolism. Most of this information comes from studies with poplar (*Populus*) since seasonal N cycling has been studied in this genus more than other species.

A number of VSPs and BSPs have been identified in plants and there does not seem to be any consistency in plants as to the class or type of proteins that serve as storage proteins. The majority of VSPs and BSPs have been identified by SDS-PAGE based on the characteristics of their accumulation and disappearance. As a consequence, little information exists about protein sequence and biochemistry. One of the earliest VSPs to be studied is in soybean where two glycosylated proteins called VSPα and VSPβ. Both VSPα and VSPβ were found to be abundant in young sink leaves but disappeared during seed fill (Wittenbach, 1982) and were shown to have acid phosphatase activity (DeWald, Mason, & Mullet, 1992). Other VSPs have also been shown to have biological properties other than their role in storage. For example, the potato storage protein patatin has lipid acyl hydrolase activity (Andrews, Beames, Summers, & Park, 1988) and the storage protein in alfalfa taproots has β-amylase activity (Gana, Kalengamaliro, Cunningham, & Volenec, 1998). Examples also exist for storage proteins that have sequence homology to active proteins but appear to have lost their activity including RNAse-like proteins in *Calystegia sepium* (Van Damme et al., 2000) and lectin-like proteins that appear to lack sugar-binding properties in *Cladrastis lutea* (Van Damme et al., 1995), and *Sambucus nigra* (Chen, Rouge, Peumans, & van Damme, 2002).

In the case of *Populus*, BSPs are members of a gene family that have homology to nucleoside phosphorylase-like (NP-like) proteins (Pettengill, Pettengill, & Coleman, 2013). Bacterial and mammalian nucleoside phosphorylases are involved in nucleoside catabolism through cleavage of the glycosidic bond of (deoxy-) ribonucleosides in the presence of inorganic phosphate (Pi) to produce (deoxy-) ribose-1-phosphate and a nucleobase (Lewkowicz & Iribarren, 2006; Pugmire & Ealick, 2002). The released nucleobase can be either used for synthesis of other metabolites or degraded while the (deoxy-) ribose-1-phosphate can be used by the Pentose Phosphate Pathway and glycolysis (Pugmire & Ealick, 2002). Purine salvaging and degradation in plants could contribute to NUE through the retention and remobilization of N (Smith & Atkins, 2002; Werner & Witte, 2011). However, there is no evidence that nucleoside phosphorylases are involved in purine nucleoside salvage in plants but instead hydrolysis of plant nucleosides seems to involve nucleosidases (EC 3.2.2x) (Guranowski, 1982; Jung, Hoffmann, & Möhlmann, 2011; Katahira & Ashihara, 2006a). Since genes in the purine salvage and degradation pathways in plants are induced by abscisic acid, drought, wounding and dark induced senescence suggests a role for purine salvage in N remobilization and NUE (Brychkova, Alikulov, Fluhr, & Sagi, 2008; Hesberg, Hänsch, Mendel, & Bittner, 2004; Katahira & Ashihara, 2006b) yet a role for nucleoside phosphorylase-like proteins in plant nucleoside salvaging and NUE are not known. Moreover, attempts in our lab to show nucleoside catabolism related activity of the nucleoside phosphorylase-like proteins of *Populus* have not been successful, so the question still remains if members of this gene family either retain or possess catalytic activity.

Thus, it appears that a number of proteins have been recruited to serve as storage proteins and there does not seem to be any consistency as to the type of proteins used for vegetative N storage. This presents an intriguing question about the evolution of N storage and the proteins used for storage. Although storage of N involving VSP seems widespread yet the proteins used for storage appears to be diverse. In poplar VSPs including BSPs are highly homologous to seed storage proteins (Beardmore, Wetzel, Burgess, & Charest, 1996) suggesting that VSPs and BSPs may have been recruited from seed storage proteins but it is still unclear why nucleoside phosphorylase-like proteins would have been used for N storage. It would be interesting to know if this relationship or relatedness between seed storage proteins and VSPs holds for other species that utilize VSPs for vegetative N storage.

3.2 *Populus* BSP and seasonal N cycling

In *Populus*, BSPs comprises the major VSPs that are involved in seasonal N cycling. During autumn and winter BSPs accumulate in *Populus* bark accounting for approximately 62% of total soluble bark protein during dormancy (Coleman, Chen, Ernst, & Fuchigami, 1991). Over the past 20 years, BSPs and the genes that encode them been described for a number of tree species (Table 1). Among these the BSPs of *Populus* are the most extensively studied in trees. Early studies using SDS-PAGE and immunological analyses of *Populus* BSP revealed multiple isoelectric isoforms with molecular masses between 32 and 38 kDa and a 32 kDa BSP was also observed in closely

Table 1 Bark storage proteins in trees and their corresponding gene names.

Name	Organisms	Tissue localization	References
BSP	*Acer*, *Salix* and *Populus* spp.	Bark	Wetzel et al. (1989)
BSP	*Populus deltoides*	Bark	Coleman et al. (1991)
BSP	*Betula papyrifera*, *Salix microstachya*	Bark	Wetzel and Greenwood (1991a)
WIN4	*Populus trichocarpa* × *P. deltoides*	Leaf and bark	Lawrence, Greenwood, Korhnak, and Davis (1997)
PNI288	*Populus trichocarpa* × *P. deltoides*	Leaf and bark	Lawrence, Cooke, Greenwood, Korhnak, and Davis (2001) and Pettengill et al. (2013)
CiVSP	*Cichorium intybus*	Taproot and leaf	Richard-Molard, Brugière, Moille, Carrayol, and Limami (2004)
VSP	*Populus canadensis*	New shoot	Tian et al. (2005)
VSP	*Litchi chinensis*	Leaf, bark and root	Tian et al. (2007)
CCA	*Castanea crenata*	Leaf and stem	Nomura, Ikegami, Koide, and Yagi (2008)
BSP A	*Populus trichocarpa*	Shoot tip and bark	Pettengill et al. (2013)
BSP B	*Populus trichocarpa*	Shoot tip and bark	Pettengill et al. (2013)
BSP C	*Populus trichocarpa*	Shoot tip, young leaf and bark	Pettengill et al. (2013)

related *Salix* (Coleman et al., 1991; Langheinrich & Tischner, 1991; Sauter, van Cleve, & Apel, 1988; Wetzel et al., 1989). Further biochemical analyses also demonstrated that both the *Populus* and *Salix* BSPs are glycosylated and this posttranslational modification appears to be the basis for the iso-electric isoforms that have been detected (Wetzel et al., 1989; Wetzel & Greenwood, 1991b).

The availability of the poplar genome (Tuskan et al., 2006) greatly facilitated our understanding of the structure and organization of the poplar VSP gene family which consists of 13 members with homology to NP-like proteins (Pettengill et al., 2013). Phylogenetic analysis of NP-like genes in plants revealed two major clusters. One cluster is characterized by proteins encoded by genes found in a wide range of plant taxa and includes four genes in *Populus*, while the second cluster consists primarily of proteins encoded by species of the order *Malpighiales*, including nine genes encoding *Populus* BSP and WIN4-like proteins. Expression analysis showed that all 13 members of *Populus* NP-like gene family are expressed in various tissues in both long-day (LD) and short-day (SD) conditions (Pettengill et al., 2013). The evolution of the *Populus* gene family involved duplication events and divergence in expression may have recently occurred (Pettengill et al., 2013).

Populus BSP accumulates in protein storage vacuoles of inner bark parenchyma and xylem ray cells during fall (Sauter, van Cleve, & Wellenkamp, 1989; van Cleve, Clausen, & Sauter, 1988; Wetzel et al., 1989). Three genes (*BSPA*, *BSPB*, *BSPC*) encode poplar BSPs and their expression is regulated by environmental factors such as SD photoperiods, N availability and low temperature (Coleman, Banados, & Chen, 1994; Coleman et al., 1991; Pettengill et al., 2013; Zhu & Coleman, 2001a). Of these factors, SD photoperiod is the main environmental cue that induces BSP gene expression and N storage. Photoperiod perception is mediated by the phytochrome photoreceptors and play an essential role in *Populus* growth cessation and bud set (Howe, Gardner, Hackett, & Furnier, 1996; Olsen et al., 1997). Like other dormancy-associated processes, phytochrome-mediated perception of SD is also involved in the induction of BSP gene expression and involves both low fluence and very low fluence pathways (Zhu & Coleman, 2001a). It has yet to be determined if phytochrome mediates BSP gene expression directly or indirectly through altered growth resulting in altered C and N partitioning. Using tobacco transformed with the *Populus BSPA* promoter fused to GUS it was found that the transcriptional activation from the *BSPA* promoter occurred in developing floral tissues and in seeds during early stages of embryogenesis and seed germination (Zhu & Coleman, 2001b).

These results are consistent with the concept of BSP gene expression being associated with N sink status of tissues and is consistent with a role for N partitioning in regulating BSP gene expression.

A lingering question concerning seasonal N cycling and storage in *Populus* is centred around the mechanism(s) that regulate BSP expression. This is fundamental to N sink establishment in bark which is an essential process for seasonal N storage. The red and far-red light photoreceptor phytochrome mediates diverse developmental and photomorphogenic responses in plants, including SD-induced BSP gene expression, through the action of the reversible inactive Pr and active Pfr forms (Quail, 2002). Phytochrome interacting factors (PIFs) are basic helix-loop-helix (bHLH) transcription factors that play roles in the early phases of phytochrome signalling (Leivar & Quail, 2011). Activated phytochrome translocates to the nucleus and directly interact with PIFs (Quail, 2002) resulting in phosphorylation and degradation of the targeted PIFs, and transcriptional changes in target genes (Al-Sady, Ni, Kircher, Schäfer, & Quail, 2006; Shen, Khanna, Carle, & Quail, 2007). PIFs are also involved in hormonal control of light signalling (Chaiwanon, Wang, Zhu, Oh, & Wang, 2016; de Lucas & Prat, 2014) including ethylene production and signalling (Finlayson, Lee, & Morgan, 1998; Finlayson, Lee, Mullet, & Morgan, 1999; Shi et al., 2016; Zdarska et al., 2015). A family of endoplasmic reticulum-located receptors perceive ethylene that is then transduced through a multistep signal cascade to the Ethylene-Insensitive 3 (EIN3) and EIN3-Like1 (EIL1) transcription factors (Alonso, Hirayama, Roman, Nourizadeh, & Ecker, 1999; Chang & Shockey, 1999). The F box proteins, EBF1 and EBF2, target EIN3/EIL1 for proteasome-mediated degradation through the SCR E3 ligase complex (Binder et al., 2007; Gagne et al., 2004; Guo & Ecker, 2003; Potuschak et al., 2003). Ethylene stabilizes EIN3/EIL1 by inhibition of EBF1/EBF2 through regulation of transcription and protein stability (An et al., 2010; Merchante et al., 2015; Qiao et al., 2012). PhyB has also been shown to directly interact with EIN3 and enhance interactions between PhyB, EIN3 and EBF1/EBF2 which modulates EIN3 degradation via the SCR E3 ligase complex (Shi et al., 2016). Ongoing research in our lab is directed at determining if an interaction between phytochrome and ethylene mediate SD-induced BSP gene expression in *Populus*.

Nitrogen remobilization from senescing leaves to storage tissues during growth cessation and the onset of dormancy is an essential process in seasonal N cycling and is the overriding factor that contributes to NUE. During the N storage phase of seasonal N cycling, senescing leaves transition to sources of N that are translocated to bark prior to leaf abscission. Consequently, both leaves and bark function as either an N sink or an N source depending on the season and so the nature of the interactions between these organs shifts seasonally. Leaf senescence is a complex developmental process utilized to remobilize nutrients and contributes to NUE (Avila-Ospina, Moison, Yoshimoto, & Masclaux-Daubresse, 2014; Kim, Woo, & Nam, 2016; Schippers, 2015). Autumn leaf senescence in *Populus* is triggered by shortening photoperiod (SD) and once initiated is accelerated by low temperature (Fracheboud et al., 2009). Growth cessation and bud set is also triggered by shortening photoperiods, but the critical photoperiods for bud set and leaf senescence appear to differ (Fracheboud et al., 2009). At a cellular level, autumn senescence of *Populus* leaves occurs in an ordered fashion involving chlorophyll degradation, chloroplast conversion to gerontoplasts, carotenoid degradation and anthocyanin accumulation, and protein degradation (Keskitalo, Bergquist, Gardeström, & Jansson, 2005). Autumn leaf senescence also involves extensive changes in gene expression of which cysteine proteases and metallothioneins are highly enriched (Andersson et al., 2004; Bhalerao, 2003).

Glutamine (Gln) is the major amino acid that accumulates in senescing *Populus* leaves and is also the major amino acid transported from senescing leaves to bark via the phloem during N remobilization and storage. The cationic amino acid transporter CAT11 is one of the few amino acid transporters characterized in *Populus* and it appears to play a role in Gln transport from leaf to bark (Couturier et al., 2010). Little is known about other amino acid transporters or facilitators that might be involved in the unloading of amino acids from phloem to bark. Based on data from DNA microarrays it was recently reported that a large number of amino acid transporters are expressed in poplar bark during SD-induced N storage (Babst & Coleman, 2018). This includes genes belonging to the ATF-AAT and NPF families and it was suggested that some of transporters encoded by these genes might be involved in phloem unloading and influx of amino acids into parenchyma storage tissues (Babst & Coleman, 2018). In *Arabidopsis* an amino acid transporter was recently identified that is responsible for loading Gln into phloem (Santiago & Tegeder, 2016) which provides some insight into this important transport step. However, the mechanism for loading and unloading Gln from xylem and phloem are not known in *Populus*. While passive symplastic phloem loading of carbohydrates via plasmodesmata is common and has been demonstrated in *Populus* (Zhang et al., 2014) metabolite un/loading has not been studied in senescing leaves. Although some modest progress has been made in deciphering the transporters involved in amino acid transport during seasonal N storage we only have a rudimentary understanding of this critical process that is likely to govern N partitioning between N source and sink tissues.

The majority of the N released during autumn leaf senescence is translocated to bark and used in the biosynthesis of BSP. Transient increases in stem Gln levels peak during autumn leaf senescence and then again during spring N remobilization, correlating with major developmental processes during the annual growth cycle of *Populus* (Sauter & van Cleve, 1992; Wildhagen, Durr, Ehlting, & Rennenberg, 2010). The first peak of bark Gln levels coincides with fall leaf senescence, whereas the second peak coincides with spring bud expansion and growth (Wildhagen et al., 2010). Bark Arg levels also increase during autumn leaf senescence and correlate with increased BSP gene expression (Couturier et al., 2010; Wildhagen et al., 2010). However, unlike Gln, increased bark Arg levels have one seasonal peak. Arginine levels begin to accumulate during leaf senescence, remain elevated during dormancy, and then decline when spring bud expansion and growth begins (Wildhagen et al., 2010). BSP gene expression can also be rapidly induced in poplar bark by feeding Gln to *Populus* stems (Zhu & Coleman, 2001a). Thus, in addition to its role as a transport amino acid and as a source of N for incorporation into BSP, Gln also appears to serve as a signal molecule that regulates BSP gene expression. Thus, Gln transport from senescing leaves (source) to bark (sink) would have dual roles in *Populus* N metabolism: (1) as a N source for BSP biosynthesis and (2) a signal for metabolic regulation of seasonal N cycling. These roles suggest a link between bark BSP accumulation and bark sink development with leaf source status.

Since induction of BSP expression and N storage appears to be coupled to changes in the N status of bark mediated by Gln transport and/or accumulation this metabolic change or increased N influx must somehow be sensed. Nutrient sensing is an important component of how organisms coordinate growth and development and multiple nutrient sensing mechanisms have evolved in both prokaryotes and eukaryotes (Chantranupong, Wolfson, & Sabatini, 2015). Plants have components of the (1) P_{II}, (2) GCN2 (general amino acid control non-derepressible 2) and (3) TOR (target of rapamycin) kinase nitrogen and amino acid sensing systems, but the role of these systems in plants is not well understood compared to yeast and mammals. Both GCN2 and TOR kinases are involved in sensing intracellular amino acid levels particularly under limiting or starvation conditions (Chantranupong et al., 2015). In plants, GCN2 has been shown to interact with uncharged tRNA and have a role in eIF2a phosphorylation that may play a role in stress and herbicide responses (Faus et al., 2015; Lageix et al., 2008; Zhang et al., 2008). GCN2 is unlikely to be important in seasonal N cycling since N becomes abundant in sink tissues as opposed to N-starvation conditions.

TOR consists of two distinct complexes; TORC1 being sensitive to rapamycin and TORC2 which is insensitive (Wullschleger, Loewith, & Hall, 2006). Plants possess all of the TORC1 components and this pathway has been shown to play roles in protein synthesis and metabolism, cell proliferation, cell size, autophagy and the auxin signalling pathway (Bögre, Henriques, & Magyar, 2013; Caldana et al., 2013; Dobrenel et al., 2011; Henriques, Bögre, Horváth, & Magyar, 2014; Leiber et al., 2010; Mahfouz, Kim, Delauney, & Verma, 2006; Menand et al., 2002; Ren et al., 2012; Robaglia, Thomas, & Meyer, 2012), but it is unknown if this pathway has any role in sensing changes in Gln levels during seasonal N storage.

The P_{II} family of conserved nutritional signalling proteins occurs in bacteria, archaea and plants (Chellamuthu, Alva, & Forchhammer, 2013; Leigh & Dodsworth, 2007; Sant'Anna et al., 2009; Uhrig, Ng, & Moorhead, 2009). P_{II} signalling proteins play roles in sensing cellular carbon and N balance and are involved in protein interactions with metabolic enzymes, signal transduction proteins, transcription factors and metabolite transporters (Leigh & Dodsworth, 2007; Osanai & Tanaka, 2007; Uhrig et al., 2009). Plant P_{II} are conserved ~15 kDa proteins that form a ~45 kDa homotrimeric complex (Mizuno, Berenger, Moorhead, & Ng, 2007). A hallmark of this sensing mechanism is ATP and 2-oxoglutarate binding to P_{II} homotrimers to produce distinct P_{II} conformations (Fokina, Chellamuthu, Forchhammer, & Zeth, 2010; Forchhammer, 2008; Huergo, Chandra, & Merrick, 2013; Zeth, Fokina, & Forchhammer, 2014). Ligand binding and receptor interaction involve a protruding T loop for each subunit (Forchhammer, 2008; Huergo et al., 2013). High-resolution structural analysis on P_{II} proteins from a variety of species have revealed that trimeric P_{II} sensors interact with hexameric NAGK to alleviate feedback inhibition by Arg (Chellamuthu et al., 2014; Llácer et al., 2007; Llácer, Fita, & Rubio, 2008; Mizuno, Berenger, et al., 2007; Mizuno, Moorhead, & Ng, 2007; Ramón-Maiques et al., 2006). Interestingly, (most) plant P_{II} proteins possess a C-terminal extension that is absent in archaea, bacteria and red algae (Uhrig et al., 2009), but which is required for Gln binding. This so-called Q-loop facilitates the interaction of trimeric P_{II} with hexameric NAGK to alleviate feedback inhibition by Arg (Chellamuthu et al., 2014). On the other hand, it is intriguing that the *Brassicaceae* P_{II} sensors, including that from *Arabidopsis*, lack part of the plant-specific C-terminal extension that comprises the Gln binding loop found in plants (including poplar) (Chellamuthu et al., 2014), yet the *A. thaliana* P_{II} retains the ability to bind NAGK and regulate enzyme activity.

Gln may serve multiple roles in *Populus* seasonal N cycling including N transport from leaves to bark during SD-mediated N storage, an N source for BSP biosynthesis, and as a metabolic signal of bark N influx. The increased influx of Gln to bark may alter C/N ratios and then be sensed through P_{II}-Gln binding which facilitates P_{II}-NAGK interactions that up-regulate the Arg pathway and Arg accumulation. Increases in bark Arg may function in N storage but may also regulate the expression and accumulation of BSP through unknown mechanisms. On the other hand, if the Arg pathway beyond the P_{II}-NADK step is unaffected, then it is possible that the P_{II}-Gln interaction regulates *BSP* expression independent of Arg. Since P_{II} has been shown to interact with other proteins in *Arabidopsis* (biotin carboxyl carrier protein) (Bourrellier et al., 2010) as well as transporters and transcriptional regulators in bacteria and archaea (Amon, Titgemeyer, & Burkovski, 2010; Herrero, Muro-Pastor, & Flores, 2001; Huergo et al., 2013; Leigh & Dodsworth, 2007; Merrick & Edwards, 1995; Sonenshein, 2007) it may be premature to conclude that other P_{II} interactions are not involved in the regulation of *BSP* expression.

3.3 Other VSP in *Populus*: WIN4

Besides BSP the *Populus* genome also contains a BSP-like subfamily consisting of 5 genes that have been termed the WIN4-subfamy and consists of WIN4, VSP 425, VSP XIII, VSP 87A and VSP 840 (Pettengill et al., 2013). WIN4, first identified as a wound-inducible gene in *Populus*, shares 75% cDNA sequence identity to BSP and WIN4 also shares common expression patterns, e.g., nitrogen- and wound-induction (Coleman et al., 1994; Davis et al., 1993; Lawrence et al., 1997). WIN4 transcripts accumulate in leaves and stems, while BSP accumulates in stem phloem parenchyma and xylem ray cells. In contrast to BSP, WIN4 is not induced in bark under either SD and low temperature (Davis et al., 1993; Lawrence et al., 2001; Rennenberg, Wildhagen, & Ehlting, 2010; Wildhagen, Bilela, & Rennenberg, 2013), but is induced in young leaves with increased

N availability and repressed by SD treatment (Coleman et al., 1994; Pettengill et al., 2013). It has been suggested that WIN4 may play a role in short-term N storage and mobilization in leaves during growth while BSP is involved in seasonal N storage and storage during periods of excess nitrogen availability (Coleman et al., 1994). Recently, *Populus* WIN4 (PtdWIN4) was reported to be involved in herbivore resistance since PtdWIN4 transgenic *Arabidopsis* exhibited high anti-herbivore activity (Hu, Wang, Ji, Song, & Yang, 2012). Methyl jasmonate has also been reported to regulate the accumulation of both BSP and WIN4 in *Populus* (Beardmore, Wetzel, & Kalous, 2000; Zhu & Coleman, 2001a), as well as VSP accumulation in soybean (Anderson, 1991; Mason & Mullet, 1990; Staswick, 1994), alfalfa (Meuriot et al., 2004) and Arabidopsis (Berger, Bell, Sadka, & Mullet, 1995) which may be indicative of possible role in herbivore resistance.

3.4 Other VSP in poplar: PNI288

PNI288 is another of the NP-like family of poplar VSPs, and is expressed in conditions similar to WIN4, supporting a possible role as a short-term storage protein (Lawrence et al., 2001). For example, like WIN4, PNI288 is also induced by both wounding and increased nitrogen availability, whereas its mRNA abundances decreased when exposed to either SD or low temperatures (Cooke, Brown, Wu, & Davis, 2003; Lawrence et al., 2001; Pettengill et al., 2013; Rennenberg et al., 2010; Wildhagen et al., 2013). Similar to WIN4, PNI288 expression was also detected in shoot tips, leaves and bark (Pettengill et al., 2013). To date, studies on PNI288 are only limited to *Populus* (Cooke et al., 2003; Lawrence et al., 2001; Pettengill et al., 2013), and future research needs explore the role of PNI288 in N storage and response to environmental stresses in other woody plants.

4. Nitrogen remobilization

In temperate deciduous trees seasonal N cycling involves two phases of N remobilization. N is mobilized from senescing leaves in the fall where it is transported to and stored in perennial tissues during winter. In the spring, as growth commences, this stored N is remobilized once again and partitioned to growing tissues. Although progress has been made towards understanding the molecular and physiological processes involved in autumn N remobilization from leaves and storage in perennial tissues such as bark, much less is known about the physiology and regulation of

N remobilization during spring growth. The role of spring N remobilization to plant growth is mostly circumstantial and is based on the seasonal dynamics of BSP accumulation and disappearance. Few studies have directly assessed the contribution of stored N to spring growth. The importance of spring N remobilization from storage pools to N sinks is likely to be related to reduced N acquisition from soil in spring when soil temperatures are cold (Rennenberg et al., 2010). For example, ^{15}N soil feeding experiments that assayed changes in xylem ^{15}N content during the spring showed that ^{15}N enrichment of xylem sap did not occur until 40 days after bud burst in poplar and 21 days after bud burst in cherry (Millard et al., 2006). Since ^{15}N in xylem sap was dependent upon the uptake of soil applied ^{15}N, this delay in the acquisition of soil N implies that initial spring growth relies upon N remobilization from stored N to developing N sinks such as expanding reproductive and vegetative buds and shoots.

Spring N remobilization involves the catabolism of stored BSP. Furthermore, since Gln is the major amino acid present in xylem sap during N remobilization then spring N remobilization must also involve metabolic processes involved in amino acid catabolism and biosynthesis, and amino acid transport to N sinks. Currently little is known about the physiology and regulation of these processes. In poplar, BSP catabolism and N remobilization is dependent upon the development of N sinks that occurs during bud break and shoot growth (Coleman, Englert, Chen, & Fuchigami, 1993). Further, the absence of expanding buds inhibits BSP catabolism even in plants with sufficient chilling to break dormancy, suggesting that N source (bark) and sink (expanding buds and young leaves) communication is an important aspect that coordinates BSP catabolism and N remobilization (Coleman et al., 1993). Because BSP catabolism depends upon expanding buds and the development of N sinks, it is likely that a signal originates in expanding buds and shoots and subsequently transported to storage tissues, such as bark phloem parenchyma, which in turn triggers BSP catabolism and N remobilization. Ongoing research in our lab is targeted towards identifying this signal and the molecular processes associated with BSP catabolism and N remobilization.

One of the key processes involved in spring N remobilization is the catabolism of BSP which is likely to involve the action of proteolytic enzymes. In apple, Kang and Titus (1980b) partially purified an endopeptidase from dormant bark that appeared to be involved in protein catabolism. Moreover, the levels of exo- and endopeptidase activity in shoot bark coincided with ABA levels of bark where increased ABA levels seemed to suppress protease synthesis or activation (Mousdale, 1982, 1983). Thus, it was proposed that in the case of apple that storage protein catabolism is inhibited by increases in ABA that occur in bark during the winter.

Proteolysis is a complex process involving multifunctional proteases important to different stages in the plant life (van der Hoorn, 2008). The genome of *Populus* is predicted to contain 955 genes that encode proteases (García-Lorenzo, Sjödin, Jansson, & Funk, 2006) making the identification and study on proteases involved in N remobilization challenging. In an attempt to identify proteases that could play a role in poplar BSP catabolism during spring N remobilization, Islam et al. (2015) used multidimensional protein identification technology (MudPIT) to identify bark protein changes that occur during BSP catabolism. Over 4000 proteins were identified in bark prior to and during BSP catabolism with the majority of the proteins associated with BSP catabolism. Thus, the bark proteome undergoes extensive changes as bark transitions to a N source in the spring. These extensive changes in the proteome of bark also included at least 30 proteases that increased in abundance during BSP catabolism and N remobilization (Islam et al., 2015). The identified proteases spanned a range of families including members of the papain-like cysteine proteases, serine carboxypeptidases and aspartyl proteases (Islam et al., 2015). Of these 30 proteases, 16 were predicted to be targeted to vacuole or secretory system which is likely to be necessary if they are involved in BSP catabolism. Half (8) of the proteases predicted to be targeted to the vacuole or secretory systems were classified as papain-like cysteine proteases (PLCPs; C1A), 3 as serine carboxypeptidases (S10), and 3 as eukaryotic aspartyl proteases (A1A and A1B) (Islam et al., 2015). Although the identified proteases may be involved in other process, such a xylem development (Funk, Kositsup, Zhao, & Beers, 2002), it is likely that many are involved in BSP catabolism, especially given their possible location to vacuoles. It is also likely that multiple proteases, including exo- and endoproteases are required for BSP catabolism which would require coordination of protease synthesis or activation. Cysteine proteases also have a role in the mobilization of seed storage proteins that involves both the activation of proteases as well as de novo biosynthesis (Müntz, 2007; Müntz, Belozersky, Dunaevsky, Schlereth, & Tiedemann, 2001; Shutov, Bäumlein, Blattner, & Müntz, 2003; Tiedemann, Schlereth, & Müntz, 2001). It is possible that similarities exist between BSP and seed storage protein catabolism and research directed at uncovering these similarities would be informative towards uncovering conserved features of N remobilization in plants. Additional studies are required to thoroughly identify the role of C1A cysteine proteases during N remobilization in *Populus*, which could provide a basis to develop strategies to manipulate N remobilization in trees.

During N remobilization in the spring, glutamine is the predominate amino acid present in xylem sap of poplar, cherry and red beech (Millard et al., 2006; Sauter & van Cleve, 1992; Schneider, Kreuzwieser, Schupp, Sauter, & Rennenberg, 1994; Wildhagen et al., 2010). Some tree species appear to use different amino acids or compounds in N remobilization including citrulline in *Betula pendula* (Millard, Wendler, Hepburn, & Smith, 1998), arginine in *Juglans nigra* × *regia* (Frak, Millard, Roux, Guillaumie, & Wendler, 2002) and asparagine in *Acer pseudoplatanus* (Millard & Grelet, 2010) and *M. domestica* (Guak, Neilsen, Millard, Wendler, & Neilsen, 2003; Malaguti, Millard, Wendler, Hepburn, & Tagliavini, 2001). Irrespective of the chemical nature of N transport, catabolism of the respective storage proteins in different species requires catabolism of the amino acids produced from storage protein catabolism and the biosynthesis of the transport form of N. As an example, *Populus* BSP is rich in serine, leucine, phenylalanine and lysine, but not glutamine (Coleman, Chen, & Fuchigami, 1992), yet glutamine is the major amino acid found in xylem sap during spring N remobilization. This implies that the breakdown of poplar BSP provides the C and N used in glutamine biosynthesis via glutamine synthetase (GS) before transport to developing buds and shoots. Thus, functional studies on specific glutamine transporters and GSs will advance our understanding of N source-sink mechanisms of seasonal N cycling.

5. Future directions

Nutrient accumulation, including N accumulation, is important to the productivity of both horticultural and forest trees. N fertilizers are an integral part of US and global agricultural systems with over 90 million metric tons (MMt) of N fertilizer applied to soils globally (Frink, Waggoner, & Ausubel, 1999) but global nitrogen use-efficiency (NUE) is only 42% (Zhang et al., 2015) and it is estimated that 50–70% of N fertilizers are lost following application (Good, Shrawat, & Muench, 2004). Further, N lost from fertilization has major environmental costs including increased atmospheric N deposition in forest and other natural ecosystems, increased leaching and surface runoff leading to the onset of algal blooms, hypoxia and fish kills (Robertson & Vitousek, 2009). Thus, any gains in enhancing plant NUE will lead to lower N inputs and would directly impact long-term improvement and sustainability of U.S. agriculture and food and fibre systems. Because seasonal N cycling is the major factor that contributes the NUE in trees, deciphering this process would reveal strategies used by perennial plants to conserve and reuse N. Although progress has been made towards understanding the mechanisms of seasonal N cycling in trees there are still substantial knowledge gaps. The molecular signalling and regulatory pathways involved in seasonal N accumulation and the factors involved in N transport for N sources to N sinks are poorly understood in trees. Understanding this process would also provide insight into long-distance source-sink communication in plants.

Acknowledgements

Research in the authors lab was supported in part by the National Science Foundation Grant IOS0922650 and by Tier 1 funds from the Vice President of Research, University of Maryland.

References

Aerts, R. (1990). Nutrient use efficiency in evergreen and deciduous species from heathlands. *Oecologia, 84*, 391–397.

Alonso, J. M., Hirayama, T., Roman, G., Nourizadeh, S., & Ecker, J. R. (1999). EIN2, a bifunctional transducer of ethylene and stress responses in Arabidopsis. *Science, 284*, 2148–2152.

Al-Sady, B., Ni, W., Kircher, S., Schäfer, E., & Quail, P. H. (2006). Photoactivated phytochrome induces rapid PIF3 phosphorylation prior to proteasome-mediated degradation. *Molecular Cell, 23*, 439–446.

Amon, J., Titgemeyer, F., & Burkovski, A. (2010). Common patterns—Unique features: Nitrogen metabolism and regulation in Gram-positive bacteria. *FEMS Microbiology Reviews, 34*, 588–605.

An, F., Zhao, Q., Ji, Y., Li, W., Jiang, Z., Yu, X., et al. (2010). Ethylene-induced stabilization of ETHYLENE INSENSITIVE3 and EIN3-LIKE1 is mediated by proteasomal degradation of EIN3 binding F-box 1 and 2 that requires EIN2 in Arabidopsis. *Plant Cell, 22*, 2384–2401.

Anderson, J. M. (1991). Jasmonic acid-dependent increase in vegetative storage protein in soybean tissue cultures. *Journal of Plant Growth Regulation, 10*, 5.

Andersson, A., Keskitalo, J., Sjödin, A., Bhalerao, R., Sterky, F., Wissel, K., et al. (2004). A transcriptional timetable of autumn senescence. *Genome Biology, 5*, R24.

Andrews, D. L., Beames, B., Summers, M. D., & Park, W. D. (1988). Characterization of the lipid acyl hydrolase activity of the major potato (*Solanum tuberosum*) tuber protein, patatin, by cloning and abundant expression in a baculovirus vector. *The Biochemical Journal, 252*, 199–206.

Avila-Ospina, L., Moison, M., Yoshimoto, K., & Masclaux-Daubresse, C. (2014). Autophagy, plant senescence, and nutrient recycling. *Journal of Experimental Botany, 65*, 3799–3811.

Babst, B. A., & Coleman, G. D. (2018). Seasonal nitrogen cycling in temperate trees: Transport and regulatory mechanisms are key missing links. *Plant Science, 270*, 268–277.

Beardmore, T., Wetzel, S., Burgess, D., & Charest, P. J. (1996). Characterization of seed storage proteins in Populus and their homology with Populus vegetative storage proteins. *Tree Physiology*, *16*, 833–840.

Beardmore, T., Wetzel, S., & Kalous, M. (2000). Interactions of airborne methyl jasmonate with vegetative storage protein gene and protein accumulation and biomass partitioning in Populus plants. *Canadian Journal of Forest Research*, *30*, 1106–1113.

Berger, S., Bell, E., Sadka, A., & Mullet, J. E. (1995). *Arabidopsis thaliana Atvsp* is homologous to soybean *VspA* and *VspB*, genes encoding vegetative storage protein acid phosphatases, and is regulated similarly by methyl jasmonate, wounding, sugars, light and phosphate. *Plant Molecular Biology*, *27*, 933–942.

Bhalerao, R. (2003). Gene expression in autumn leaves. *Plant Physiology*, *131*, 430–442.

Binder, B. M., Walker, J. M., Gagne, J. M., Emborg, T. J., Hemmann, G., Bleecker, A. B., et al. (2007). The Arabidopsis EIN3 binding F-box proteins EBF1 and EBF2 have distinct but overlapping roles in ethylene signaling. *Plant Cell*, *19*, 509–523.

Birk, E. M., & Vitousek, P. M. (1986). Nitrogen availability and nitrogen use efficiency in loblolly pine stands. *Ecology*, *67*, 69–79.

Boerner, R. (1984). Foliar nutrient dynamics and nutrient use efficiency of four deciduous tree species in relation to site fertility. *Journal of Applied Ecology*, *21*, 1029–1040.

Bögre, L., Henriques, R., & Magyar, Z. (2013). TOR tour to auxin. *The EMBO Journal*, *32*, 1069–1071.

Bourrellier, A. B. F., Valot, B., Guillot, A., Ambard-Bretteville, F., Vidal, J., & Hodges, M. (2010). Chloroplast acetyl-CoA carboxylase activity is 2-oxoglutarate–regulated by interaction of PII with the biotin carboxyl carrier subunit. *Proceedings of the National Academy of Sciences of the United States of America*, *107*, 502–507.

Bravo, K., Marcolini, G., Sorrenti, G., Baldi, E., Quartieri, M., & Toselli, M. (2017). Effect of time of application on nitrogen uptake, partitioning, and remobilization in walnut trees. *Journal of Plant Nutrition*, *40*, 719–725.

Brychkova, G., Alikulov, Z., Fluhr, R., & Sagi, M. (2008). A critical role for ureides in dark and senescence-induced purine remobilization is unmasked in the Atxdh1 Arabidopsis mutant. *The Plant Journal*, *54*, 496–509.

Caldana, C., Li, Y., Leisse, A., Zhang, Y., Bartholomaeus, L., Fernie, A. R., et al. (2013). Systemic analysis of inducible target of rapamycin mutants reveal a general metabolic switch controlling growth in Arabidopsis thaliana. *The Plant Journal*, *73*, 897–909.

Chaiwanon, J., Wang, W., Zhu, J.-Y., Oh, E., & Wang, Z.-Y. (2016). Information integration and communication in plant growth regulation. *Cell*, *164*, 1257–1268.

Chang, C., & Shockey, J. A. (1999). The ethylene-response pathway: Signal perception to gene regulation. *Current Opinion in Plant Biology*, *2*, 352–358.

Chantranupong, L., Wolfson, R. L., & Sabatini, D. M. (2015). Nutrient-sensing mechanisms across evolution. *Cell*, *161*, 67–83.

Chapin, F. S., & Kedrowski, R. A. (1983). Seasonal changes in nitrogen and phosphorus fractions and autumn retranslocation in evergreen and deciduous taiga trees. *Ecology*, *64*, 376–391.

Chapin, F. S., Schulze, E. D., & Mooney, H. A. (1990). The ecology and economics of storage in plants. *Annual Review of Ecology and Systematics*, *21*, 423–447.

Chellamuthu, V. R., Alva, V., & Forchhammer, K. (2013). From cyanobacteria to plants: Conservation of PII functions during plastid evolution. *Planta*, *237*, 451–462.

Chellamuthu, V.-R., Ermilova, E., Lapina, T., Lüddecke, J., Minaeva, E., Herrmann, C., et al. (2014). A widespread glutamine-sensing mechanism in the plant kingdom. *Cell*, *159*, 1188–1199.

Chen, Y., Rouge, P., Peumans, W. J., & van Damme, E. J. M. (2002). Mutational analysis of the carbohydrate-binding activity of the NeuAc(alpha-2,6)Gal/GalNAc-specific type 2 ribosome-inactivating protein from elderberry (*Sambucus nigra*) fruits. *The Biochemical Journal*, *364*, 587–592.

Cheng, L., Ma, F., & Ranwala, D. (2004). Nitrogen storage and its interaction with carbohydrates of young apple trees in response to nitrogen supply. *Tree Physiology*, *24*, 91–98.
Coleman, G. D., Banados, M. P., & Chen, T. H. (1994). Poplar bark storage protein and a related wound-induced gene are differentially induced by nitrogen. *Plant Physiology*, *106*, 211–215.
Coleman, G. D., Chen, T. H. H., Ernst, S. G., & Fuchigami, L. (1991). Photoperiod control of poplar bark storage protein accumulation. *Plant Physiology*, *96*, 686–692.
Coleman, G. D., Chen, T. H. H., & Fuchigami, L. H. (1992). Complementary DNA cloning of poplar bark storage protein and control of its expression by photoperiod. *Plant Physiology*, *98*, 687–693.
Coleman, G. D., Englert, J. M., Chen, T. H., & Fuchigami, L. H. (1993). Physiological and environmental requirements for poplar (*Populus deltoides*) bark storage protein degradation. *Plant Physiology*, *102*, 53–59.
Cooke, J. E. K., Brown, K. A., Wu, R., & Davis, J. M. (2003). Gene expression associated with N-induced shifts in resource allocation in poplar. *Plant, Cell & Environment*, *26*, 757–770.
Côté, B., & Dawson, J. O. (1986). Autumnal changes in total nitrogen, salt-extractable proteins and amino acids in leaves and adjacent bark of black alder, eastern cottonwood and white basswood. *Physiologia Plantarum*, *67*, 102–108.
Côté, B., Vogel, C. S., & Dawson, J. O. (1989). Autumnal changes in tissue nitrogen of autumn olive, black alder and eastern cottonwood. *Plant and Soil*, *118*, 23–32.
Couturier, J., Doidy, J., Guinet, F., Wipf, D., Blaudez, D., & Chalot, M. (2010). Glutamine, arginine and the amino acid transporter Pt-CAT11 play important roles during senescence in poplar. *Annals of Botany*, *105*, 1159–1169.
Crowther, T. W., Glick, H. B., Covey, K. R., Bettigole, C., Maynard, D. S., Thomas, S. M., et al. (2015). Mapping tree density at a global scale. *Nature*, *525*, 201–205.
Davis, J. M., Egelkrout, E. E., Coleman, G. D., Chen, T. H., Haissig, B. E., Riemenschneider, D. E., et al. (1993). A family of wound-induced genes in Populus shares common features with genes encoding vegetative storage proteins. *Plant Molecular Biology*, *23*, 135–143.
de Lucas, M., & Prat, S. (2014). PIFs get BRright: PHYTOCHROME INTERACTING FACTORs as integrators of light and hormonal signals. *The New Phytologist*, *202*, 1126–1141.
DeWald, D. B., Mason, H. S., & Mullet, J. E. (1992). The soybean vegetative storage proteins VSP alpha and VSP beta are acid phosphatases active on polyphosphates. *The Journal of Biological Chemistry*, *267*, 15958–15964.
Dobrenel, T., Marchive, C., Sormani, R., Moreau, M., Mozzo, M., Montané, M., et al. (2011). Regulation of plant growth and metabolism by the TOR kinase. *Biochemical Society Transactions*, *39*, 477–481.
Eckstein, R. L., Karlsson, P. S., & Weih, M. (1999). Leaf life span and nutrient resorption as determinants of plant nutrient conservation in temperate-arctic regions. *The New Phytologist*, *143*, 177–189.
Faus, I., Zabalza, A., Santiago, J., Nebauer, S. G., Royuela, M., Serrano, R., et al. (2015). Protein kinase GCN2 mediates responses to glyphosate in Arabidopsis. *BMC Plant Biology*, *15*, 14.
Ferguson, C. W. (1968). Bristlecone pine: Science and esthetics: A 7100-year tree-ring chronology aids scientists; old trees draw visitors to California mountains. *Science*, *159*, 839–846.
Fife, D. N., & Nambiar, E. K. S. (1982). Accumulation and retranslocation of mineral nutrients in developing needles in relation to seasonal growth of young Radiata pine trees. *Annals of Botany*, *50*, 817–829.

Finlayson, S. A., Lee, I.-J., & Morgan, P. W. (1998). Phytochrome B and the regulation of circadian ethylene production in sorghum. *Plant Physiology, 116*, 17–25.
Finlayson, S. A., Lee, I.-J., Mullet, J. E., & Morgan, P. W. (1999). The mechanism of rhythmic ethylene production in sorghum. The role of phytochrome B and simulated shading. *Plant Physiology, 119*, 1083–1090.
Fokina, O., Chellamuthu, V.-R., Forchhammer, K., & Zeth, K. (2010). Mechanism of 2-oxoglutarate signaling by the *Synechococcus elongatus* PII signal transduction protein. *Proceedings of the National Academy of Sciences of the United States of America, 107*, 19760–19765.
Forchhammer, K. (2008). PII signal transducers: Novel functional and structural insights. *Trends in Microbiology, 16*, 65–72.
Fracheboud, Y., Luquez, V., Bjorken, L., Sjodin, A., Tuominen, H., & Jansson, S. (2009). The control of autumn senescence in European aspen. *Plant Physiology, 149*, 1982–1991.
Frak, E., Millard, P., Roux, X. L., Guillaumie, S., & Wendler, R. (2002). Coupling sap flow velocity and amino acid concentrations as an alternative method to ^{15}N labeling for quantifying nitrogen remobilization by walnut trees. *Plant Physiology, 130*, 1043–1053.
Frink, C. R., Waggoner, P. E., & Ausubel, J. H. (1999). Nitrogen fertilizer: Retrospect and prospect. *Proceedings of the National Academy of Sciences of the United States of America, 96*, 1175–1180.
Funk, V., Kositsup, B., Zhao, C., & Beers, E. P. (2002). The Arabidopsis xylem peptidase XCP1 is a tracheary element vacuolar protein that may be a papain ortholog. *Plant Physiology, 128*, 84–94.
Gagne, J. M., Smalle, J., Gingerich, D. J., Walker, J. M., Yoo, S.-D., Yanagisawa, S., et al. (2004). Arabidopsis EIN3-binding F-box 1 and 2 form ubiquitin-protein ligases that repress ethylene action and promote growth by directing EIN3 degradation. *Proceedings of the National Academy of Sciences of the United States of America, 101*, 6803–6808.
Gana, J. A., Kalengamaliro, N. E., Cunningham, S. M., & Volenec, J. J. (1998). Expression of beta-amylase from alfalfa taproots. *Plant Physiology, 118*, 1495–1506.
García-Lorenzo, M., Sjödin, A., Jansson, S., & Funk, C. (2006). Protease gene families in Populus and Arabidopsis. *BMC Plant Biology, 6*, 30.
Gérant, D., Pluchon, M., Mareschal, L., Koutika, L. S., & Epron, D. (2017). Seasonality of nitrogen partitioning (non-structural vs structural) in the leaves and woody tissues of tropical eucalypts experiencing a marked dry season. *Tree Physiology, 37*, 790–798.
Godfray, H. C. J., Beddington, J. R., Crute, I. R., Haddad, L., Lawrence, D., Muir, J. F., et al. (2010). Food security: The challenge of feeding 9 billion people. *Science, 327*, 812–818.
Good, A. G., Shrawat, A. K., & Muench, D. G. (2004). Can less yield more? Is reducing nutrient input into the environment compatible with maintaining crop production? *Trends in Plant Science, 9*, 597–605.
Guak, S., Neilsen, D., Millard, P., Wendler, R., & Neilsen, G. H. (2003). Determining the role of N remobilization for growth of apple (*Malus domestica* Borkh.) trees by measuring xylem-sap N flux. *Journal of Experimental Botany, 54*, 2121–2131.
Guo, H., & Ecker, J. R. (2003). Plant responses to ethylene gas are mediated by SCF(EBF1/EBF2)-dependent proteolysis of EIN3 transcription factor. *Cell, 115*, 667–677.
Guranowski, A. (1982). Purine catabolism in plants. *Plant Physiology, 70*, 344–349.
Harms, U., & Sauter, J. J. (1991). Storage proteins in the wood of *Taxodiaceae* and of *Taxus*. *Journal of Plant Physiology, 138*, 497–499.
Henriques, R., Bögre, L., Horváth, B., & Magyar, Z. (2014). Balancing act: Matching growth with environment by the TOR signalling pathway. *Journal of Experimental Botany, 65*, 2691–2701.

Herman, E. M., & Larkins, B. A. (1999). Protein storage bodies and vacuoles. *Plant Cell, 11*, 601–613.

Hermans, C., Hammond, J. P., White, P. J., & Verbruggen, N. (2006). How do plants respond to nutrient shortage by biomass allocation? *Trends in Plant Science, 11*, 610–617.

Herrero, A., Muro-Pastor, A. M., & Flores, E. (2001). Nitrogen control in cyanobacteria. *Journal of Bacteriology, 183*, 411–425.

Hesberg, C., Hänsch, R., Mendel, R. R., & Bittner, F. (2004). Tandem orientation of duplicated xanthine dehydrogenase genes from *Arabidopsis thaliana*: Differential gene expression and enzyme activities. *The Journal of Biological Chemistry, 279*, 13547–13554.

Howe, G. T., Gardner, G., Hackett, W. P., & Furnier, G. R. (1996). Phytochrome control of short-day-induced bud set in black cottonwood. *Physiologia Plantarum, 97*, 95–103.

Hu, R., Wang, J., Ji, Y., Song, Y., & Yang, S. (2012). Overexpression of poplar wounding-inducible genes in Arabidopsis caused improved resistance against *Helicoverpa armigera* (Hübner) larvae. *Breeding Science, 62*, 288–291.

Huergo, L. F., Chandra, G., & Merrick, M. (2013). PII signal transduction proteins: Nitrogen regulation and beyond. *FEMS Microbiology Reviews, 37*, 251–283.

Islam, N., Li, G., Garrett, W. M., Lin, R., Sriram, G., Cooper, B., et al. (2015). Proteomics of nitrogen remobilization in poplar bark. *Journal of Proteome Research, 14*, 1112–1126.

Jung, B., Hoffmann, C., & Möhlmann, T. (2011). Arabidopsis nucleoside hydrolases involved in intracellular and extracellular degradation of purines. *The Plant Journal, 65*, 703–711.

Kang, S.-M., & Titus, J. S. (1980a). Qualitative and quantitative changes in nitrogenous compounds in senescing leaf and bark tissues of the apple. *Physiologia Plantarum, 50*, 285–290.

Kang, S.-M., & Titus, J. S. (1980b). Isolation and partial characterization of an acid endoprotease present in dormant apple shoot bark. *Plant Physiology, 66*, 984–989.

Katahira, R., & Ashihara, H. (2006a). Profiles of purine biosynthesis, salvage and degradation in disks of potato (*Solanum tuberosum* L.) tubers. *Planta, 225*, 115–126.

Katahira, R., & Ashihara, H. (2006b). Role of adenosine salvage in wound-induced adenylate biosynthesis in potato tuber slices. *Plant Physiology and Biochemistry, 44*, 551–555.

Kato, T., Yamagata, M., Tsukahara, S., & Shikoku N.A.E.S (1984). Storage forms and reservoirs of nitrogen used for new shoot development in satsuma mandarin tree. *Journal of the Japanese Society for Horticultural Science, 52*, 393–398.

Keskitalo, J., Bergquist, G., Gardeström, P., & Jansson, S. (2005). A cellular timetable of autumn senescence. *Plant Physiology, 139*, 1635–1648.

Kim, J., Woo, H. R., & Nam, H. G. (2016). Toward systems understanding of leaf senescence: An integrated multi-omics perspective on leaf senescence research. *Molecular Plant, 9*, 813–825.

Kost, J. A., & Boerner, R. E. J. (1985). Foliar nutrient dynamics and nutrient use efficiency in *Cornus florida*. *Oecologia, 66*, 602–606.

Lageix, S., Lanet, E., Pouch-Pélissier, M.-N., Espagnol, M.-C., Robaglia, C., Deragon, J.-M., et al. (2008). Arabidopsis eIF2α kinase GCN2 is essential for growth in stress conditions and is activated by wounding. *BMC Plant Biology, 8*, 134.

Langheinrich, U., & Tischner, R. (1991). Vegetative storage proteins in poplar induction and characterization of a 32-and a 36-kilodalton polypeptide. *Plant Physiology, 97*, 1017–1025.

Lawrence, S. D., Cooke, J. E. K., Greenwood, J. S., Korhnak, T. E., & Davis, J. M. (2001). Vegetative storage protein expression during terminal bud formation in poplar. *Canadian Journal of Forest Research, 31*, 1098–1103.

Lawrence, S. D., Greenwood, J. S., Korhnak, T. E., & Davis, J. M. (1997). A vegetative storage protein homolog is expressed in the growing shoot apex of hybrid poplar. *Planta, 203*, 237–244.

Leiber, R.-M., John, F., Verhertbruggen, Y., Diet, A., Knox, J. P., & Ringli, C. (2010). The TOR pathway modulates the structure of cell walls in Arabidopsis. *Plant Cell, 22*, 1898–1908.
Leigh, J. A., & Dodsworth, J. A. (2007). Nitrogen regulation in bacteria and archaea. *Annual Review of Microbiology, 61*, 349–377.
Leivar, P., & Quail, P. H. (2011). PIFs: Pivotal components in a cellular signaling hub. *Trends in Plant Science, 16*, 19–28.
Lewkowicz, E. S., & Iribarren, A. M. (2006). Nucleoside phosphorylases. *Current Organic Chemistry, 10*, 1197–1215.
Llácer, J. L., Contreras, A., Forchhammer, K., Marco-Marín, C., Gil-Ortiz, F., Maldonado, R., et al. (2007). The crystal structure of the complex of PII and acetylglutamate kinase reveals how PII controls the storage of nitrogen as arginine. *Proceedings of the National Academy of Sciences of the United States of America, 104*, 17644–17649.
Llácer, J. L., Fita, I., & Rubio, V. (2008). Arginine and nitrogen storage. *Current Opinion in Structural Biology, 18*, 673–681.
Mahfouz, M. M., Kim, S., Delauney, A. J., & Verma, D. P. S. (2006). Arabidopsis TARGET OF RAPAMYCIN interacts with RAPTOR, which regulates the activity of S6 kinase in response to osmotic stress signals. *Plant Cell, 18*, 477–490.
Malaguti, D., Millard, P., Wendler, R., Hepburn, A., & Tagliavini, M. (2001). Translocation of amino acids in the xylem of apple (Malus domestica Borkh.) trees in spring as a consequence of both N remobilization and root uptake. *Journal of Experimental Botany, 52*, 1665–1671.
Mason, H. S., & Mullet, J. E. (1990). Expression of two soybean vegetative storage protein genes during development and in response to water deficit, wounding, and jasmonic acid. *Plant Cell, 2*, 569–579.
Mason, A. C., & Whitfield, A. B. (1960). Seasonal changes in the uptake and distribution of mineral elements in apple trees. *Journal of Horticultural Science, 35*, 34–55.
Menand, B., Desnos, T., Nussaume, L., Berger, F., Bouchez, D., Meyer, C., et al. (2002). Expression and disruption of the Arabidopsis TOR (target of rapamycin) gene. *Proceedings of the National Academy of Sciences of the United States of America, 99*, 6422–6427.
Merchante, C., Brumos, J., Yun, J., Hu, Q., Spencer, K. R., Enríquez, P., et al. (2015). Gene-specific translation regulation mediated by the hormone-signaling molecule EIN2. *Cell, 163*, 684–697.
Merrick, M. J., & Edwards, R. A. (1995). Nitrogen control in bacteria. *Microbiological Reviews, 59*, 604–622.
Meuriot, F., Noquet, C., Avice, J.-C., Volenec, J. J., Cunningham, S. M., Sors, T. G., et al. (2004). Methyl jasmonate alters N partitioning, N reserves accumulation and induces gene expression of a 32-kDa vegetative storage protein that possesses chitinase activity in *Medicago sativa* taproots. *Physiologia Plantarum, 120*, 113–123.
Millard, P., & Grelet, G. (2010). Nitrogen storage and remobilization by trees: Ecophysiological relevance in a changing world. *Tree Physiology, 30*, 1083–1095.
Millard, P., & Thomson, C. M. (1989). The effect of the autumn senescence of leaves on the internal cycling of nitrogen for the spring growth of apple trees. *Journal of Experimental Botany, 40*, 1285–1289.
Millard, P., Wendler, R., Grassi, G., Grelet, G. A., & Tagliavini, M. (2006). Translocation of nitrogen in the xylem of field-grown cherry and poplar trees during remobilization. *Tree Physiology, 26*, 527–536.
Millard, P., Wendler, R., Hepburn, A., & Smith, A. (1998). Variations in the amino acid composition of xylem sap of *Betula pendula* Roth. Trees due to remobilization of stored N in the spring. *Plant, Cell & Environment, 21*, 715–722.
Mizuno, Y., Berenger, B., Moorhead, G. B. G., & Ng, K. K.-S. (2007). Crystal structure of Arabidopsis PII reveals novel structural elements unique to plants. *Biochemistry, 46*, 1477–1483.

Mizuno, Y., Moorhead, G. B. G., & Ng, K. K.-S. (2007). Structural basis for the regulation of N-acetylglutamate kinase by PII in *Arabidopsis thaliana*. *The Journal of Biological Chemistry*, *282*, 35733–35740.

Mousdale, D. M. (1982). Endogenous abscisic acid and the regulation of protein metabolism in shoot bark of apple (*Malus domestics* Borkh.). *Biochemie und Physiologie der Pflanzen*, *177*, 601–610.

Mousdale, D. M. (1983). Endogenous abscisic acid and the regulation of proteolytic activities in shoot bark of apple (*Malus domestica* Borkh.). *Biochemie und Physiologie der Pflanzen*, *178*, 373–379.

Muñoz, N., Guerri, J., Legaz, F., & Primo-Millo, E. (1993). Seasonal uptake of 15N-nitrate and distribution of absorbed nitrogen in peach trees. *Plant and Soil*, *150*, 263–269.

Müntz, K. (1998). Deposition of storage proteins. *Plant Molecular Biology*, *38*, 77–99.

Müntz, K. (2007). Protein dynamics and proteolysis in plant vacuoles. *Journal of Experimental Botany*, *58*, 2391–2407.

Müntz, K., Belozersky, M. A., Dunaevsky, Y. E., Schlereth, A., & Tiedemann, J. (2001). Stored proteinases and the initiation of storage protein mobilization in seeds during germination and seedling growth. *Journal of Experimental Botany*, *52*, 1741–1752.

Murneek, A. E., & Logan, J. C. (1932). *Autumnal migration of nitrogen and carbohydrates in the apple tree*. Columbia, Missouri: University of Missouri, Agricultural Experiment Station.

Nambiar, E. K. S., & Fife, D. N. (1991). Nutrient retranslocation in temperate conifers. *Tree Physiology*, *9*, 185–207.

Nomura, K., Ikegami, A., Koide, A., & Yagi, F. (2008). Japanese chestnut (*Castanea crenata*) agglutinin may have a role as vegetative storage protein. *Journal of Plant Physiology*, *165*, 671–675.

O'Kennedy, B. T., Hennerty, M. J., & Titus, J. S. (1975). The effects of autumn foliar urea sprays on storage forms of nitrogen extracted from bark and wood of apple shoots. *Journal of Horticultural Science*, *50*, 331–338.

Oland, K. (1959). Nitrogenous reserves of apple trees. *Physiologia Plantarum*, *12*, 594–648.

Olsen, J. E., Junttila, O., Nilsen, J., Eriksson, M. E., Martinussen, I., Olsson, O., et al. (1997). Ectopic expression of oat phytochrome A in hybrid aspen changes critical daylength for growth and prevents cold acclimatization. *The Plant Journal*, *12*, 1339–1350.

Osanai, T., & Tanaka, K. (2007). Keeping in touch with PII: PII-interacting proteins in unicellular cyanobacteria. *Plant & Cell Physiology*, *48*, 908–914.

Pettengill, E. A., Pettengill, J. B., & Coleman, G. D. (2013). Elucidating the evolutionary history and expression patterns of nucleoside phosphorylase paralogs (vegetative storage proteins) in *Populus* and the plant kingdom. *BMC Plant Biology*, *13*, 118.

Potuschak, T., Lechner, E., Parmentier, Y., Yanagisawa, S., Grava, S., Koncz, C., et al. (2003). EIN3-dependent regulation of plant ethylene hormone signaling by two Arabidopsis F box proteins: EBF1 and EBF2. *Cell*, *115*, 679–689.

Pugmire, M. J., & Ealick, S. E. (2002). Structural analyses reveal two distinct families of nucleoside phosphorylases. *The Biochemical Journal*, *361*, 1–25.

Qiao, H., Shen, Z., Huang, S. C., Schmitz, R. J., Urich, M. A., Briggs, S. P., et al. (2012). Processing and subcellular trafficking of ER-tethered EIN2 control response to ethylene gas. *Science*, *338*, 390–393.

Quail, P. H. (2002). Phytochrome photosensory signalling networks. *Nature Reviews. Molecular Cell Biology*, *3*, 85–93.

Quartieri, M., Millard, P., & Tagliavini, M. (2002). Storage and remobilisation of nitrogen by pear (*Pyrus communis* L.) trees as affected by timing of N supply. *European Journal of Agronomy*, *17*, 105–110.

Ramón-Maiques, S., Fernández-Murga, M. L., Gil-Ortiz, F., Vagin, A., Fita, I., & Rubio, V. (2006). Structural bases of feed-back control of arginine biosynthesis, revealed by the structures of two hexameric n-acetylglutamate kinases, from *Thermotoga maritima* and *Pseudomonas aeruginosa*. *Journal of Molecular Biology, 356*, 695–713.

Ren, M., Venglat, P., Qiu, S., Feng, L., Cao, Y., Wang, E., et al. (2012). Target of rapamycin signaling regulates metabolism, growth, and life span in Arabidopsis. *Plant Cell, 24*, 4850–4874.

Rennenberg, H., & Schmidt, S. (2010). Perennial lifestyle—An adaptation to nutrient limitation? *Tree Physiology, 30*, 1047–1049.

Rennenberg, H., Wildhagen, H., & Ehlting, B. (2010). Nitrogen nutrition of poplar trees: Nitrogen nutrition of poplar trees. *Plant Biology, 12*, 275–291.

Richard-Molard, C., Brugière, N., Moille, M., Carrayol, E., & Limami, A. M. (2004). Molecular characterization of a gene encoding a vegetative storage protein (CiVSP) from *Cichorium intybus* and its expression in the root and shoot in relation to nitrogen status and pathogen resistance. *Physiologia Plantarum, 121*, 568–577.

Robaglia, C., Thomas, M., & Meyer, C. (2012). Sensing nutrient and energy status by SnRK1 and TOR kinases. *Current Opinion in Plant Biology, 15*, 301–307.

Roberts, R. H. (1921). Nitrogen reserve in apple trees. *Proceedings of the American Society for Horticultural Science, 18*, 143–145.

Robertson, G. P., & Vitousek, P. M. (2009). Nitrogen in agriculture: Balancing the cost of an essential resource. *Annual Review of Environment and Resources, 34*, 97–125.

Ryan, D., & Bormann, F. (1982). Nutrient resorption in northern hardwood forests. *Bioscience, 32*, 29–32.

Salzer, M. W., Hughes, M. K., Bunn, A. G., & Kipfmueller, K. F. (2009). Recent unprecedented tree-ring growth in bristlecone pine at the highest elevations and possible causes. *Proceedings of the National Academy of Sciences of the United States of America, 106*, 20348–20353.

Sant'Anna, F. H., Trentini, D. B., de Souto Weber, S., Cecagno, R., da Silva, S. C., & Schrank, I. S. (2009). The PII superfamily revised: A novel group and evolutionary insights. *Journal of Molecular Evolution, 68*, 322–336.

Santiago, J. P., & Tegeder, M. (2016). Connecting source with sink: The role of Arabidopsis AAP8 in phloem loading of amino acids. *Plant Physiology, 171*, 508–521. pp.00244.2016.

Sauter, J. J., & van Cleve, B. (1992). Seasonal variation of amino acids in the xylem sap of "Populus × canadensis" and its relation to protein body mobilization. *Trees, 7*, 26–32.

Sauter, J. J., van Cleve, B., & Apel, K. (1988). Protein bodies in ray cells of *Populus x canadensis* Moench 'robusta'. *Planta, 173*, 31–34.

Sauter, J. J., van Cleve, B., & Wellenkamp, S. (1989). Ultrastructural and biochemical results on the localization and distribution of storage proteins in a poplar tree and in twigs of other tree species. *Holzforschung, 43*, 1–6.

Schippers, J. H. (2015). Transcriptional networks in leaf senescence. *Current Opinion in Plant Biology, 27*, 77–83.

Schmidt, S., & Stewart, G. R. (1998). Transport, storage and mobilization of nitrogen by trees and shrubs in the wet/dry tropics of northern Australia. *Tree Physiology, 18*, 403–410.

Schneider, A., Kreuzwieser, J., Schupp, R., Sauter, J. J., & Rennenberg, H. (1994). Thiol and amino acid composition of the xylem sap of poplar trees (*Populus × canadensis* 'robusta'). *Canadian Journal of Botany, 72*, 347–351.

Shen, Y., Khanna, R., Carle, C. M., & Quail, P. H. (2007). Phytochrome induces rapid PIF5 phosphorylation and degradation in response to red-light activation. *Plant Physiology, 145*, 1043–1051.

Shi, H., Shen, X., Liu, R., Xue, C., Wei, N., Deng, X. W., et al. (2016). The red light receptor phytochrome B directly enhances substrate-E3 ligase interactions to attenuate ethylene responses. *Developmental Cell, 39*, 597–610.

Shutov, A. D., Bäumlein, H., Blattner, F. R., & Müntz, K. (2003). Storage and mobilization as antagonistic functional constraints on seed storage globulin evolution. *Journal of Experimental Botany*, *54*, 1645–1654.
Smith, P. M. C., & Atkins, C. A. (2002). Purine biosynthesis. Big in cell division, even bigger in nitrogen assimilation. *Plant Physiology*, *128*, 793–802.
Sonenshein, A. L. (2007). Control of key metabolic intersections in *Bacillus subtilis*. *Nature Reviews. Microbiology*, *5*, 917–927.
Staswick, P. E. (1994). Storage proteins of vegetative plant tissues. *Annual Review of Plant Biology*, *45*, 303–322.
Taylor, B. K. (1967). The nitrogen nutrition of the peach tree I. seasonal changes in nitrogenous constituents in mature trees. *Australian Journal of Biological Sciences*, *20*, 379–388.
Taylor, B. K., & May, L. H. (1967). The nitrogen nutrition of the peach tree II. Storage and mobilization of nitrogen in young trees. *Australian Journal of Biological Sciences*, *20*, 389–412.
Taylor, B. K., & van den Ende, B. (1969). The nitrogen nutrition of the peach tree. IV. Storage and mobilization of nitrogen in mature trees. *Australian Journal of Agricultural Research*, *20*, 869–881.
Taylor, B. K., & van den Ende, B. (1970). The nitrogen nutrition of the peach tree. VI. Influence of autumn nitrogen applications on the accumulation of nitrogen, carbohydrate, and macroelements in 1-year-old peach trees. *Australian Journal of Agricultural Research*, *21*, 693–698.
Taylor, B. K., van den Ende, B., & Canterford, R. L. (1975). Effects of rate and timing of nitrogen applications on the performance and chemical composition of young pear trees, cv Williams' Bon Chretien. *Journal of Horticultural Science*, *50*, 29–40.
Thomas, W. (1927). Nitrogenous metabolism of *Pyrus malus* L. III. The partition of nitrogen in the leaves, one and two year branch growth and non-bearing spurs throughout a year's cycle. *Plant Physiology*, *2*, 109–137.
Tian, W.-M., Peng, S.-Q., Hao, B.-Z., Wu, J.-L., Hu, Z.-H., & Cui, K.-M. (2005). Poplar trees (*Populus canadensis* Moench) initiate vegetative storage protein accumulation during new shoot development in spring. *Journal of Integrative Plant Biology*, *47*, 717–725.
Tian, W.-M., Peng, S.-Q., Wang, X.-C., Shi, M.-J., Chen, Y.-Y., & Hu, Z.-H. (2007). Vegetative storage protein in *Litchi chinensis*, a subtropical evergreen fruit tree, possesses trypsin inhibitor activity. *Annals of Botany*, *100*, 1199–1208.
Tiedemann, J., Schlereth, A., & Müntz, K. (2001). Differential tissue-specific expression of cysteine proteinases forms the basis for the fine-tuned mobilization of storage globulin during and after germination in legume seeds. *Planta*, *212*, 728–738.
Titus, J. S., & Kang, S.-M. (1982). Nitrogen metabolism, translocation, and recycling in apple trees. In J. Janick (Ed.), *Horticultural reviews* (pp. 204–246): John Wiley & Sons, Inc.
Tuskan, G. A., et al. (2006). The genome of black cottonwood, *Populus trichocarpa* (Torr. & Gray). *Science*, *313*, 1596–1604.
Uhrig, R. G., Ng, K. K. S., & Moorhead, G. B. G. (2009). PII in higher plants: A modern role for an ancient protein. *Trends in Plant Science*, *14*, 505–511.
van Cleve, B., Clausen, S., & Sauter, J. J. (1988). Immunochemical localization of a storage protein in poplar wood. *Journal of Plant Physiology*, *133*, 371–374.
Van Damme, E. J., Barre, A., Bemer, V., Rougé, P., Van Leuven, F., & Peumans, W. J. (1995). A lectin and a lectin-related protein are the two most prominent proteins in the bark of yellow wood (*Cladrastis lutea*). *Plant Molecular Biology*, *29*, 579–598.
Van Damme, E. J. M., Hao, Q., Barre, A., Rougé, P., Van Leuven, F., & Peumans, W. J. (2000). Major protein of resting rhizomes of *Calystegia sepium* (hedge bindweed) closely resembles plant RNases but has no enzymatic activity. *Plant Physiology*, *122*, 433–446.
van der Hoorn, R. A. L. (2008). Plant proteases: From phenotypes to molecular mechanisms. *Annual Review of Plant Biology*, *59*, 191–223.

Vitousek, P. (1982). Nutrient cycling and nutrient use efficiency. *The American Naturalist, 119*, 553–572.
Werner, A. K., & Witte, C.-P. (2011). The biochemistry of nitrogen mobilization: Purine ring catabolism. *Trends in Plant Science, 16*, 381–387.
Wetzel, S., Demmers, C., & Greenwood, J. S. (1989). Seasonally fluctuating bark proteins are a potential form of nitrogen storage in three temperate hardwoods. *Planta, 178*, 275–281.
Wetzel, S., & Greenwood, J. S. (1989). Proteins as a potential nitrogen storage compound in bark and leaves of several softwoods. *Trees, 3*, 149–153.
Wetzel, S., & Greenwood, J. S. (1991a). A survey of seasonal bark proteins in eight temperate hardwoods. *Trees, 5*, 153–157.
Wetzel, S., & Greenwood, J. S. (1991b). The 32-kilodalton vegetative storage protein of *Salix microstachya* Turz characterization and immunolocalization. *Plant Physiology, 97*, 771–777.
Wildhagen, H., Bilela, S., & Rennenberg, H. (2013). Low temperatures counteract short-day induced nitrogen storage, but not accumulation of bark storage protein transcripts in bark of grey poplar (*Populus* × *canescens*) trees. *Plant Biology, 15*, 44–56.
Wildhagen, H., Durr, J., Ehlting, B., & Rennenberg, H. (2010). Seasonal nitrogen cycling in the bark of field-grown Grey poplar is correlated with meteorological factors and gene expression of bark storage proteins. *Tree Physiology, 30*, 1096–1110.
Wittenbach, V. A. (1982). Effect of pod removal on leaf senescence in soybeans. *Plant Physiology, 70*, 1544–1548.
Wullschleger, S., Loewith, R., & Hall, M. N. (2006). TOR signaling in growth and metabolism. *Cell, 124*, 471–484.
Xu, G., Fan, X., & Miller, A. J. (2012). Plant nitrogen assimilation and use efficiency. *Annual Review of Plant Biology, 63*, 153–182.
Zdarska, M., Dobisová, T., Gelová, Z., Pernisová, M., Dabravolski, S., & Hejátko, J. (2015). Illuminating light, cytokinin, and ethylene signalling crosstalk in plant development. *Journal of Experimental Botany, 66*, 4913–4931.
Zeth, K., Fokina, O., & Forchhammer, K. (2014). Structural basis and target-specific modulation of ADP sensing by the *Synechococcus elongatus* PII signaling protein. *The Journal of Biological Chemistry, 289*, 8960–8972.
Zhang, X., Davidson, E. A., Mauzerall, D. L., Searchinger, T. D., Dumas, P., & Shen, Y. (2015). Managing nitrogen for sustainable development. *Nature, 528*, 51–59.
Zhang, C., Han, L., Slewinski, T. L., Sun, J., Zhang, J., Wang, Z.-Y., et al. (2014). Symplastic phloem loading in poplar. *Plant Physiology, 166*, 306–313.
Zhang, Y., Wang, Y., Kanyuka, K., Parry, M. A. J., Powers, S. J., & Halford, N. G. (2008). GCN2-dependent phosphorylation of eukaryotic translation initiation factor-2α in Arabidopsis. *Journal of Experimental Botany, 59*, 3131–3141.
Zhu, B., & Coleman, G. D. (2001a). Phytochrome-mediated photoperiod perception, shoot growth, glutamine, calcium, and protein phosphorylation influence the activity of the poplar bark storage protein gene promoter (bspA). *Plant Physiology, 126*, 342–351.
Zhu, B., & Coleman, G. D. (2001b). The poplar bark storage protein gene (Bspa) promoter is responsive to photoperiod and nitrogen in transgenic poplar and active in floral tissues, immature seeds and germinating seeds of transgenic tobacco. *Plant Molecular Biology, 46*, 383–394.

CHAPTER FIVE

Embryology in conifers

Sara von Arnold[a,*], David Clapham[a], Malin Abrahamsson[b]
[a]Department of Plant Biology, Uppsala BioCenter, Swedish University of Agricultural Sciences, Uppsala, Sweden
[b]SweTree Technologies AB, Umeå, Sweden
*Corresponding author: e-mail address: sara.von.arnold@slu.se

Contents

Abstract

Conifers, which grow in all climate zones, are important both ecologically and economically. An understanding of the regulation of the development of coniferous trees from single cells to mature trees is critical for scientific and biotechnological applications. Early embryogenesis is a crucial developmental phase during which the basic features of the plant body are established: the apical-basal axis of polarity, the various tissue layers, and the root and the shoot poles. Conifer somatic embryogenesis is of special interest both as a model system for zygotic embryogenesis and because of its potential for mass propagation of selected genotypes. The cell and molecular biology of higher plant embryogenesis has been most studied in the angiosperm Arabidopsis. Like seeds in Arabidopsis, conifer seeds contain a simple mature embryo. However, the patterning of cell division and cell differentiation leading to the final shape are dramatically different in the two groups of species. This is not surprising since the common ancestor of gymnosperms and angiosperms dates from *ca.* 300 Myr ago. The aim of this review is to

Advances in Botanical Research, Volume 89
ISSN 0065-2296
https://doi.org/10.1016/bs.abr.2018.11.005

give an updated survey of the developmental patterning and its regulation during zygotic and somatic embryogenesis in conifers, particularly the genera *Picea* and *Pinus*. Somatic embryogenesis biotechnology is much affected by the occurrence of cleavage polyembryony in *Pinus* and its absence in *Picea*.

1. Introduction

The conifers are by far the largest and most diverse gymnosperm group. Conifers grow in all climate zones, but the most extensive conifer stands are found at the higher latitudes in the Northern Hemisphere through the boreal forests of North America, Asia and Europe.

Conifer trees are significant both ecologically and economically. In many parts of the Northern Hemisphere, conifers make up the majority of the biomass, and are important sources of the primary production in several ecosystems. Some conifer forests support complex ecosystems with very high levels of biodiversity. Conifers are also very important economically because they provide wood and wood products that are used, e.g., to make buildings, furniture, paper and chemicals.

Embryogenesis is a critical stage of the sporophytic life cycle during which the basic features of the plant body are established: embryo polarity and the primary meristems. After the establishment of meristems for the root and shoot systems in the embryo, root and shoot meristems are maintained throughout the life of the organism, and produce most of the plant body postembryonically. Thus all of the rudimentary structures and information necessary for the development of a mature tree are present in the embryo. Knowledge about the molecular regulation of embryonic pattern formation in plants has to a large extent been derived from studies of embryo-defective mutants in the angiosperm model plant Arabidopsis (*Arabidopsis thaliana*). However, embryogenesis in Arabidopsis is not representative for all plant species. Molecular data suggest that extant seed plants (gymnosperms and angiosperms) shared a final common ancestor about 300 Myr ago (Smith, Beaulieu, & Donoghue, 2010) and several important embryonal patterning processes differ between the two groups of seed plants. The regulation of gymnosperm embryo development is therefore interesting from an evolutionary point of view. Another reason to study gymnosperms, and especially conifers, is their great commercial importance.

2. Zygotic embryo development

The reproductive cycle in a conifer is schematically shown in Fig. 1. Most conifers are pollinated and fertilized during the same season, but in *Pinus* species, a period of 12–14 months intervenes between pollination of the ovules and fertilization (Gifford & Foster, 1989). Furthermore, it has to be kept in mind that there are differences between species, for example

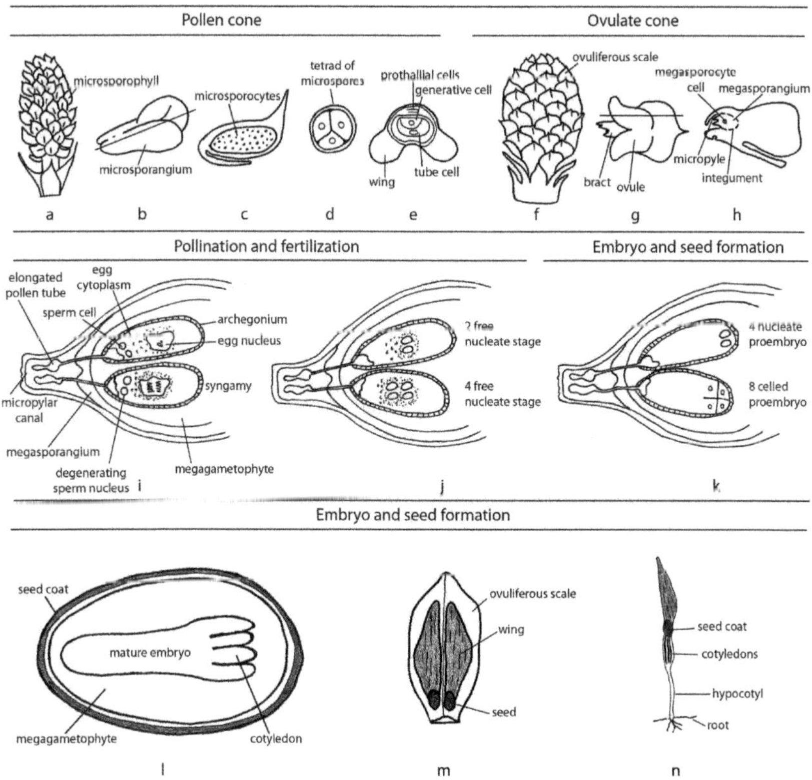

Fig. 1 Reproductive cycle of Pinaceae. (a) A pollen cone consisting of many small spirally arranged scales. Each scale is called microsporophyll. (b, c) Two microsporangia, located on the underside of each microsporophyll, contain massive amounts of spore-producing cells called microsporocytes. (d) In early spring, the nucleus in each microsporocyte divides by meiosis to produce four microspores. (e) Each microspore develops two gas-filled wings. Successively three mitotic divisions take place, creating two prothallial cells, one generative cell and one tube cell. (f) The ovulate cone consists of numerous spirally arranged ovuliferous scales. (g) Each ovuliferous scale has a small *(Continued)*

in the development and structure of the male gametophyte, the types of pollination mechanisms, the development and behaviour of the proembryo, and the occurrence and type of polyembryony (Gifford & Foster, 1989).

The multicellular megagametophyte originates from the megaspore mother cell and develops before fertilization into a haploid tissue (Fig. 1). Conifers, like all gymnosperms, display a single fertilization event when only the larger male gamete migrates through the egg cytoplasm and fuses with the egg nucleus, while the smaller male gamete degenerates (Singh, 1978). The cytoplasmic inheritance is biparental, with paternal inheritance of chloroplasts and maternal inheritance of mitochondia (Neale & Sederoff, 1989).

The zygotic embryo develops within the corrosion cavity within the megagametophyte (Fig. 1). The (haploid) megagametophyte, which functions as storage tissue analogous to the (triploid) endosperm of angiosperms, starts to accumulate storage material (lipids, proteins and starch) when the early embryo grows into the megagametophyte (Hakman, 1993).

Fig. 1—Cont'd bract fused to the lower surface, and two ovules located on the upper surface. (h) Each ovule consists of a megasporangium surrounded by an outer protective layer called the integument. An opening in the integument, the micropyle, consists of a canal covered by a sticky secretion. Each megasporangium contains a megasporocyte cell that, after pollination, undergoes meiosis giving rise to four megaspores. Only one of these megaspores will develop into the megagametophyte, and the rest will soon degenerate. (i) After pollination the male gametophyte continues the development. The generative cell of the pollen grain undergoes division, giving rise to a sterile cell and a spermatogenous cell. Simultaneously, at the micropylar end of the megagametophyte, several pores start to form of which each leads to an archegonium, containing a single egg cell. The pollen tube grows through the megasporangium to the megagametophyte. As a pollen tube approaches an archegonium, the spermatogenous cell divides producing two sperm cells. The tube nucleus and the two sperm cells and the sterile cell are discharged into the egg cytoplasm. One sperm nucleus unites with the egg nucleus (syngamy) to create a diploid zygote, and the other degenerates. (j–k) Protected inside the ovule the fertilized eggs are ready to form embryos. The development of the embryo can be divided into three phases: (i) proembryogeny, (ii) early embryogeny, and (iii) late embryogeny. A detailed description of the embryo development will follow in Fig. 2. (l) At seed maturation, the embryo desiccates and enters dormancy. The integuments are transformed into a tough protective seed coat. (m) The seed and surface tissue are separated from the cone scale and form a winged seed. (n) The seed can germinate in the spring the year after the seed is released from the cone. *After Owens, J. N., & Morris, S. J. (1988). An ultrastructural study of fertilization in Douglas fir (*Pseudotsuga menziesii *(Mirb.) Franco). In M. Cresti, et al. (Eds.),* Sexual reproduction in higher plants. *Berlin Heidelberg: Springer-Verlag; Raven, P. H., Evert, R. F., & Eichorn, S. E. (1999).* Biology of plants *(6th ed.) New York, USA: W.H. Freeman and Company.*

The megagametophyte also plays an important role in regulating embryo development. Damage to, or misdevelopment of, the megagametophyte can lead to embryo degradation (Dogra, 1967).

2.1 From proembryo to cotyledonary embryo

Embryo development can be divided into two main steps: (i) embryogenesis sensu stricto and (ii) maturation of the embryo followed by germination (Dodeman, Ducreux, & Kreis, 1997). The sequence of embryo development in gymnosperms can be divided into three phases: (i) proembryogeny—all stages before elongation of the suspensor; (ii) early embryogeny—all stages after elongation of the suspensor and before establishment of the root meristem; (iii) late embryogeny—intensive histogenesis including the establishment of the root and shoot meristems (Singh, 1978). There are four types of embryogeny in gymnosperms, of which the conifer type is the most common, and in the following presentation, focus will be on conifer embryogeny (Dogra, 1967; Gifford & Foster, 1989; Singh, 1978).

2.1.1 Proembryogeny

Proembryogeny starts with a free nuclear stage (Fig. 2a–h). Initially the zygote undergoes several rounds of nuclear duplication that are not followed by cytokinesis. The number of free nuclei differs among species, and in the conifer type of embryogeny up to 64 free nuclei can be present (Chowdhury, 1962). In Pinaceae, the fertilized egg nucleus divides into four free nuclei that are arranged in a tier. After division, two tiers of four nuclei each are formed, the primary embryonal tier and the primary upper tier, which go through cellularization. Internal divisions of these two tiers produce four tiers of which the lowest tier will form the embryonal mass and the second lowest tier will elongate to form the embryonal suspensor. The phase of proembryogeny ends when the suspensor cells elongate and force the embryonal tier into the megagametophyte.

2.1.2 Early embryogeny

During early embryogeny, the primary body plan is established together with the apical-basal symmetry and the radial symmetry (Fig. 2i and j).

The cells of the lower tier divide and form an embryonal mass. The basal cells of the embryonal mass have been suggested to be distal stem cells that by asymmetric division give rise to both apical meristematic daughter cells in the embryonal mass and basal vacuolated suspensor cells (Zhu, Moschou, Alvarez, Sohlberg, & von Arnold, 2016). The embryonal mass displays a distinct protoderm with a smooth surface. The suspensor cells are committed to

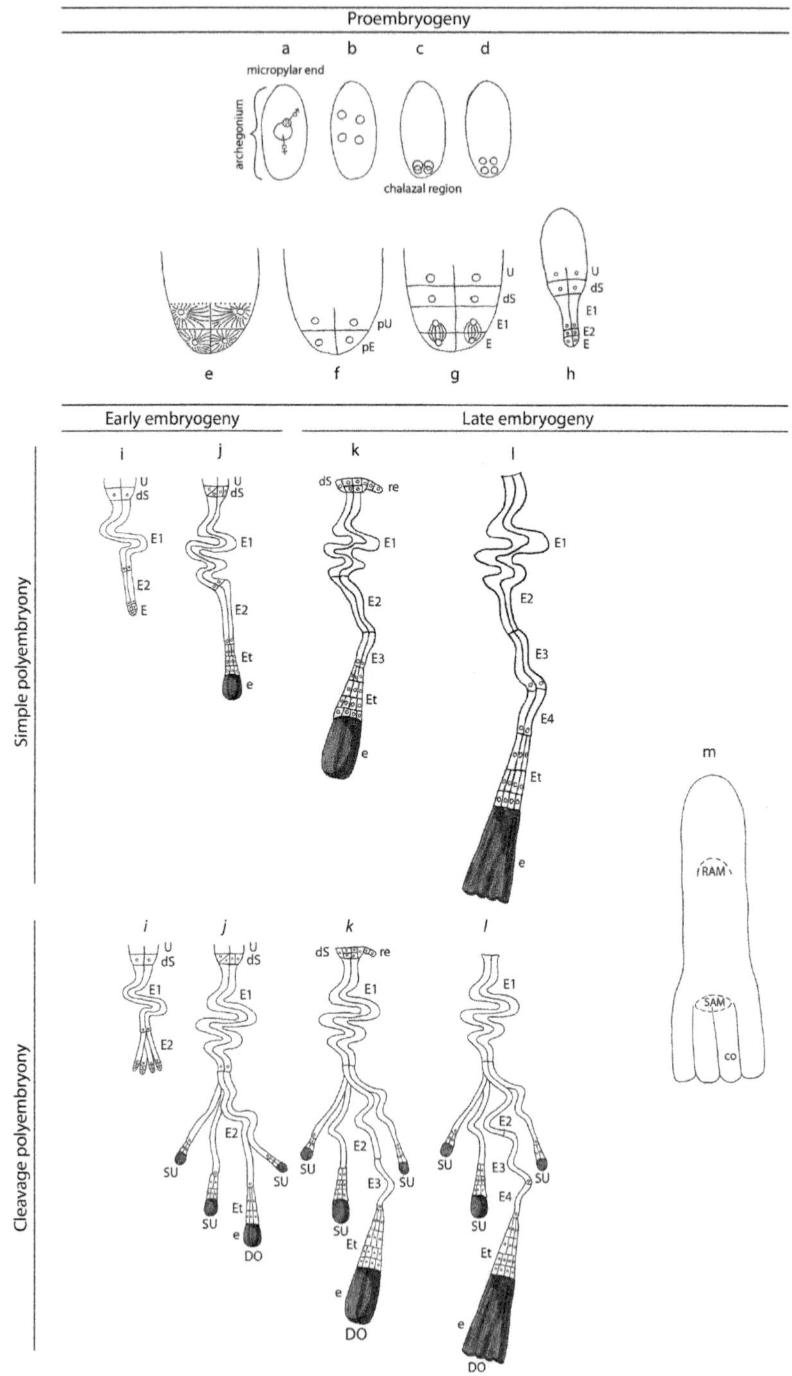

Fig. 2 See legend on next page

programmed cell death (PCD) as soon as they are formed, and the oldest part of the suspensor system continuously degenerates (Filonova, Bozhkov, Brukhin, et al., 2000; Filonova, Bozhkov, & von Arnold, 2000). The suspensor cells closest to the embryonal mass have been named embryonal tube cells. The suspensor cells form the thick secondary suspensor, which consists of several files of terminally differentiated non-dividing cells. The enlarging

Fig. 2 Schematic drawing of zygotic embryo development in Pinaceae. (a–h) Proembryogeny. (i–l) Simple polyembryony. (i–l) Cleavage polyembryony. (a) The male and female nuclei fuse. The fusing nuclei are large and full of dense nucleoplasm. After the nuclei (including the nucleoplasms) have merged, separate chromatin networks are formed. The first mitosis is intranuclear and the resulting nuclei are formed within the nucleoplasm of the zygote. The nuclear membrane of the zygote disappears at the end of the mitosis. (b) Two free nuclear divisions take place resulting in four free nuclei. (c) The four free nuclei move towards the chalazal portion of the zygote, i.e., the base of the archegonium, where they become arranged in one layer. (d) The four nuclei divide with vertical spindles to give rise to eight nuclei arranged in two layers of four each. (e, f) Secondary spindles develop and wall formation results in two tiers: the primary upper (or open) tier (pU) and the primary embryonal tier (pE), each comprising four cells. The pU tier is open towards the micropylar end, i.e., the neck of the archegonium, and is continuous with the general cytoplasm of the proembryo. (g) The pU tier and the pE tier go through internal divisions to form four tiers. First the pU tier divides into an upper (or open) tier (U) and into a dysfunctional suspensor tier (dS). However, it should be noted that the dS tier, in some literature, is also called the rosette tier. The U tier remains open towards the archegonium. Then the pE tier divides into an embryonal tier (E) and into an embryonal tier 1 (E1). The E1 tier is often also referred to as the primary suspensor. (h) The upper four cells of the E tier (E1) elongate to function as a suspensor, and the lower four cells of the E tier will form the embryonal mass. (i) The cells in the E tier divide and form a separate embryo. The U tier has started to degenerate. (j) The basal cells of the embryonal mass divide asymmetrically and give rise to meristematic cells in the embryonal mass and elongated suspensor cells. The suspensor cells closest to the embryonal mass have been named embryonal tube cells (Et). (k) Note the developing rosette embryo (re) from the dS tier. (l) The embryo has started to differentiate cotyledons. (m) A mature cotyledonary embryo with two polar meristems (SAM and RAM) and a whorl of cotyledons (co). (i) The four apical cells in the proembryo develop into four filamentous embryos. (j–l) One of the four embryos become the dominant embryo (DO), while the remaining subordinate embryos (SU) stop developing. (m) The dominant embryo develop into a mature cotyledonary embryo. *After Dogra, P. D. (1967). Seed sterility and disturbances in embryogeny in conifers with particular reference to seed testing and tree breeding in Pinaceae.* Studia Forestalia Suecica 45, *Royal College of Forestry, Stockholm; Singh, H. (1978). Embryology of gymnosperms. In W. Zimmermann, Z. Carlquist, P. Ozenda, & H. D. Wulff (Eds.),* Handbuch der Pflanzenanatomie *(pp 187–241). Berlin: Gebrüder Borntrager; Gifford, E. M., & Foster, A. S. (1989).* Morphology and evolution of vascular plants *(pp. 401–453). New York: W.H. Freeman Company.*

suspensor pushes the embryo out of its archegonial jacket into the nutritive reserves of the megagametophyte. The suspensor also acts to transport nutrients and hormones to the embryo.

Differentiation of a protoderm is the earliest event of radial pattern formation. Cells in the outer layer of the embryonal mass divide mainly anticlinally, but also periclinally, giving rise to additional internal layers (Singh, 1978), unlike angiosperms where only anticlinal cell divisions take place. Nevertheless, the outer cell layer in the embryonal mass in conifers defines a functional protoderm (Ingouff, Farbos, Lagercrantz, & von Arnold, 2001).

2.1.3 Late embryogeny

Late embryogeny is a period of histogenesis and organogenesis (Fig. 2k and l). In the beginning of late embryogeny the root and shoot apical meristems (RAM and SAM) are delineated and the plant axis established (von Arnold, Sabala, Bozkov, Dyachok, & Filonova, 2002). The first notable sign of RAM and SAM development in zygotic embryos of white spruce is the appearance of starch at the respective poles of the embryo (Yeung, Stasolla, & Kong, 1998). Later the root organizing centre is formed near the centre of the embryo. The surface layer of cells of the future SAM originates at the distal part of the embryonal mass. Cotyledonary primordia arise in a ring around the SAM. The number of cotyledons varies between species from 2 to 24 (Chandler, 2008). The variation in cotyledon number correlates with the size of the embryo, which varies from year to year (Butts & Buchholtz, 1940).

2.1.4 Cleavage polyembryony

Multiple embryos are formed in all conifers by the fertilization of eggs in more than one archegonium, simple polyembryony (Owens & Blake, 1985). Multifertilizations usually occur about the same time, and resultant proembryos having different genotypes develop at a similar rate. Multiple embryos can also be formed in conifers by the cleavage of embryos, cleavage polyembryony (Singh, 1978). Cleavage polyembryony has been reported from four genera in the Pinaceae family, *Cedrus*, *Keteleeria*, *Pinus* and *Tsuga*, (Dogra, 1967). Rarely does more than one proembryo develop further into an embryo.

During cleavage polyembryony, the four apical cells and the suspensor network in the proembryo separate into four filamentous embryos (Fig. 2i–l). The four embryos, which arise from the separated tiers, begin their development by apical cell growth (Buchholtz, 1920). The apical cell disappears as the embryonal mass starts to form. Initially the four embryos are of equal size, but successively the embryos start to compete for nutrition and space to grow and survive. It has been suggested that the embryo with the best physiological constitution situated in the most suitable environment becomes the dominant embryo, which usually develops to maturity (Buchholtz, 1926). The dominant embryo suppresses the growth and development of the subordinate embryos mechanically by means of its suspensor system, and thereby causes starvation. It has also been suggested that the dominant embryo secretes growth-inhibiting factors that suppress the subordinate embryos (Dogra, 1967). The subordinate embryos are degraded by PCD (Filonova, von Arnold, Daniel, & Bozhkov, 2002), and the dominant embryo follows a similar developmental pattern as embryos derived after single polyembryony (described above). Various hypotheses have been raised about the significance of cleavage polyembryony, to date inconclusively.

The cleavage process can be disturbed in trees exposed to very harsh conditions, for example in artic regions. In these seeds, all embryos develop further and no dominant embryo is formed. This phenomenon has been named stress-induced persistent polyembryony (Dogra, 1967). If all embryos survive to maturity, they are malformed with increased width and lack of apical-basal polarization.

2.1.5 Maturation

During embryo maturation the developmental programme switches from pattern formation to cell expansion and storage products deposition. The maturation programme is responsible for: (i) synthesis of large amounts of storage products; (ii) inducing water loss; (iii) preventing premature germination, and (iv) establishing a state of dormancy (Goldberg, de Paiva, & Yadegari, 1994).

Storage proteins, lipid bodies and starch grains accumulate both in the megagametophyte and the maturing embryo (Hakman, 1993). The accumulated storage proteins belong to the globulin (legumin and vicilin) and albumin families (Trontin, Klimaszewska, Morel, Hargreaves, & Lelu-Walter, 2016, with references). Water stress, imposed naturally on orthodox seeds during later stages of maturation, allows the embryo to survive severe dehydration when moisture content can drop as low as 5%. After imbibition, the seeds can germinate. Seed germination is recognized by the emergence of the radicle from the seed.

The mature embryo is comprised of cotyledons, SAM, hypocotyl, RAM and root cap. The SAM, localized at the base of the cotyledons, is a small dome of inactive cells (Yeung et al., 1998), which are activated upon germination. The innermost tissue of the hypocotyl is the pith, which is surrounded by the procambium containing the first elements of the xylem and phloem. Outside the procambium is the cortex, which is covered by the protodermis. The protodermis provides protection and reduces water loss. The RAM, localized just below the centre of the embryo, is protected by a large root cap consisting of parenchyma cells that have accumulated starch.

3. Somatic embryo development

Somatic embryogenesis in conifers was first reported for Norway spruce (Chalupa, 1985; Hakman, Fowke, von Arnold, & Erikson, 1985) and European larch (*Larix decidua*) (Nagmani & Bonga, 1985), and subsequently for many other coniferous species, a large majority of them belonging to Pinaceae, and only a few to the Cupressaceae, Taxaceae, Cephalotaxaceae, and Araucariaceae families.

Somatic embryos can be used for large-scale vegetative propagation, and have a great potential to capture the genetic gain in breeding programmes (Appendix). In addition, somatic embryos are used for the conservation of valuable genotypes by cryopreservation, and in basic research for studying the regulation of embryo development.

The process to regenerate conifers via somatic embryogenesis is comprised of a sequence of steps including initiation, proliferation, early embryo formation, maturation, desiccation and germination. A number of experimental approaches, mainly based on trial and error strategies, have been taken to improve the protocols for initiation of embryogenic tissue and subsequent development of cotyledonary embryos (Häggman, Vuosku, Sarjala, Jokeal, & Niemi, 2006; Klimaszewska, Hargreaves, Lelu-Walter, & Trontin, 2016; Klimaszewska et al., 2007; Lelu-Walter, Teyssier, Guérin, & Pâques, 2016; Pullman & Bucalo, 2014; von Aderkas, Kong, & Prior, 2016; von Arnold et al., 2005; von Arnold & Clapham, 2008). Based on these studies it has been possible to increase the yield of high quality somatic embryo plants in conifers.

In this review we will focus on the developmental pattern of somatic embryos, without discussing the various methods under development to optimize the process.

3.1 Initiation and proliferation of embryogenic tissue

Embryogenic tissue is initiated from zygotic embryos of conifers. From species where the zygotic embryo has simple polyembryony, mature cotyledonary embryos are used as primary explants. Mature cotyledonary embryos from both fresh and stored seeds are suitable, although the initiation frequency is higher if fresh seeds before desiccation are used (S. von Arnold, unpublished). In contrast, in species having cleavage polyembryony, embryogenic tissue is commonly initiated from immature zygotic embryos prior to cotyledon development and often at the cleavage stage (Burg, Helmersson, Bozhkov, & von Arnold, 2007; Klimaszewska et al., 2007, with references). A strong limitation when establishing embryogenic cultures from immature zygotic embryos is that the time period when the zygotic embryos are at the critical developmental stage is short, usually only a few weeks. Furthermore, the initiation frequency is also greatly dependent on the genotypes of the parents (MacKay, Becwar, Park, Cordero, & Pullman, 2006).

Initiation of embryogenic tissue is stimulated by treating the primary explants with plant growth regulators (PGRs), usually auxin (2,4-dichlorophenoxyacetic acid; 2.4-D) and cytokinin (N^6-benzyladenine; BA). During initiation of embryogenic tissue from cotyledonary embryos, differentiated cells in the zygotic embryo are first stimulated to dedifferentiate and then to give rise to proliferating proembryogenic masses (PEMs) (Mo & von Arnold, 1991). When initiating embryogenic tissue from immature embryos, whole megagametophytes are most commonly used as primary explants (Klimaszewska et al., 2007). One or more zygotic embryos protrude from the megagametophyte at the micropylar end. In Scots pine (*Pinus sylvestris*), the protruding embryos initially degenerate before new embryogenic cell aggregates are formed (Abrahamsson, Valladares, Merino, Larsson, & von Arnold, 2017), and in loblolly pine (*Pinus taeda*) translucent and mucilaginous tissue differentiate from the primary explants before somatic embryos are formed (Becwar, Nagmani, & Wann, 1990). This suggests that the initiation of embryogenic tissue from immature embryos is not a direct continuation of the cleavage process.

The initiated embryogenic tissue continues to proliferate under similar culture conditions as during the initiation process. Embryogenic cultures are composed of two cell types: rounded, densely cytoplasmic cells and highly vacuolated cells. The majority of the cells are clustered into cell aggregates. In Norway spruce (*Picea abies*) the cell aggregates are designated PEMs, which pass through a series of three stages distinguished by cellular organization and cell numbers (stages I, II and III) (Filonova, Bozhkov, & von Arnold, 2000).

3.2 From proliferating embryogenic cells to cotyledonary embryos

The developmental patterning during differentiation and development of somatic embryos has been described in several coniferous species (Abrahamsson, Valladares, Larsson, Clapham, & von Arnold, 2012; Cairney & Pullman, 2007; Klimaszewska et al., 2007; von Arnold et al., 2002, 2016). Although differences have been observed in different species and among different cell lines within the same species, the general developmental pathway is similar in most conifers, and also similar to the development of zygotic embryos.

In Norway spruce, early somatic embryos differentiate from PEM III after withdrawal of PGRs (Filonova, Bozhkov, & von Arnold, 2000). The time for the transition from proliferation to embryo differentiation varies both among species and cell lines from 2 days to 2 weeks. Early embryos are polarized with a distinct border between the compact globular embryonal mass covered by a protoderm in the apical part and vacuolated suspensor cells in the basal part (Zhu et al., 2016). The suspensor cells do not divide but are committed to PCD as soon as they are formed (Bozhkov et al., 2004). The basal cells of the embryonal mass divide asymmetrically and give rise to meristematic daughter cells in the embryonal mass and vacuolated suspensor cells (Zhu, Moschou, Alvarez, Sohlberg, & von Arnold, 2014). Inhibition of suspensor differentiation and death blocks the development of somatic embryos (Bozhkov et al., 2004; Helmersson, von Arnold, & Bozhkov, 2008). Vacuolar PCD is essential for differentiation and subsequent elimination of the suspensor, thus contributing to the establishment of embryo polarity (Smertenko & Bozhkov, 2014). The inability of a cell line to form normal cotyledonary somatic embryos is to a large extent associated with a disturbed transition from proliferating embryogenic cells to differentiation of early embryos; for example if the degradation of the suspensor cells is delayed the embryos will have an intensive radial growth during the whole maturation process (Filonova, Bozhkov, & von Arnold, 2000).

For further development, the early embryos are transferred to a maturation medium that includes abscisic acid (ABA), and in some cases high molecular weight osmotic agents such as polyethylene glycol. Late embryos usually develop after a couple of weeks on maturation medium. RAM,

SAM and procambium are delineated during late embryogeny (Filonova, Bozhkov, & von Arnold, 2000). The establishment of the SAM occurs concomitantly with the formation of a crown of cotyledons surrounding the meristem.

Maturation includes accumulation of storage reserves, growth arrest, and acquisition of desiccation tolerance, which is induced by ABA and increased osmotic pressure. Similar storage compounds are accumulated in mature somatic embryos as in zygotic embryos, although the amount and the timing can vary depending on the concentration of ABA and the supplement of various osmotic compounds. When the somatic embryos resemble mature zygotic embryos (usually after 6–12 weeks on maturation medium, depending on the species), they are harvested and partially desiccated before germination. Only those embryos that have accumulated enough storage compounds and have acquired desiccation tolerance will develop into normal plants. Maturation is regarded as a major bottleneck in species recalcitrant to somatic embryogenesis, but in most cases the problem to stimulate the embryos to mature is a consequence of disturbed patterning during early and late embryogeny.

Generally speaking, embryogenic tissue, from which high quality mature somatic embryos can routinely be regenerated, can be initiated from a high frequency of cotyledonary zygotic embryos (Klimaszewska et al., 2007; Lelu-Walter et al., 2013). However, in most *Pinus* species, where embryogenic cultures are initiated from immature zygotic embryos during the cleavage process, both the initiation frequency and the regeneration rate are low (Abrahamsson et al., 2017; Klimaszewska et al., 2007). Thus, further research is needed to elucidate how to regain the embryogenic potential in mature zygotic embryos in species having cleavage polyembryony.

4. Molecular regulation of zygotic and somatic embryo development

The regulation of the development of zygotic embryos is difficult to study in conifers owing to the inaccessability of the embryo inside the ovule and it is very difficult to sample sufficient quantities of early zygotic embryos at specific developmental stages. In contrast, the development of somatic embryos can be synchronized by specific treatments, making it possible to collect a large number of embryos at specific developmental stages. In angiosperms it has been shown that the key genes controlling zygotic embryogenesis perform similar roles during somatic embryogenesis

(Mordhorst, Hartog, El Tamer, Laux, & de Vries, 2002) and based on what has been reported so far this also seems to be true for conifers. Therefore, somatic embryogenesis is considered as a model system to study the molecular regulation of embryogenesis in conifers. However, not all embryogenic cell lines of conifers develop normal somatic embryos. Identification of differences in gene expression during normal and abnormal development of somatic embryos could enable new approaches to improve somatic embryogenesis in recalcitrant species. In addition, molecular studies open up ways to identify molecular makers for specific processes and patterning, and in the future perhaps also for determining the quality of somatic embryos.

Microarray and RNA sequencing methods have provided advances for the genome-wide profiling of gene expression, and in conifers transcriptomics is developing rapidly (reviewed in Plomion et al., 2016). Genome-wide profiling (transcriptomics, proteomics, metabolomics) have shown that the molecular regulation from early to late embryogenesis in conifers mainly arises from spatiotemporal modulation of auxin-, gibberellin- and ABA-mediated responses, in a similar way as has been reported for angiosperms (reviewed in Trontin et al., 2016). The maintenance of cellular redox homeostasis might be a generic sensor for controlling embryo development (Stasolla et al., 2004). Transcripts related to embryogenic competence, cell wall modifications, cell division pattern, axis specification and response to hormones and stress are highly abundant and differentially expressed at different developmental stages during early zygotic embryo development in Scots pine (Merino et al., 2016).

4.1 Initiation and proliferation of embryogenic tissue

The earliest developmental stages are different in zygotic and somatic embryos. Zygotic embryogenesis starts from the fertilized egg cell, while somatic embryogenesis starts from somatic cells (single cells or groups of cells), which are stimulated to dedifferentiate and proliferate as embryogenic tissue. Owing to these differences, it is not possible to compare the regulation of the earliest events during zygotic and somatic embryogenesis.

Initiation of embryogenic tissue from differentiated somatic cells requires an induction signal that causes the somatic cell to change its identity and become embryogenic and also that the somatic cell is competent to respond to the signal. There may be an association between biotic defence elicitation and suppression of embryogenic potential, at least when initiating embryogenic tissue from buds of white spruce (*Picea glauca*) (Rutledge et al., 2013).

However, it has also been suggested that the jasmonic acid pathway, involved in biotic defence elicitation, is activated during initiation of embryogenic tissue (Elhiti, Stasolla, & Wang, 2013). In addition, stress-related processes and auxin-mediated processes were associated with early somatic embryo development in Norway spruce (Vestman et al., 2011). Based on changes in gene expression during initiation of embryogenic tissue, candidate genes associated with embryogenic competence have been identified (reviewed by Miguel, Rupps, Raschke, Rodrigues, & Trontin, 2016). The authors specifically lifted up *LEAFY COTYLEDON (LEC)* and *WUSCHEL* (*WUS*) genes as potential makers for early somatic embryogenesis in plants. They pointed out that *SOMATIC EMBRYOGENESIS RECEPTOR-LIKE KINASE (SERK)* is less reliable as a maker since it can also be expressed in non-embryogenic cultures. To date, the mechanisms underlying the initiation of embryogenic tissue are still poorly understood.

The transcript abundance of putative homologues to *ARABIDOPSIS THALIANA HOMEOBOX PROTEIN 22/MATERNAL EFFECT EMBRYO ARREST 68* (*ATHB22/MEE68*), *MEE49* and *NAC-REGULATED SEED MORPHOLOGY 2* (*NARS2*) increases during the differentiation of somatic embryos of Norway spruce (Vestman et al., 2011). The corresponding Arabidopsis genes, which are expressed in the female gametophyte (Pagnussat et al., 2005) and in the integument (Kunieda et al., 2008), have been shown to be important for the progression of embryogenesis. Furthermore, *Chia4-Pa*, the gene encoding chitinase 4, is expressed in the layer of single cells surrounding the corrosion cavity in seeds of Norway spruce (Wiweger, Farbos, Ingouff, Lagercrantz, & von Arnold, 2003). In embryogenic cultures of Norway spruce, *Chia4-Pa* is expressed in a sub-population of cells close to the developing embryo but not in the embryo proper itself. It was suggested that cells expressing *Chia4-Pa* have a megagametophyte signalling function and that CHIA4-Pa promotes the transition from PEM to somatic embryo (Wiweger et al., 2003). In addition, extracellular arabinogalactan proteins (AGPs) and lipo-chitooligosaccharides (LCOs) stimulate development of somatic embryos in Norway spruce (Egertsdotter & von Arnold, 1995; Dyachok, Wiweger, Kenne, & von Arnold, 2002). In carrot (*Daucus carrota*), chitinase-mediated cleavage of AGPs generates AGPs with an increased capacity to promote somatic embryogenesis (van Hegel, van Kammen, & de Vries, 2002). It is tempting to assume that CHIA4-Pa is involved in the cleavage of compounds like AGPs and LCOs. These examples indicate that similar signalling molecules are involved in early differentiation of zygotic and somatic embryos both in conifers and angiosperms.

4.2 Early and late embryogeny

Early embryos have a polar structure with an embryonal mass in the apical part and a suspensor in the basal part. Embryo polarization is established through polar auxin transport (PAT) (Larsson, Sitbon, Ljung, & von Arnold, 2008; Palovaara, Hallberg, Stasolla, Luit, & Hakman, 2010). When PAT is blocked, after treatment with an auxin transport inhibitor, the amount of endogenous auxin (IAA) increases in the early embryos concomitantly with a decrease in PCD and abnormal differentiation of the suspensor, including supernumerary suspensor cells and meristematic cells in the suspensor (Larsson, Sitbon, Ljung, et al., 2008; Larsson, Sitbon, & von Arnold, 2008). Furthermore, the mature embryos have fused cotyledons, lack an organized SAM, and have irregular cell divisions in the area of the root meristem. The auxin responses during embryo development were analyzed in transgenic cell lines of Norway spruce expressing the GUS reporter gene under an auxin responsive promoter (Larsson, 2011; Larsson, Sitbon, & von Arnold, 2008; von Arnold et al., 2016). During early embryogeny and in the beginning of late embryogeny an auxin response maximum was generated in the stem cells in the basal part of the embryonal mass, indicating that auxin might be involved in the specification of the stem cells. Analyses of the effects of blocked PAT on auxin response and embryo development suggested that auxin is transported from the suspensor to the embryonal mass during early embryogeny, and that this transport is crucial for the differentiation of suspensor cells. Thus, PAT is essential for apical-basal patterning and related stem niche positioning during early embryogenesis, and disturbed PAT causes aberrant embryo development.

During differentiation of early and late embryos in Norway spruce, *PaWOX8/9*, a member of the *WUSCHEL-RELATED HOMEOBOX* (*WOX*) gene family, is important for the correct orientation of the cell division plane and cell fate determination of the stem cells in the basal part of the embryonal mass (Zhu et al., 2014). *PaWOX8/9* is active upstream of several important cell-cycle regulators. Down-regulation of *PaWOX8/9* results in embryos having an aberrant morphology caused by disturbed orientation of the cell division plane in the stem cells (Zhu et al., 2014). In addition to the normal anticlinal division, the stem cells also divide periclinally and incline, which retains both daughter cells in the embryonal mass and results in a radial growth of the embryonal mass. Thus, *PaWOX8/9* acts as an important regulator for establishing the apical-basal embryo pattern. The caspase-related protease separase (*EXTRA SPINDLE POLES, ESP*) is required to sustain cell elongation (Moschou, Gutierrez-Beltran, Bozhkov, & Smertenko, 2016).

Down-regulation of *PaESP* in somatic embryos of Norway spruce affected the suspensor cells so that supernumerary, non-elongating suspensor cells were formed. The defects induced by *PaESP* deficiency resembled those caused by exposing the embryos to a PAT inhibitor (Larsson, Sitbon, Ljung, et al., 2008), indicating that *PaESP* perturbs auxin signalling. Embryo polarization is also dependent on a gradient of cells at different stages of PCD along its apical-basal axis. The molecular regulation of PCD of the suspensor has been reviewed by Smertenko and Bozhkov (2014). The proper timing of PCD in the suspensor cells, which starts as soon as the early embryos start to differentiate, is crucial for normal embryo development, and dysfunction of PCD leads to embryonic aberrations (Bozhkov et al., 2004; Smertenko, Bozhkov, Filonova, von Arnold, & Hussey, 2003).

The embryonal mass in early somatic embryos is covered by a protoderm. In Norway spruce, *PaWOX2* is important for specification of the protoderm (Zhu et al., 2016). Down-regulation of *PaWOX2* during development of early embryos results in aberrant early embryos, which fail to form a protoderm. It has also been shown that a protoderm can only differentiate when *PICEA ABIES HOMEOBOX1* (*PaHB1*) is specifically expressed in the outer cell layer of the early embryo (Ingouff et al., 2001). These results suggest that differentiation of a functional protoderm is conserved between angiosperms and conifers, despite differences in the division pattern for protodermal cells.

The establishment of stem cell niches in shoot and root meristems take place during late embryogeny. The expression pattern of genes related to *ZWILLE*, *FIDDLEHEAD*, *FUSCA* and *SCARECROW* during somatic embryo development in white spruce suggest that they are involved in controlling the formation of shoot and root meristems (Stasolla et al., 2004). In addition, a transcript encoding a PINHEAD/ZWILLE-like protein was shown to be up-regulated during initiation of adventitious roots in contorta pine (*Pinus contorta*) (Brinker et al., 2004). Four class 1 *KNOTTED-LIKE HOMEOBOX* (*KNOX1*) genes are expressed during somatic embryogenesis in Norway spruce (Larsson, Sitbon, & von Arnold, 2012). The expression of *HBK2* and *HBK4* was shown to be correlated with SAM differentiation. Furthermore, overexpression of *HBK3* leads to embryos with increased SAM, and down-regulation inhibits embryogenesis (Belmonte, Tahir, Schroeder, & Stasolla, 2007). During white spruce embryo development, an Argonaut family member, *PgAGO*, is required for proper RAM and SAM differentiation (Tahir, Law, & Stasolla, 2006). The expression of a *GLUTAMINE SYNTHETASE* gene (*GS1b*) is induced in procambial cells in zygotic and somatic embryos of maritime pine (*Pinus pinaster*) and Scots pine very early in vascular development, even before the differentiation of mature vascular elements (Rodríguez et al., 2005). The importance of the expression of *GS1b* during early embryogeny was strengthened by the fact that *GS1b* transcripts could hardly be detected in abnormal somatic embryos. However, the functions of these genes during meristem differentiation in conifers have so far not been shown.

In most conifers, the cotyledons develop as a crown surrounding the incipient crown and the radial symmetry is retained in the seedlings. In Norway spruce embryos, a PIN-FORMED (PIN) auxin efflux carrier, encoded by *PIN1*, is involved in the localization of auxin in the epidermis, which is important for correct cotyledon formation (Palovaara et al., 2010). Two *CUP-SHAPED COTYLEDON* (*CUC*) genes have been described in Norway spruce, *PaNAC01* and *PaNAC02* (Larsson, Sundström, Sitbon, & von Arnold, 2012). The expression of *PaNAC01* is PAT-regulated, and *PaNAC01* can rescue the Arabidopsis mutant *cuc1cuc2*, having partially fused cotyledons or cup-shaped cotyledons, indicating that *PaNAC01* is involved in the formation of cotyledon boundaries. Another gene, also important for differentiation of cotyledons in Norway spruce, is *PaWOX3* (Alvarez et al., 2015). *PaWOX3* is highly expressed in mature somatic embryos at the base of each cotyledon and in the lateral margins of cotyledons, separating them into adaxial and abaxial side. Down-regulation of *PaWOX3* results in embryos having short and thick cotyledons, lacking lateral margin outgrowth. Despite the diversity of cotyledon morphology between gymnosperms and angiosperms, at least some cotyledon development genes have been shown to be conserved.

4.3 Transition from the morphogenic phase to the maturation phase

During the maturation phase embryo morphogenesis is arrested and the embryo increases in size by cell elongation. Storage compounds are accumulated at the end of the maturation phase. The transcript accumulation of several genes changes during the developmental switch from the morphogenic phase to the maturation phase (Vega-Bartol et al., 2013; Vestman et al., 2011).

In zygotic and somatic embryos of Norway spruce and Scots pine the developmental switch coincides with a decreased accumulation of transcripts related to a *LEC-like* gene (*PaHAP3A*/*PsHAP3A*) and an increased accumulation of transcripts related to a *Viviparous-1* gene (*PaVP1*/*PsVP1*)

(Uddenberg et al., 2011). Furthermore, when embryogenic cultures are treated with the histone deacetylase inhibitor trichostatin A (TSA) during embryo maturation, the maturation process is arrested and the accumulation of *PaHAP3A* remains high while the accumulation of *PaVP1* remains low. This suggests that the switch in the expression of *PaHAP3A* and *PaVP1* is related to the transition from the morphogenic phase to the maturation phase, and also that the shift in the expression level of *HAP3A* and *VP1* provides a possible link between chromatin structure and expression of embryogenesis-related genes (Uddenberg et al., 2011).

Overexpression of a *HAP3A* gene in somatic embryos of white spruce (Klimaszewska, Pelletier, & Overton, 2010) and of *PaHAP3A* in somatic embryos of Norway spruce (Uddenberg, Abrahamsson, & von Arnold, 2016) is neither sufficient to affect the maturation process nor sufficient to alter the embryogenic potential after germination. In germinated somatic embryos of Norway spruce, embryogenic tissue differentiates from localized regions in the upper part of the hypocotyl and in the cotyledons (Uddenberg et al., 2011). In order to localize the expression of *PaHAP3A* during initiation of embryogenic tissue in germinated somatic embryos, embryogenic cell lines were transformed with *pPaHAP3A:GUS* (Uddenberg et al., 2016). Patchy patterns of GUS staining were detected in the upper part of the hypocotyl and along the cotyledons. The patchy expression pattern of *PaHAP3A* coincides with the place where embryogenic tissue differentiates. The authors suggested that *PaHAP3A* is important for the embryogenic potential, although it is not by itself enough to stimulate initiation of embryogenic tissue.

4.4 Cleavage polyembryony

Four filamentous embryos are formed during the cleavage process, of which one becomes the dominant embryo (Chowdhury, 1962). In maritime pine it was found that epigenetic regulation and transcriptional control related to auxin transport and responses are critical during the differentiation of the dominant zygotic embryos (Vega-Bartol et al., 2013). By performing transcriptomic analyses of early stages during zygotic embryo development in Scots pine it was possible to identify differentially accumulated transcripts, which might be involved in the cleavage process and the development of a dominant embryo (Merino et al., 2016, with references). For example, transcripts related to beta-expansin (*EXPB1*) and *SERK1* were highly abundant in proembryos but decreased dramatically during the cleavage process.

It was suggested that *EXPB1* is important for loosening the cell walls in the proembryo to allow separation of the four filamentous embryos. Furthermore, the high expression of *SERK1* in proembryos might be important for stimulating the four apical cells to differentiate into separate embryos, but thereafter the expression must be down-regulated to allow the four embryos to develop further and not go through a second round of cleavage. *ROP-INTERACTIVE CRIB MOTIF-CONTAINING PROTEIN 3* (*RIC3*)-related transcripts were abundant in proembryos and during the cleavage process, but not during later developmental stages, which might reflect that the early embryos develop by apical cell growth. Several transcripts related to embryo polarization were more abundant in dominant embryos than in subordinate embryos, including *NAC09* and *FAMA*, which in angiosperms regulate cell division and cell differentiation, and *PLANTACYANIN* (*ARPN*), involved in PCD. Furthermore, transcripts related to *AINTEGUMENTA-like 5* (*AIL5*), which is important to maintain the embryonic identity in Arabidopsis, were highly expressed in subordinate embryos but not in dominant embryos.

Based on processes identified in GO term enrichment analyses and the expression of selected transcripts, Merino et al. (2016) suggested that (i) processes related to embryogenic competence and cell wall loosening are involved in activating the cleavage process; (ii) apical-basal polarization is strictly regulated in dominant embryos but not in subordinate embryos; (iii) the transition from the morphogenic phase to the maturation phase is not completed in subordinate embryos.

4.5 Maturation

Low global DNA methylation in embryogenic cultures of black spruce (*Picea nigra*) was shown to be associated with higher maturation ability (Noceda et al., 2009). Furthermore, based on transcriptomic and proteomic profiling during maturation of somatic embryos of maritime pine, novel predictive makers for monitoring adaptive molecular response to favourable maturation conditions were identified (Morel et al., 2014; Teyssier et al., 2014).

During embryo maturation in conifers, several genes encoding late embryogenesis abundant (LEA) proteins and heat-shock proteins (HSPs) are up-regulated (reviewed by Miguel et al., 2016). Both LEAs and HSPs are important for protecting cellular components from severe dehydration, which is important for the acquisition of embryo desiccation tolerance.

Members of these protein families were also shown to be up-regulated in mature somatic embryos of Norway spruce (Businge, Bygdell, Wingsle, Moritz, & Egertsdotter, 2013).

Proteins belonging to the albumin and globulin (legumin and vicilin) families are the major storage proteins in conifers (e.g. Morel et al., 2014). In loblolly pine, legumin –/vicilin-like transcripts showed a similar accumulation during maturation of somatic and zygotic embryos (Brownfield, Todd, Stone, Deyholos, & Gifford, 2007; Lara-Chavez, Egertsdotter, & Flinn, 2012).

5. Conclusion

Embryogenesis in conifers starts with a single fertilization event in which the larger male gamete fuses with the egg cell. The zygote, which develops within the haploid megagametophyte, undergoes initially several rounds of duplication before cellularization. The proembryo develops into an early embryo characterized by a distinct embryonal mass in the apical part and terminally differentiated suspensor cells in the basal part. Radial patterning in the early embryo starts with the differentiation of a protoderm. In the beginning of late embryogeny the RAM and the SAM are delineated and the plant axis established. Cotyledonary primordia arise in a ring around the SAM. In some genera, for example *Pinus*, the proembryo goes through a cleavage process when the four apical cells and the suspensor network separate into four filamentous embryos. One of the embryos, the dominant embryo, develops further, while the three others, the subordinate embryos, are degraded by PCD. During the maturation phase the developmental programme switches from pattern formation to cell expansion and deposition of storage products.

Somatic embryogenesis in conifers is used as a model system for studying the regulation of embryo development. The developmental pathway from early embryos to mature cotyledonary embryos is very similar during zygotic and somatic embryogenesis. Although the earliest developmental stages are different for zygotic and somatic embryos, it has been shown that similar signalling molecules are involved in both processes. More recently, transcriptomic analyses have made it possible to identify genes and processes important for the progression of embryo development, and functional studies have been performed for some candidate genes. During the cleavage process *SERK1* might stimulate the four apical cells in the proembryo to differentiate into new embryos, and *EXPB1* to loosen up the cell walls for facilitating the separation of the embryos. Apical-basal polarization is established in early embryos through PAT, which generates an auxin response maximum in the stem cells in the basal part of the embryonal mass. *WOX8/9* is important for the correct orientation of the cell division plane and cell fate determination in the stem cells. The stem cells divide anticlinally, giving rise to meristematic cells in the embryonal mass and terminally differentiated, vacuolated suspensor cells. The suspensor cells are degenerated by PCD. *WOX2* is important for specification of protodermal cells. Based on the expression pattern and transgenic analyses of *KNOX1* genes (*HBK2*, *HBK3* and *HBK4*) it is likely that they directly or indirectly are involved in SAM differentiation. Differentiation and development of cotyledons are regulated by *NAC01* and *WOX3*, respectively. *HAP3A* and *VP1* are involved in the transition from the morphogenic phase to the maturation phase. Storage products are accumulated during the maturation phase and the embryos become desiccation tolerant, where genes encoding for HSPs probably play an important function. Despite the developmental pattern during embryogenesis having evolved differently in conifers and angiosperms, several processes are conserved in higher plants.

In addition to their value for studying the molecular regulation of embryo development in conifers, somatic embryos are also used for large-scale vegetative propagation. However, the presence or absence of cleavage polyembryony greatly influences the course of somatic embryogenesis and its practical application, which highlights the importance of learning more about the regulation of the cleavage process.

Appendix. Breeding of conifers

Breeding coniferous species is a slow process in which economically important traits are continuously improved. Progenies from controlled crosses are grown in field trials at various locations. After about 10–15 years, superior genotypes are selected. The selected genotypes are used as parents in the next breeding generation. Furthermore, material from selected trees is taken for grafts and used for establishing seed orchards for large-scale propagation. A limitation is that parts of the genetic gain achieved in the breeding programme are lost owing to sexual recombination, and contamination from foreign pollen. If the selected genotypes could be propagated vegetatively, it would be possible to capture the genetic gain obtained in the breeding programme. In principle, embryogenic cultures can be initiated from selected elite families and cryopreserved. Plants regenerated from each cell-line are tested in field trials and, after 10–20 years, the top clones are identified and mass propagated from the cryopreserved embryogenic cultures (Klimaszewska et al., 2016; Lelu-Walter et al., 2013), removing the need for seed orchards. The practical objections to this procedure are: (1) the great expense of carrying out the clonal tests, and the need to wait 10–20 years before learning the results of the tests; (2) the difficulty of identifying absolutely top clones because of relatively modest juvenile-mature correlations for economically important characters even after 10–20 years, compounded with sampling errors arising from the small number of individuals from each clone it is feasible to test; (3) the environmental and economic objections to planting a stand with only one (supposedly) top clone. An alternative procedure is 'family forestry' (Lindgren, 2008): at each stand plant about 25 untested embryogenic clones from about six families already tested and selected in the traditional breeding programme. This procedure exploits the genetic gains from the traditional breeding programme without the losses from pollen contamination in seed orchards and without the expense and time associated with clonal testing, and with sufficient genetic variation to override the environmental objections to clonal forestry. Another potentially important application of SE to mass propagation is the limited cloning of selected genotypes for subsequent incorporation as grafts in traditional seed orchards (e.g. Lindgren, 2008).

References

Abrahamsson, M., Valladares, S., Larsson, E., Clapham, D., & von Arnold, S. (2012). Patterning during somatic embryogenesis in Scots pine in relation to polar auxin transport and programmed cell death. *Plant Cell Tissue and Organ Culture*, *109*, 391–400.

Abrahamsson, M., Valladares, S., Merino, I., Larsson, E., & von Arnold, S. (2017). Degeneration patterning in somatic embryos of *Pinus sylvestris* L. *In Vitro Cellular & Developmental Biology*, *53*, 86–96.

Alvarez, J., Sohlberg, J., Engström, P., Zhu, T., Englund, M., Moschou, P. N., et al. (2015). The WUSCHEL-RELATED HOMEOBOX 3 gene PaWOX3 regulates lateral organ formation in Norway spruce. *New Phytologist*, *208*, 1078–1088.

Becwar, M. R., Nagmani, R., & Wann, S. R. (1990). Initiation of embryogenic cultures and somatic embryo development in loblolly pine (*Pinus taeda*). *Canadian Journal of Forest Research*, *20*, 775–780.

Belmonte, M., Tahir, M., Schroeder, D., & Stasolla, C. (2007). Overexpression of HBK3, a class 1 KNOX homeobox gene, improves the development of Norway spruce (*Picea abies*) somatic embryos. *Journal of Experimental Botany*, *58*, 2851–2861.

Bozhkov, P., Filonova, L., Suarez, M., Helmersson, A., Smertenko, P., Zhivotovsky, B., et al. (2004). VEIDase is a principal caspase-like activity involved in plant programmed cell death and essential for embryonic pattern formation. *Cell Death and Differentiation*, *11*, 175–182.

Brinker, M., van Zyl, L., Liu, W., Craig, D., Sederoff, R., Clapham, D., et al. (2004). Microarray analyses of gene expression during adventitious root development in *Pinus contorta*. *Plant Physoiology*, *135*, 1526–1539.
Brownfield, D., Todd, C., Stone, S., Deyholos, M., & Gifford, D. (2007). Patterns of storage protein and triacylglycerol accumulation during loblolly pine somatic embryo maturation. *Plant Cell Tissue and Organ Culture*, *88*, 217–223.
Buchholtz, J. T. (1920). Embryo development and polyembryony in relation to phylogeny of conifers. *American Journal of Botany*, 7(4), 125–145.
Buchholtz, J. T. (1926). Origin of cleavage polyembryony in conifers. *Botanical Gazette*, *81*(1), 55–71.
Burg, K., Helmersson, A., Bozhkov, P., & von Arnold, S. (2007). Developmental and genetic variation in nuclear microsatellite instability during pine embryogenesis. *Journal of Experimental Botany*, *58*, 687–698.
Businge, E., Bygdell, J., Wingsle, G., Moritz, T., & Egertsdotter, U. (2013). The effect of carbohydrates and osmoticum on storage reserve accumulation and germination of Norway spruce somatic embryos. *Physiologia Plantarum*, *149*, 273–285.
Butts, D., & Buchholtz, J. T. (1940). Cotyledon numbers in conifers. *Transaction of the Illinois State Academy of Science*, *33*(2), 58–62.
Cairney, J., & Pullman, G. S. (2007). The cellular and molecular biology of conifer embryogenesis. *New Phytologist*, *176*(3), 511–536.
Chalupa, V. (1985). Somatic embryogenesis and plantlet regeneration from cultured immature and mature embryos of Picea abies (L) karst. *Communicationes Instituti Forestalis Czechosloveniae*, *14*, 57–63.
Chandler, J. (2008). Cotyledon organogenesis. *Journal of Experimental Botany*, *59*, 2917–2931.
Chowdhury, C. R. (1962). The embryogeny of conifers: A review. *Phytomorphology*, *12*, 313–338.
Dodeman, V. L., Ducreux, G., & Kreis, M. (1997). Zygotic embryogenesis versus somatic embryogenesis. *Journal of Experimental Botany*, *48*(313), 1493–1509.
Dogra, P. D. (1967). Seed sterility and disturbances in embryogeny in conifers with particular reference to seed testing and tree breeding in Pinaceae. *Studia Forestalia Suecica*: *Vol. 45*. Stockholm: Royal College of Forestry.
Dyachok, J., Wiweger, M., Kenne, L., & von Arnold, S. (2002). Endogenous nod-factor-like signal molecules suppress cell death and promote early somatic embryo development in Norway spruce. *Plant Physiology*, *128*, 523–533.
Egertsdotter, U., & von Arnold, S. (1995). Importance of arabinogalactan proteins for the development of somatic embryos in Norway spruce. *Physiologia Plantarum*, *93*, 334–345.
Elhiti, M., Stasolla, C., & Wang, A. (2013). Molecular regulation of plant somatic embryogenesis. *In Vitro Cellular & Developmental Biology. Plant*, *49*, 631–642.
Filonova, L., Bozhkov, P., Brukhin, V., Daniel, G., Zhivotovsky, B., & von Arnold, S. (2000). Developmental programmed cell death in plant embryogenesis: Exploring a model system of Norway spruce somatic embryogenesis. *Journal of Cell Science*, *113*, 4399–4411.
Filonova, L., Bozhkov, P., & von Arnold, S. (2000). Developmental pathway of somatic embryogenesis in *Picea abies* as revealed by time-laps tracking. *Journal of Experimental Botany*, *51*(343), 249–264.
Filonova, L. H., von Arnold, S., Daniel, G., & Bozhkov, P. V. (2002). Programmed cell death eliminates all but one embryo in a polyembryonic plant seed. *Cell Death and Differentiation*, *9*, 1057–1062.
Gifford, E. M., & Foster, A. S. (1989). *Morphology and evolution of vascular plants* (pp. 401–453). New York: WH Freeman Company.
Goldberg, R. B., de Paiva, G., & Yadegari, R. (1994). Plant embryogenesis: Zygote to seed. *Science*, *266*, 605–614.

Häggman, H., Vuosku, J., Sarjala, T., Jokeal, A., & Niemi, K. (2006). Somatic embryogenesis of pine species: From functional genomics to plantation forestry. In A. Mujib & J. Šamaj (Eds.), *Somatic embryogenesis: Vol. 2. Plant cell monographs* (pp. 119–140). Berlin: Springer.

Hakman, I. (1993). Embryology in Norway spruce (*Picea abies*). Immunochemical studies on transport of a seed storage protein. *Physiologia Plantarum, 88*, 427–433.

Hakman, I., Fowke, L. C., von Arnold, S., & Erikson, T. (1985). The development of somatic embryos in tissue cultures initiated from immature embryos of *Picea abies* (Norway spruce). *Plant Science, 38*, 53–59.

Helmersson, A., von Arnold, S., & Bozhkov, P. (2008). The level of free intracellular zink mediates programmed cell death/cell survival decisions in plant embryos. *Plant Physiology, 147*, 1158–1167.

Ingouff, M., Farbos, I., Lagercrantz, U., & von Arnold, S. (2001). *PaHB1* is an evolutionary conserved HD-GL2 homeobox gene defining the protoderm during Norway spruce embryo development. *Genesis, 30*, 220–230.

Klimaszewska, K., Hargreaves, C., Lelu-Walter, M.-A., & Trontin, J.-F. (2016). Advances in conifer somatic embryogenesis since year 2000. *Methods in Molecular Biology, 1359*, 131–166. https://doi.org/10.1007/978-1-4939-3061-6_7.

Klimaszewska, K., Pelletier, G., & Overton, C. (2010). Hormonally regulated overexpression of *Arabidopsis WUS* and conifer *LEC1* (*CHAP3A*) in transgenic white spruce: Implications for somatic embryo development and somatic seedling growth. *Plant Cell Reports, 29*, 723–734.

Klimaszewska, K., Trontin, J.-F., Becwar, M. R., Devillard, C., Park, Y.-S., & Lelu-Walter, M.-A. (2007). Recent progress in somatic embryogenesis of four *Pinus* spp. *Tree and Forestry Science and Biotechnology, 1*, 11–25.

Kunieda, T., Mitsuda, N., Ohme-Takagi, M., Takeda, S., Aida, M., Tasaka, M., et al. (2008). NAC family proteins NARS1/NAC2 and NARS2/NAM in the outer integument regulate embryogenesis in Arabidopsis. *The Plant Cell, 20*, 2631–2642.

Lara-Chavez, A., Egertsdotter, U., & Flinn, B. S. (2012). Comparison of gene expression makers during zygotic and somatic embryogenesis in pine. *In Vitro Cellular & Developmental Biology – Plant, 48*, 341–354.

Larsson, E. (2011). Molecular regulation of embryo development in Norway spruce—Polar auxin transport and transcription factors. Doctoral thesis *Vol. 2011* (p. 68). Swedish University of Agricultural Sciences. ISBN 978-91-576-7612-2.

Larsson, E., Sitbon, F., Ljung, K., & von Arnold, S. (2008). Inhibited polar auxin transport results in aberrant embryo development in Norway spruce. *New Phytologist, 177*, 356–366.

Larsson, E., Sitbon, F., & von Arnold, S. (2008). Polar auxin transport controls suspensor fate. *Plant Signaling & Behavior, 3*, 469–470.

Larsson, E., Sitbon, F., & von Arnold, S. (2012). Differential regulation of *Knotted1-like* genes during establishment of the shoot apical meristem in Norway spruce (*Picea abies*). *Plant Cell Reports, 31*, 1053–1060.

Larsson, E., Sundström, J., Sitbon, F., & von Arnold, S. (2012). Expression of PaNAC01, a Picea abies CUP-SHAPED COTYLEDON orthologue, is regulated by polar auxin transport and associated with differentiation of the shoot apical meristem and formation of separated cotyledons. *Annals of Botany, 110*, 923–934.

Lelu-Walter, M.-A., Teyssier, C., Guérin, V., & Pâques, M. (2016). Vegetative propagation of larch species: Somatic embryogenesis improvement towards its integration in breeding programs. In Y.-S. Park, J. M. Bonga, & H.-K. Moo (Eds.), *Vegetative propagation of forest trees* (pp. 551–571). Republic of Korea: National Institute of Forest Science. ISBN 978-89-8176-064-9.

Lelu-Walter, M.-A., Thompson, D., Harvengt, L., Sanchez, L., Toribio, M., & Paques, L. E. (2013). Somatic embryogenesis in forestry with a focus on Europe: State-of-the-art, benefits, challenges and future direction. *Tree Genetics & Genomes, 9*, 883–899.

Lindgren, D. (2008). *A way to utilize the advantages of clonal forestry for Norway spruce? Vol. 114* (pp. 8–15). Working papers of the Finnish Forest Research Institute.

MacKay, J. J., Becwar, M. R., Park, Y. S., Cordero, J. P., & Pullman, G. S. (2006). Genetic control of somatic embryogenesis initiation in loblolly pine and implications for breeding. *Tree Genetics & Genomes, 2*, 1–9.

Merino, I., Abrahamsson, M., Sterck, L., Craven-Bartle, B., Canovas, F., & von Arnold, S. (2016). Transcript profiling for early stages during embryo development in Scots pine. *BMC Plant Biology, 16*, 255. https://doi.org/10.1186/s12870-016-0939-5.

Miguel, C. M., Rupps, A., Raschke, J., Rodrigues, A. S., & Trontin, J.-F. (2016). Impact of molecular studies on somatic embryogenesis development for implementation in conifer multi-varietal forestry. In Y.-S. Park, J. M. Bonga, & H.-K. Moo (Eds.), *Vegetative propagation of forest trees* (pp. 373–421). Republic of Korea: National Institute of Forest Science. ISBN 978-89-8176-064-9.

Mo, L. H., & von Arnold, S. (1991). Origin and development of embryogenic cultures from seedlings of Norway spruce (*Picea abies*). *Journal of Plant Physiology, 138*, 223–230.

Mordhorst, A. P., Hartog, M. V., El Tamer, M. K., Laux, T., & de Vries, S. (2002). Somatic embryogenesis from Arabidopsis shoot apical meristem mutants. *Planta, 214*(6), 829–836.

Morel, A., Teyssier, C., Trontin, J.-F., Eliášová, K., Pesek, B., Beaufour, M., et al. (2014). Early molecular events in *Pinus pinaster* Ait. Somatic embryo development under reduced water availability: Transcriptomic and proteomic analysis. *Physiologia Plantarum, 152*, 184–201.

Moschou, P. N., Gutierrez-Beltran, E., Bozhkov, P. V., & Smertenko, A. (2016). Separase promotes microtubule polymerization by activating CENP-E-related Kinesin kin 7. *Developmental Cell, 37*, 350–361.

Nagmani, R., & Bonga, J. M. (1985). Embryogenesis in sub-cultured callus of *Larix decidua*. *Canadian Journal of Forest Research, 15*, 1088–1091.

Neale, D. B., & Sederoff, R. R. (1989). Paternal inheritance of chloroplast DNA and maternal inheritance of mitochondrial DNA in loblolly pine. *Theoretical and Applied Genetics, 77*, 212–216.

Noceda, C., Salaj, T., Pérez, M., Viejo, M., Cañal, M. J., Salaj, J., et al. (2009). DNA demethylation and decrease on free polyamines is associated with the embryogenic capacity of *Pinus nigra* Arn. Cell culture. *Trees, 23*, 1285–1293. https://doi.org/10.1007/s00468-009-0370-8.

Owens, J. N., & Blake, M. D. (1985). *Forest tree seed production. Information report PI-X-53.* Petawawa National Forestry Institute.

Pagnussat, G. C., Yu, H., Ngo, Q. A., Rajani, S., Mayalagu, S., Johnson, C. S., et al. (2005). Genetic and molecular identification of genes required for female gametophyte development and function in Arabidopsis. *Development, 132*, 603–614.

Palovaara, J., Hallberg, H., Stasolla, C., Luit, B., & Hakman, I. (2010). Expression of a gymnosperm *PIN* homologous gene correlates with auxin immunolocalization pattern at cotyledon formation and in demarcation of the procambium during *Picea abies* somatic embryo development and in seedling tissues. *Tree Physiology, 30*, 479–489.

Plomion, C., Bastien, C., Bogeat-Triboulot, M.-B., Bouffier, L., Déjardin, A., Duplessis, S., et al. (2016). Forest tree genomics: 10 achievements from the past 10 years and future prospects. *Annals of Forest Science, 73*, 77–103.

Pullman, G. S., & Bucalo, K. (2014). Pine somatic embryogenesis: Analyses of seed tissue and medium to improve protocol development. *New Forest, 45*, 353–377.

Rodríguez, M. P., Fernanda Suárez, M., Heredia, R., Concepción, A., Breton, D., Trontin, J.-F., et al. (2005). Developmental regulation of glutamine sythetase genes during embryogenesis of pine. *New Phytologist, 169*, 35–44.

Rutledge, R. G., Stewart, D., Caron, S., Overton, C., Boyle, B., MacKay, J., et al. (2013). Potential link between biotic defense activation and recalcitrance to induction of somatic embryogenesis in shoot primordia from adult trees of white spruce (*Picea glauca*). *BMC Plant Biology*, *13*, 116. https://doi.org/10.1186/1471-2229-13-116.

Singh, H. (1978). Embryology of gymnosperms. In W. Zimmermann, Z. Carlquist, P. Ozenda, & H. D. Wulff (Eds.), *Handbuch der Pflanzenanatomie* (pp. 187–241). Berlin: Gebrüder Borntrager.

Smertenko, A., & Bozhkov, P. V. (2014). Somatic embryogenesis: Life and death processes during apical-basal patterning. *Journal of Experimental Botany*, *55*, 1343–1360.

Smertenko, A. P., Bozhkov, P. V., Filonova, L. H., von Arnold, S., & Hussey, P. J. (2003). Re-organization of the cytoskeleton during developmental programmed cell death in *Picea abies* embryos. *Plant Journal*, *33*, 813–824.

Smith, S. A., Beaulieu, J. M., & Donoghue, M. J. (2010). An uncorrelated relaxed-clock analysis suggests an earlier origin for flowering plants. *Proceedings of the National Academy of Sciences of the United States of America*, *107*(13), 5897–5902.

Stasolla, C., Bozhkov, P., Chu, T.-M., van Zyl, L., Egertsdotter, U., Suarez, M., et al. (2004). Variation in transcript abundance during somatic embryogenesis in gymnosperms. *Tree Physiology*, *24*, 1073–1085.

Tahır, M., Law, D. A., & Stasolla, C. (2006). Molecular characterization of *PgAGO*, a novel conifer gene of the Argonaute family expressed in apical cells and required for somatic embryo development in spruce. *Tree Physiology*, *26*, 1257–1270.

Teyssier, C., Maury, S., Beaufour, M., Grondin, C., Delaunay, A., Le Mette, C., et al. (2014). In search of markers for somatic embryo maturation in hybrid larch (*Larix x eurolepis*): Global DNA methylation and proteomic analyses. *Physiologia Plantarum*, *150*, 271–291.

Trontin, J.-F., Klimaszewska, K., Morel, A., Hargreaves, C., & Lelu-Walter, M.-A. (2016). Molecular aspects of conifer zygotic and somatic embryo development: A review of genome-wide approaches and recent insights. In M. A. Germanà & M. Lambardi (Eds.), *Vol. 1359. In vitro plant embryogenesis in higher plants, methods in molecular biology* (pp. 167–208): Springer Humana Press.

Uddenberg, D., Abrahamsson, M., & von Arnold, S. (2016). Overexpression of *PaHAP3A* stimulates differentiation of ectopic embryos on maturing somatic embryos of Norway spruce. *Tree Genetics & Genomes*, *12*, 18. https://doi.org/10.1007/s11295-016-0971-2.

Uddenberg, D., Valladares, S., Abrahamsson, M., Sundström, J., Sundås-Larsson, A., & von Arnold, S. (2011). Embryogenic potential and expression of embryogenesis-related genes in conifers are affected by treatment with a histone deacetylase inhibitor. *Planta*, *234*, 527–539.

van Hegel, A. J., van Kammen, A., & de Vries, S. (2002). A relationship between seed development and arabinogalactan-proteins (AGPs) and the AGP mediated promotion of somatic embryogenesis. *Physiologia Plantarum*, *114*, 637–644.

Vega-Bartol, J. J., Simoes, M., Lorenz, W. W., Rodrigues, A. S., Alba, R., Dean, J. F. D., et al. (2013). Transcriptomic analysis highlights epigenetic and transcriptional regulation during zygotic embryo development of *Pinus pinaster*. *BMC Plant Biology*, *13*, 123. https://doi.org/10.1186/1471-2229-13-123.

Vestman, D., Larsson, E., Uddenberg, D., Cairney, J., Clapham, D., & von Arnold, S. (2011). Important processes during differentiation and early development of somatic embryos of Norway spruce as revealed by changes in global gene expression. *Tree Genetics & Genomes*, *7*, 347–362.

von Aderkas, P., Kong, L., & Prior, N. A. (2016). In vitro techniques for conifer embryogenesis. In Y.-S. Park, J. M. Bonga, & H.-K. Moo (Eds.), *Vegetative propagation of forest trees* (pp. 335–350). Republic of Korea: Institute of Forest Science. ISBN 978-89-8176-064-9.

von Arnold, S., Bozhkov, P., Clapham, D., Dyachok, J., Filonova, L., Högberg, K.-A., et al. (2005). Propagation of Norway spruce via somatic embryogenesis. *Plant Cell Tissue and Organ Culture*, *81*, 323–329.

von Arnold, S., & Clapham, D. (2008). Spruce embryogenesis. In M. F. Suárez & P. V. Bozhkov (Eds.), *Vol. 427. Plant embryogenesis, methods in molecular biology* (pp. 31–47): Springer Humana Press. ISBN 978-1-58829-931-4.

von Arnold, S., Larsson, E., Moschou, P., Zhu, T., Uddenberg, D., & Bozhkov, P. (2016). Norway spruce as a model for studying regulation of somatic embryo development in conifers. In Y.-S. Park, J. M. Bonga, & H.-K. Moo (Eds.), *Vegetative propagation of forest trees* (pp. 351–372). Republic of Korea: National Institute of Forest Science. ISBN 978-89-8176-064-9.

von Arnold, S., Sabala, I., Bozkov, P., Dyachok, J., & Filonova, L. (2002). Developmental pathways of somatic embryogenesis. *Plant Cell Tissue and Organ Culture*, *69*, 233–249.

Wiweger, M., Farbos, I., Ingouff, M., Lagercrantz, U., & von Arnold, S. (2003). Expression of Chia4-Pa chitinase genes during somatic and zygotic embryo development in Norway spruce (*Picea abies*): Similarities and differences between gymnosperm and angiosperm class IV chitinases. *Journal of Experimental Botany*, *54*(393), 2691–2699.

Yeung, E. C., Stasolla, C., & Kong, L. S. (1998). Apical meristem formation during zygotic embryo development of white spruce. *Canadian Journal of Botany*, *76*, 751–761.

Zhu, T., Moschou, P. N., Alvarez, J. M., Sohlberg, J., & von Arnold, S. (2014). Wuschel-related homeobox 8/9 is important for proper embryo patterning in the gymnosperm Norway spruce. *Journal of Experimental Botany*, *65*, 6543–6552.

Zhu, T., Moschou, P. N., Alvarez, J. M., Sohlberg, J., & von Arnold, S. (2016). WUSCHEL-RELATED HOMEOBOX 2 is important for protoderm and suspensor development in the gymnosperm Norway spruce. *BMC Plant Biology*, *16*, 19.

CHAPTER SIX

Activity of the shoot apical and cambial meristems: Coordination and responses to environmental signals

Rita Teresa Teixeira, Xiaoyan Sheng, Amy M. Brunner*
Department of Forest Resources and Environmental Conservation, Virginia Tech, Blacksburg, VA, United States
*Corresponding author: e-mail address: abrunner@vt.edu

Contents

Abstract

A tree shoot includes multiple developmental gradients that reflect the relationships of the shoot apical meristem, the cambial meristem and the tissues and organs they produce. The adaptation of trees to temperate and boreal climates depends on the ability of these two meristems to appropriately respond to environmental signals that are markers of seasonal changes. The aim of this review is to present a more holistic view of the tree shoot, including the developmental changes associated with seasonally regulated growth transitions. We consider the role of long-distance communication in coordinating leaf development, the transition to secondary growth and meristem activity. Using *Populus* as the model, we discuss the roles of signalling mediated by hormones, sugar and *FLOWERING LOCUS T2* (*FT2*) in coordinating tree shoot developmental processes.

1. Introduction

Shoot growth and tree form rely on the initiation, maintenance and activities of shoot apical, axillary and cambial meristems. The shoot apical meristem (SAM) produces phytomers consisting of internode, a node bearing

Advances in Botanical Research, Volume 89
ISSN 0065-2296
https://doi.org/10.1016/bs.abr.2018.11.006

a leaf and an axillary meristem. When, where and to what extent axillary meristems outgrow has a major role in determining shoot architecture. As the source of cells that differentiate into wood, the role of the vascular cambium (VC) is most pronounced in trees. Management of meristem activity over extended periods of time and in response to both seasonal and episodic environmental changes is central to tree productivity, adaptation and longevity. In this review, we focus on coordination of the SAM and VC in trees and related shoot developmental transitions, particularly synchronization of leaf development and the stem transition from primary to secondary growth.

2. Coordination during growth

The SAM contains three functional domains. In the central zone (CZ), slowly dividing stem cells maintain the CZ stem cell population as well as provide daughter cells that are displaced to the peripheral zone (PZ) on either side or to the rib meristem (RM) directly below (Laux, 2003). Lateral organ primordia initiate in the PZ and the RM provides the cells of the stem, including the vasculature of both the stem and lateral organs. In the stem, the procambium originating from the RM gives rise to a ring of vascular bundles where the fascicular cambium produces primary phloem and xylem. The subsequent development of interfascicular cambium completes the continuous ring of VC to initiate secondary growth and increase stem diameter.

Tree species exhibit a range of shoot growth patterns (Barthelemy & Caraglio, 2007); however, only monopodial shoots of free-growing (shoots capable of continuous growth under favourable conditions) taxa such as *Populus* are described here as they display well-defined developmental gradients. An actively growing shoot displays an ontogenic series of leaf development. After leaves initiate, they are displaced further below the SAM by the production of new phytomers. As leaves expand, they undergo spatiotemporally regulated anatomical and physiological changes that enable their transition from sink to source (Kalve, De Vos, & Beemster, 2014). Leaf maturation proceeds from tip to base; thus, the apical part of a transitional leaf acts as source, whereas the basal region is still importing photosynthate. Leaf plastochron index (LPI) based on leaf length is widely used to describe developing shoots (Larson & Isebrands, 1971). The apical to basipetal leaf development zone also encompasses changes in stem anatomy and growth direction. As an internode is displaced from the apex, its elongation is first driven by cell division and then by cell expansion. An internode ceases elongation after the VC begins to produce secondary xylem and phloem. However, similar to leaf sink to source transitions, stem growth transitions are not abrupt; transitional internodes are still elongating as secondary growth commences. The exact relationships among leaf maturation stages and primary-secondary growth transitions vary within and between taxa, with environmental conditions, growth rate and shoot size. For example, in *Populus deltoides* plants grown from seed, both the primary to secondary growth transition zone and leaf maturation occurred at LPI3 for small plants, but for plants with 60 PIs, the stem growth transition occurred at LPI10 and leaf maturation at LPI13 (Larson, 1980). In vegetatively propagated *Populus tremula x P. alba* clone 717-1B of around 20 PIs, the VC generally forms at LPIs where leaves have begun the transition to a source organ, but have not fully matured (Fig. 1). Note that what is referred to as a recently mature leaf is not necessarily a fully expanded leaf; for example, Larson (1980) defined a poplar leaf as mature when lamina growth was not more than 1 mm on two consecutive days.

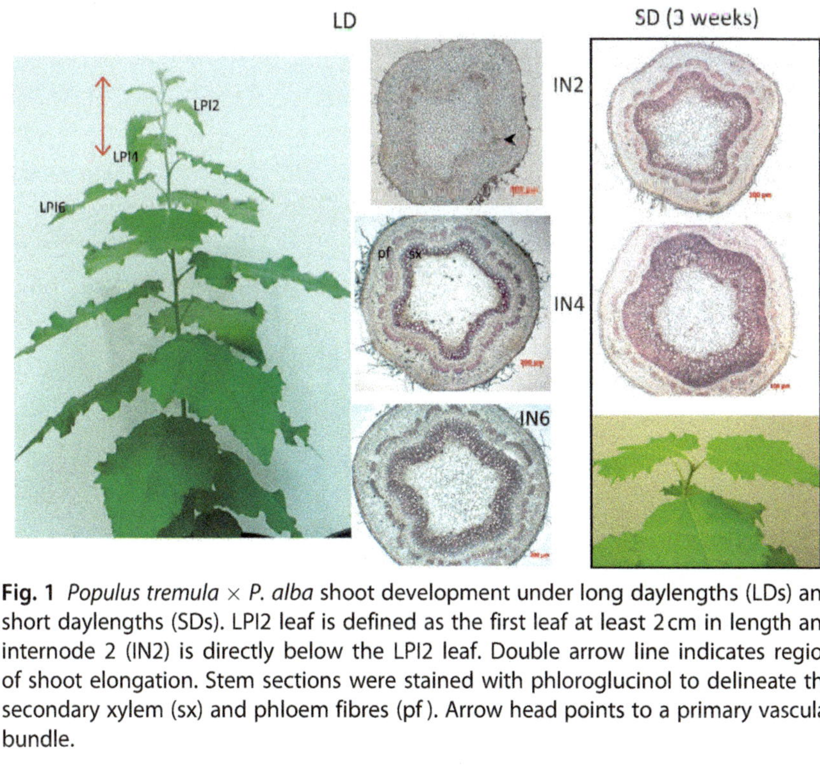

Fig. 1 *Populus tremula* × *P. alba* shoot development under long daylengths (LDs) and short daylengths (SDs). LPI2 leaf is defined as the first leaf at least 2 cm in length and internode 2 (IN2) is directly below the LPI2 leaf. Double arrow line indicates region of shoot elongation. Stem sections were stained with phloroglucinol to delineate the secondary xylem (sx) and phloem fibres (pf). Arrow head points to a primary vascular bundle.

2.1 Regulation of shoot development

Studies have begun to reveal regulatory aspects of SAM activity, leaf development and the formation, maintenance and activity of the VC (Campbell & Turner, 2017; Kalve et al., 2014; Ruonala, Ko, & Helariutta, 2017). However, these shoot developmental processes have seldom been studied as an integrated whole. The signalling pathways that link these are mostly unknown, but a number of studies provide clues. The formation of the VC introduces another sink into a developing shoot. Regulatory aspects similar to those revealed by studies of other competing shoot sinks could potentially apply to VC development and activity. Hormones have long been recognized to have major roles in regulating the outgrowth of axillary buds, but a more recent study in pea indicated that the initial signal for axillary bud outgrowth is an increase in sucrose (Suc) level (Mason, Ross, Babst, Wienclaw, & Beveridge, 2014). The authors hypothesized that apical dominance results from an active SAM's sink strength limiting sugar translocation to axillary buds. Trehalose 6-phosphate (T6P) levels correlate with Suc availability and a proposed model (Yadav et al., 2014) includes that T6P synthesis is positively regulated by Suc levels and T6P also functions as negative feedback regulator of Suc to maintain a T6P: sucrose ratio appropriate for the cell type, plant developmental stage and environmental conditions. T6P functions in both source and sink organs (Figueroa & Lunn, 2016). In the Arabidopsis SAM, increased production of T6P promotes flowering and it is also necessary for *FLOWERING LOCUS T* (*FT*) expression in leaves (Wahl et al., 2013). *FT* is expressed in specific phloem companion cells in leaves (Chen et al., 2018), and the FT protein is a mobile signal that acts in the SAM to promote flowering (Corbesier et al., 2007). Onset of pea axillary bud outgrowth following decapitation was also correlated with increased T6P in the bud (Fichtner et al., 2017). Whereas there is not yet direct evidence for a role for T6P in the regulation of SAM and VC activity in trees, there is for FT.

Populus has two *FT* homologues, one of these, *FT2*, is expressed in leaves, showing highest expression in the youngest mature leaf compared to very low expression in young sink leaves immediately below the shoot apex (Hsu, Liu, Luthe, & Yuceer, 2006; Ruonala, Rinne, Kangasjarvi, & van der Schoot, 2008). However, *FT2* differs from *FT* in that its major role is in vegetative development (Bohlenius et al., 2006; Hsu et al., 2011). *FT2* expression is necessary to maintain shoot growth and thus, the generation of an *FT2*-dependent mobile signal appears to be important for SAM and VC activity. In Arabidopsis, studies have identified two proteins necessary for FT transport. FT-INTERACTING PROTEIN 1 (FTIP1) interacts with FT in phloem companion cells and is necessary for its translocation from companion cell to sieve element (Liu et al., 2012). The metallochaperone-like protein SODIUM POTASSIUM ROOT DEFECTIVE 1 (NaKR1) mediates long-distance transport of FT from leaves to the shoot apex (Zhu, Liu, Shen, & Yu, 2016). Although movement of FT2 in poplar has not been shown, differential regulation of transporters could potentially direct its transport to SAM or VC depending on, for example, leaf developmental stage/position on shoot. Additional signalling pathways that integrate with or act in parallel with the long-distance *FT2* pathway as well as differences in signalling pathways for SAM and VC activity, including local regulation, are also possible.

In addition to providing carbon to support growth of sink tissues, sugars act as long distance signals (Wingler, 2018). ^{14}C tracer studies of *Populus* grown in both controlled and field environments (Dickson & Larson, 1981; Isebrands & Nelson, 1983; Larson, Dickson, & Isebrands, 1972) showed that transport patterns depend on leaf developmental stage. Developing leaves (e.g. only apical portion is mature) mostly transport carbon upward to younger leaves and the SAM, recently mature leaves transport in both directions, and older mature leaves primarily transport carbon to the lower stem and roots. Thus, sugars potentially could act as signals that connect leaf development stage with the transition to secondary growth.

Hormonal signalling has a role in all aspects of plant development and we highlight a couple examples that relate to coordination among SAM, VC and the tissues or organs they produce. Auxin is a key regulator of cell division in plants (Teale, Paponov, & Palme, 2006) and considerable evidence supports that formation of the VC depends on basipetal transport of auxin. In Arabidopsis, an in vitro assay showed that application of synthetic auxin to apical, but not basal, ends of a stem segment induced VC formation as evidenced by interfascicular cell divisions and phloem-specific marker gene expression (Agusti, Lichtenberger, Schwarz, Nehlin, & Greb, 2011). In *Populus*, study of auxin transporter gene expression in internodes and leaves along an apical-basal developmental gradient revealed differential expression (Schrader et al., 2003). In particular, the expression of specific influx and efflux transporters was highest in internodes where the VC was establishing.

Auxin regulates multiple aspects of leaf development, including a central role in leaf vascularization (Biedron & Banasiak, 2018; Kalve et al., 2014). In Arabidopsis, the efflux carrier *PIN-FORMED1* (*PIN1*) has a central role in leaf vascularization and is expressed in the first leaf. Similarly,

Schrader et al. (2003) identified a poplar *PIN1* homologue predominately expressed in the youngest leaves where vascularization and cell division occur. As this occurs at a lower LPI (closer to the apex) than VC formation, auxin from young leaves could directly (e.g. young leaves are a source of auxin for VC initiation) or indirectly (e.g. auxin-mediated vascularization of leaves enables export/import of other signalling molecules) link leaf development with VC initiation. Detailed study of GUS expression driven by the auxin-responsive DR5 promoter in developing *Populus* stems showed an auxin response was present within and between (interfascicular region) primary vascular bundles (Spicer, Tisdale-Orr, & Talavera, 2013). A continuous ring of GUS expressed was seen before and after the formation of the VC, consistent with an auxin signal initiating formation of the VC. Although the authors state that the rapidly expanding leaves above are the likely source of the auxin that results in this GUS pattern, they also note that it remains unclear to what extent basipetal or lateral transport delineates the auxin response zone.

Bioactive gibberellins (GAs) have prominent roles in stem elongation, leaf expansion and cambial activity (Bhalerao & Fischer, 2017; Hedden & Sponsel, 2015). Movement of GA from leaves to stem (Binenbaum, Weinstain, & Shani, 2018) could also contribute to the coordination among SAM and VC and primary and secondary growth. In tobacco, defoliation of all but the youngest leaves significantly reduced GA levels in the stem and resulted in both reduced internode elongation and cambial activity (Dayan et al., 2012). In Arabidopsis, a mobile shoot-derived GA signal induces xylem expansion in the hypocotyl following the transition to flowering (Ragni et al., 2011). In *Populus*, reduction of bioactive GAs in different parts of the vasculature via targeted expression of GA 2-oxidase genes reduced both stem elongation and radial expansion (Mauriat, Sandberg, & Moritz, 2011). The two promoters used directed somewhat different expression (i.e. predominate in VC/developing phloem vs. VC/xylem) within the secondary stem, and directed either high (LMP1 promoter) or very low (LMX5 promoter) expression in the shoot apex and young leaf. Transgenics with LMP1-directed GA2ox expression also had a unique phenotype: smaller leaves with a modified shape and increased sylleptic branching compared to wild-type plants. Furthermore, these transgenics showed altered auxin partitioning within the leaf and reduced expression of a *PIN1* homologue in the shoot apex, suggesting that GA affects auxin transport. Application of auxin and GA individually or in combination to decapitated poplar stems indicated that GA promoted polar auxin transport resulting in higher auxin levels in the stem (Bjorklund, Antti, Uddestrand, Moritz, & Sundberg, 2007). Whereas these studies suggest that integration of auxin and GA synthesis, transport and signalling plays a role in relating SAM, VC and shoot developmental transitions, there other hormones and other types of signalling pathways (e.g. sugar) that are also likely to have significant roles in coordinating the different aspects of woody shoot development.

3. The seasonal cycle: Responses to photoperiod and temperature signalling

Due to their sessile nature, plants have elaborate mechanisms to cope with changing seasons and sporadic adverse environmental conditions. One of the major adaptive traits of trees growing in temperate and boreal forests is the appropriate timing of developmental and physiological alterations that enable survival during winter and resumption of growth in spring. Here, we highlight aspects of the related responses of SAM and VC to environmental conditions and refer the reader to recent reviews (Brunner, Varkonyi-Gasic, & Jones, 2017; Cooke, Eriksson, & Junttila, 2012; Ding & Nilsson, 2016; Singh, Svystun, AlDahmash, Jonsson, & Bhalerao, 2017) for more comprehensive information on the molecular regulation and physiology of phenology in trees.

For many tree species adapted to temperate and boreal climates, reductions in photoperiod are the primary signal to begin entry into winter dormancy (also referred to as endodormancy). In *Populus* and other free-growing species, this is also the main signal for growth cessation and bud set; and thus, these taxa provide a useful system to manipulate growth and begin to elucidate the signals and molecular mechanisms that coordinate the activities of SAM and VC and the transitions of the developing woody plant shoot. In photoperiodic-responsive taxa, growth and dormancy transitions are initiated when daylength falls below a critical threshold. The critical daylength (CDL) is generally correlated with latitude of origin (Keller et al., 2011). Tree populations at higher latitudes receive winter earlier and therefore have to be prepared earlier than populations growing at lower latitudes, which typically have a shorter CDL. Once exposed to its CDL, a shoot's developmental gradient begins to change. As elongation growth ceases and bud set begins, leaves initiated under long day conditions expand and the internodes directly below the shoot apex transition to secondary growth (Fig. 1). Cessation of SAM and cambial activity is coordinated (Fig. 2) but the degree

Fig. 2 Coordination of shoot apex and vascular cambium. *Populus trichocarpa* shoot apices and stem anatomy during growth (A and B) and ecodormancy (C and D). (A) Apex of plants growing under long daylength regime with numerous young leaves emerging from the SAM. Bar 5 mm. (B) Internode 4 cross section of the same plant shown in (A). Note the numerous meristematic cell layers present in the vascular cambium (red line) characteristic of active growth. The lack of secondary cell wall thickening of the adjacent xylem (xy) is another indicator of active growth. (C) Ecodormant apical bud stage of plants after exposure to short-daylengths for 42 days. Bar 5 mm. (D) Cross section of the internode 4 from the same plant shown in (C). Arrow points to the greatly reduced vascular cambium. Sclereid cells (*) and thickening of xylem secondary cell walls further indicate absence of cell division. ph, phloem; vc, vascular cambium; xy, xylem.

to which the relative timing of SAM and VC quiescence varies within and between species is largely unknown. Especially in field conditions, multiple varying environmental components could result in, for example, a greater delay in VC quiescence relative to bud set.

Study of *Populus FT* homologues (Bohlenius et al., 2006) marked the beginning of a series of studies supporting that photoperiodic pathways controlling vegetative growth and flowering share common components (Ding & Nilsson, 2016). As described above, leaf-expressed *FT2* has a major role in maintaining growth in *Populus* and it is downregulated by short-daylengths (Hsu et al., 2011). Studies of other genes acting in the *FT2* pathway have focused on events at the SAM; however, *AINTEGUMENTA-LIKE* (*AIL*) genes are involved in promoting the activity of both SAM and VC. AILs are broadly conserved members of the APETALA2 transcription factor family and promote organ growth (Horstman, Willemsen, Boutilier, & Heidstra, 2014). All four *Populus AIL* genes are downregulated in the shoot apex by short daylengths (SDs) and overexpression of *AIL1* or *AIL3* delayed SD-induced bud set (Karlberg, Bako, & Bhalerao, 2011). Moreover, AIL1 was able to bind D-type cyclin (CYCD) promoters, consistent with a role in promoting cell division and thus, SAM activity. Subsequent studies suggested that a poplar member of the *APETALA1* subgroup of the MADS-box transcription factor family acts downstream of *FT2* in the SAM to promote growth via activation of the *AIL1* promoter (Azeez, Miskolczi, Tylewicz, & Bhalerao, 2014). *AILs* and *CYCDs* also appear to have conserved roles in promoting cambial activity (Randall et al., 2015). Cytokinins (CKs) promote VC activity and in *Populus;* both CK treatment and upregulation of CK synthesis in the cambial zone increased *AIL1* expression in the VC (Immanen et al., 2016; Randall et al., 2015). Considered together, these results suggest that *AIL1* has a similar function in SAM and VC, but the signalling pathways that activate *AIL* might differ between the two meristems.

SD responses also involve hormone synthesis and signalling (Ruttink et al., 2007). Poplar transgenics with reduced levels of bioactive GA via downregulation of GA20ox set bud earlier that wild-type in SDs (Eriksson, Hoffman, Kaduk, Mauriat, & Moritz, 2015). Similarly, the VC also reduced activity sooner in these transgenics, indicating that SD responses of both SAM and VC involve changes in GA metabolism. In addition, gene expression studies indicate a role for DELLA repressors of GA signalling in SD-mediated growth cessation of both SAM and VC (Druart et al., 2007; Zawaski & Busov, 2014). Moreover, upregulation of DELLA proteins or GA2ox accelerated SD-induced growth cessation. Time series transcript profiling combined with monitoring of the auxin-responsiveness of the cambium linked SD-induced cambial quiescence with a reduction in auxin response, which correlated with upregulation of repressor auxin response factors (Baba et al., 2011). A similar prominent role for auxin responses in SD-induced SAM quiescence has not been indicated, possibly reflecting differences in the SD response between the two meristems.

Following release from dormancy by extended exposure to chilling temperatures, the timing of meristem reactivation is largely determined by accumulated warm temperature units (heat sum) in most species. Whereas photoperiod is primarily perceived by leaves, requiring long-distance signalling to meristems, temperature is directly perceived by meristems (Amasino, 2010). Consistent with local temperature perception, localized heating of a poplar stem stimulated earlier activation of the VC but did not advance bud flush (Begum, Nakaba, Yamagishi, Oribe, & Funada, 2013). Are reactivation of SAM and VC only coordinated by a similar temperature requirement? If SAM and cambium are responding separately, this is likely of limited duration. Resumption of growth relies on remobilization of stored resources, especially nitrogen and carbohydrates, but maintenance of growth requires photosynthesis and water and nutrient acquisition. Furthermore, the growth promoter *FT2* is upregulated after bud flush (Hsu et al., 2011), consistent with a role in maintaining growth when photosynthetic leaves are providing carbon for growth.

Compared to photoperiod-mediated growth cessation, less is known about the molecular regulation of temperature-mediated reactivation. However, the poplar transcription factor EARLY BUD BREAK1 (EBB1) is a notable exception (Yordanov, Ma, Strauss, & Busov, 2014). *EBB1* is upregulated in shoot apices shortly before spring bud flush and promotes SAM reactivation. Further study of *EBB1* upregulated and downregulated transgenics showed that *EBB1* does not alter the timing of SD-mediated growth cessation and bud set. Transcriptomic analysis of transgenic and wild-type shoot apices coupled with identification of putative EBB1 targets suggested a role in auxin signalling. Consistent with EBB1-mediated SAM reactivation and with studies indicating that nitrogen remobilization from storage proteins is mediated by sinks (Islam et al., 2015), genes involved in nitrogen metabolism were the most enriched category among putative EBB1 targets. Though not specifically studied, *EBB1* could also have a role in the cambial reactivation as it is expressed in stem sections undergoing secondary growth.

4. Conclusions

All plants regulate growth in response to environmental signals, but tree shoot development also requires a well-orchestrated integration of SAM and VC activity to spatially and temporally balance primary and secondary growth with leaf development (Fig. 3). Moreover, SAM, VC and shoot development coordination is altered seasonally to cope with winter conditions. This likely requires multiple long-distance signals as well as local

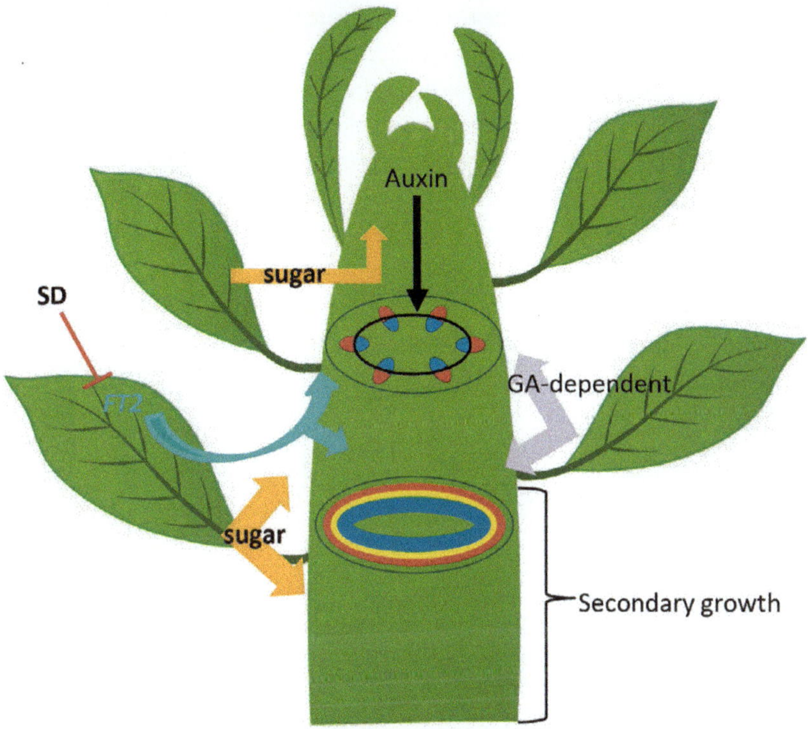

Fig. 3 Conceptual model of long-distance signals that coordinate SAM and VC activities and developmental transitions. Basipetal auxin transport is proposed to initiate a ring of auxin response which bisects the primary vascular bundles (represented as ovals) in the primary stem to specify the location and development of the VC. The direction of sugar transport from leaves changes with leaf developmental stage and could help coordinate the timing/position of the transition to secondary growth. Middle (yellow) concentric ring denotes the VC which produces secondary xylem (blue) to the inside and phloem (red) to the outside. A bidirectional GA-dependent signal from maturing/mature leaves is proposed to promote shoot elongation in the upper shoot as well as support cambial activity and production of secondary xylem. Short daylengths as well as other environmental signals (e.g. drought) that limit growth repress *FT2* expression in mature leaves. *FT2*-mediated long-distance signal promotes SAM and VC activity to maintain growth when conditions are favourable.

response pathways. In *Populus*, *FT2* expression in leaves has a major role in regulating growth in response to environmental conditions that likely extends beyond responding to seasonal changes in photoperiod as it is down-regulated by different stresses, such as drought (Hsu et al., 2011). Thus, analogous to the role of *FT* in Arabidopsis flowering time, *FT2* likely acts as an integrator of multiple environmental signals as well as endogenous pathways to ensure that growth is only maintained under conducive conditions.

Whereas there is a rich history of carbon transport studies in trees, the role of sugar as a long distance signal that coordinates leaf development with the transition to secondary growth has barely begun. Phytohormone synthesis, transport and signalling have major roles in regulating SAM and VC activity and in initiating the transition to secondary growth as well as coordinating this transition with leaf development stage. Although there is evidence for multiple long distance signals acting to coordinate aspects of tree shoot development, the signalling pathways, including long-distance signal identities and routes as well as the components of local responses in different tissues and organs, are largely unknown.

Acknowledgement

This work was supported by the Office of Biological and Environmental Research (BER) in the DOE Office of Science, DOE Office of Science (BER) Grant No. DE-SC0012574.

References

Agusti, J., Lichtenberger, R., Schwarz, M., Nehlin, L., & Greb, T. (2011). Characterization of transcriptome remodeling during cambium formation identifies MOL1 and RUL1 as opposing regulators of secondary growth. *PLoS Genetics*, 7(2), e1001312. https://doi.org/10.1371/journal.pgen.1001312.

Amasino, R. (2010). Seasonal and developmental timing of flowering. *Plant Journal*, *61*(6), 1001–1013. https://doi.org/10.1111/j.1365-313X.2010.04148.x.

Azeez, A., Miskolczi, P., Tylewicz, S., & Bhalerao, R. P. (2014). A tree ortholog of APETALA1 mediates photoperiodic control of seasonal growth. *Current Biology*, *24*(7), 717–724. https://doi.org/10.1016/j.cub.2014.02.037.

Baba, K., Karlberg, A., Schmidt, J., Schrader, J., Hvidsten, T. R., Bako, L., et al. (2011). Activity-dormancy transition in the cambial meristem involves stage-specific modulation of auxin response in hybrid aspen. *Proceedings of the National Academy of Sciences of the United States of America*, *108*(8), 3418–3423. https://doi.org/10.1073/pnas.1011506108.

Barthelemy, D., & Caraglio, Y. (2007). Plant architecture: A dynamic, multilevel and comprehensive approach to plant form, structure and ontogeny. *Annals of Botany*, *99*(3), 375–407. https://doi.org/10.1093/aob/mcl260.

Begum, S., Nakaba, S., Yamagishi, Y., Oribe, Y., & Funada, R. (2013). Regulation of cambial activity in relation to environmental conditions: Understanding the role of temperature in wood formation of trees. *Physiologia Plantarum*, *147*(1), 46–54.

Bhalerao, R. P., & Fischer, U. (2017). Environmental and hormonal control of cambial stem cell dynamics. *Journal of Experimental Botany*, *68*(1), 79–87. https://doi.org/10.1093/jxb/erw466.

Biedron, M., & Banasiak, A. (2018). Auxin-mediated regulation of vascular patterning in Arabidopsis thaliana leaves. *Plant Cell Reports*, *37*, 1215–1229. https://doi.org/10.1007/s00299-018-2319-0.

Binenbaum, J., Weinstain, R., & Shani, E. (2018). Gibberellin localization and transport in plants. *Trends in Plant Science*, *23*(5), 410–421. https://doi.org/10.1016/j.tplants.2018.02.005.

Bjorklund, S., Antti, H., Uddestrand, I., Moritz, T., & Sundberg, B. (2007). Cross-talk between gibberellin and auxin in development of Populus wood: Gibberellin stimulates polar auxin transport and has a common transcriptome with auxin. *Plant Journal, 52*(3), 499–511. https://doi.org/10.1111/j.1365-313X.2007.03250.x.

Bohlenius, H., Huang, T., Charbonnel-Campaa, L., Brunner, A. M., Jansson, S., Strauss, S. H., et al. (2006). CO/FT regulatory module controls timing of flowering and seasonal growth cessation in trees. *Science, 312*(5776), 1040–1043.

Brunner, A. M., Varkonyi-Gasic, E., & Jones, R. C. (2017). Phase change and phenology in trees. In A. Groover & Q. Cronk (Eds.), *Comparative and evolutionary genomics of angiosperm trees* (pp. 227–274). Cham, Switzerland: Springer.

Campbell, L., & Turner, S. (2017). Regulation of vascular cell division. *Journal of Experimental Botany, 68*(1), 27–43. https://doi.org/10.1093/jxb/erw448.

Chen, Q., Payyavula, R. S., Chen, L., Zhang, J., Zhang, C., & Turgeon, R. (2018). FLOWERING LOCUS T mRNA is synthesized in specialized companion cells in Arabidopsis and Maryland Mammoth tobacco leaf veins. *Proceedings of the National Academy of Sciences of the United States of America, 115*(11), 2830–2835. https://doi.org/10.1073/pnas.1719455115.

Cooke, J. E. K., Eriksson, M. E., & Junttila, O. (2012). The dynamic nature of bud dormancy in trees: Environmental control and molecular mechanisms. *Plant, Cell & Environment, 35*(10), 1707–1728.

Corbesier, L., Vincent, C., Jang, S., Fornara, F., Fan, Q., Searle, I., et al. (2007). FT protein movement contributes to long-distance signaling in floral induction of Arabidopsis. *Science, 316*(5827), 1030–1033. https://doi.org/10.1126/science.1141752.

Dayan, J., Voronin, N., Gong, F., Sun, T. P., Hedden, P., Fromm, H., et al. (2012). Leaf-induced gibberellin signaling is essential for internode elongation, cambial activity, and fiber differentiation in tobacco stems. *Plant Cell, 24*(1), 66–79. https://doi.org/10.1105/tpc.111.093096.

Dickson, R. E., & Larson, P. R. (1981). C-14 fixation, metabolic labeling patterns, and translocation profiles during leaf development in Populus deltoides. *Planta, 152*(5), 461–470. https://doi.org/10.1007/Bf00385364.

Ding, J. H., & Nilsson, O. (2016). Molecular regulation of phenology in trees—Because the seasons they are a-changin'. *Current Opinion in Plant Biology, 29*, 73–79.

Druart, N., Johansson, A., Baba, K., Schrader, J., Sjodin, A., Bhalerao, R. R., et al. (2007). Environmental and hormonal regulation of the activity-dormancy cycle in the cambial meristem involves stage-specific modulation of transcriptional and metabolic networks. *Plant Journal, 50*(4), 557–573.

Eriksson, M. E., Hoffman, D., Kaduk, M., Mauriat, M., & Moritz, T. (2015). Transgenic hybrid aspen trees with increased gibberellin (GA) concentrations suggest that GA acts in parallel with FLOWERING LOCUS T2 to control shoot elongation. *New Phytologist, 205*(3), 1288–1295.

Fichtner, F., Barbier, F. F., Feil, R., Watanabe, M., Annunziata, M. G., Chabikwa, T. G., et al. (2017). Trehalose 6-phosphate is involved in triggering axillary bud outgrowth in garden pea (Pisum sativum L.). *Plant Journal, 92*(4), 611–623. https://doi.org/10.1111/tpj.13705.

Figueroa, C. M., & Lunn, J. E. (2016). A tale of two sugars: Trehalose 6-phosphate and sucrose. *Plant Physiology, 172*(1), 7–27. https://doi.org/10.1104/pp.16.00417.

Hedden, P., & Sponsel, V. (2015). A century of gibberellin research. *Journal of Plant Growth Regulation, 34*(4), 740–760. https://doi.org/10.1007/s00344-015-9546-1.

Horstman, A., Willemsen, V., Boutilier, K., & Heidstra, R. (2014). AINTEGUMENTA-LIKE proteins: Hubs in a plethora of networks. *Trends in Plant Science, 19*(3), 146–157.

Hsu, C. Y., Adams, J. P., Kim, H. J., No, K., Ma, C. P., Strauss, S. H., et al. (2011). *FLOWERING LOCUS T* duplication coordinates reproductive and vegetative growth in perennial poplar. *Proceedings of the National Academy of Sciences of the United States of America, 108*(26), 10756–10761. https://doi.org/10.1073/pnas.1104713108.

Hsu, C. Y., Liu, Y. X., Luthe, D. S., & Yuceer, C. (2006). Poplar *FT2* shortens the juvenile phase and promotes seasonal flowering. *Plant Cell, 18*(8), 1846–1861.

Immanen, J., Nieminen, K., Smolander, O. P., Kojima, M., Alonso Serra, J., Koskinen, P., et al. (2016). Cytokinin and auxin display distinct but interconnected distribution and signaling profiles to stimulate cambial activity. *Current Biology, 26*(15), 1990–1997. https://doi.org/10.1016/j.cub.2016.05.053.

Isebrands, J. G., & Nelson, N. D. (1983). Distribution of [C-14]-labeled photosynthates within intensively cultured populus clones during the establishment year. *Physiologia Plantarum, 59*(1), 9–18. https://doi.org/10.1111/j.1399-3054.1983.tb06563.x.

Islam, N., Li, G., Garrett, W. M., Lin, R., Sriram, G., Cooper, B., et al. (2015). Proteomics of nitrogen remobilization in poplar bark. *Journal of Proteome Research, 14*(2), 1112–1126. https://doi.org/10.1021/pr501090p.

Kalve, S., De Vos, D., & Beemster, G. T. (2014). Leaf development: A cellular perspective. *Frontiers in Plant Science, 5*, 362. https://doi.org/10.3389/fpls.2014.00362.

Karlberg, A., Bako, L., & Bhalerao, R. P. (2011). Short day-mediated cessation of growth requires the downregulation of AINTEGUMENTALIKE1 transcription factor in hybrid aspen. *PLoS Genetics*, 7(11), e1002361.

Keller, S. R., Soolanayakanahally, R. Y., Guy, R. D., Silim, S. N., Olson, M. S., & Tiffin, P. (2011). Climate-driven local adaptation of ecophysiology and phenology in balsam poplar, Populus balsamifera L. (Salicaceae). *American Journal of Botany, 98*(1), 99–108. https://doi.org/10.3732/ajb.1000317.

Larson, P. R. (1980). Interrelations between phyllotaxis, leaf development and the primary-secondary vascular transition in Populus deltoides. *Annals of Botany, 46*(6), 757–769. https://doi.org/10.1093/oxfordjournals.aob.a085973.

Larson, P. R., Dickson, R. E., & Isebrands, J. G. (1972). Fixation patterns of C-14 within developing leaves of eastern cottonwood. *Planta, 107*(4), 301–314. https://doi.org/10.1007/Bf00386392.

Larson, P. R., & Isebrands, J. G. (1971). The plastochron index as applied to developmental studies of cottonwood. *Canadian Journal of Forest Research, 1*, 1–11.

Laux, T. (2003). The stem cell concept in plants: A matter of debate. *Cell, 113*(3), 281–283.

Liu, L., Liu, C., Hou, X., Xi, W., Shen, L., Tao, Z., et al. (2012). FTIP1 is an essential regulator required for florigen transport. *PLoS Biology, 10*(4), e1001313. https://doi.org/10.1371/journal.pbio.1001313.

Mason, M. G., Ross, J. J., Babst, B. A., Wienclaw, B. N., & Beveridge, C. A. (2014). Sugar demand, not auxin, is the initial regulator of apical dominance. *Proceedings of the National Academy of Sciences of the United States of America, 111*(16), 6092–6097. https://doi.org/10.1073/pnas.1322045111.

Mauriat, M., Sandberg, L. G., & Moritz, T. (2011). Proper gibberellin localization in vascular tissue is required to control auxin-dependent leaf development and bud outgrowth in hybrid aspen. *Plant Journal, 67*(5), 805–816. https://doi.org/10.1111/j.1365-313X.2011.04635.x.

Ragni, L., Nieminen, K., Pacheco-Villalobos, D., Sibout, R., Schwechheimer, C., & Hardtke, C. S. (2011). Mobile gibberellin directly stimulates Arabidopsis hypocotyl xylem expansion. *Plant Cell, 23*(4), 1322–1336. https://doi.org/10.1105/tpc.111.084020.

Randall, R. S., Miyashima, S., Blomster, T., Zhang, J., Elo, A., Karlberg, A., et al. (2015). AINTEGUMENTA and the D-type cyclin CYCD3;1 regulate root secondary growth and respond to cytokinins. *Biology Open, 4*(10), 1229–1236. https://doi.org/10.1242/bio.013128.

Ruonala, R., Ko, D., & Helariutta, Y. (2017). Genetic networks in plant vascular development. *Annual Review of Genetics*, *51*(51), 335–359.

Ruonala, R., Rinne, P. L., Kangasjarvi, J., & van der Schoot, C. (2008). CENL1 expression in the rib meristem affects stem elongation and the transition to dormancy in Populus. *Plant Cell*, *20*(1), 59–74. https://doi.org/10.1105/tpc.107.056721.

Ruttink, T., Arend, M., Morreel, K., Storme, V., Rombauts, S., Fromm, J., et al. (2007). A molecular timetable for apical bud formation and dormancy induction in poplar. *Plant Cell*, *19*(8), 2370–2390.

Schrader, J., Baba, K., May, S. T., Palme, K., Bennett, M., Bhalerao, R. P., et al. (2003). Polar auxin transport in the wood-forming tissues of hybrid aspen is under simultaneous control of developmental and environmental signals. *Proceedings of the National Academy of Sciences of the United States of America*, *100*(17), 10096–10101. https://doi.org/10.1073/pnas.1633693100.

Singh, R. K., Svystun, T., AlDahmash, B., Jonsson, A. M., & Bhalerao, R. P. (2017). Photoperiod- and temperature-mediated control of phenology in trees—A molecular perspective. *The New Phytologist*, *213*(2), 511–524. https://doi.org/10.1111/nph.14346.

Spicer, R., Tisdale-Orr, T., & Talavera, C. (2013). Auxin-responsive DR5 promoter coupled with transport assays suggest separate but linked routes of auxin transport during woody stem development in Populus. *PLoS One*, *8*(8), e72499. https://doi.org/10.1371/journal.pone.0072499.

Teale, W. D., Paponov, I. A., & Palme, K. (2006). Auxin in action: Signalling, transport and the control of plant growth and development. *Nature Reviews. Molecular Cell Biology*, 7(11), 847–859. https://doi.org/10.1038/nrm2020.

Wahl, V., Ponnu, J., Schlereth, A., Arrivault, S., Langenecker, T., Franke, A., et al. (2013). Regulation of flowering by trehalose-6-phosphate signaling in Arabidopsis thaliana. *Science*, *339*(6120), 704–707. https://doi.org/10.1126/science.1230406.

Wingler, A. (2018). Transitioning to the next phase: The role of sugar signaling throughout the plant life cycle. *Plant Physiology*, *176*(2), 1075–1084. https://doi.org/10.1104/pp.17.01229.

Yadav, U. P., Ivakov, A., Feil, R., Duan, G. Y., Walther, D., Giavalisco, P., et al. (2014). The sucrose-trehalose 6-phosphate (Tre6P) nexus: Specificity and mechanisms of sucrose signalling by Tre6P. *Journal of Experimental Botany*, *65*(4), 1051–1068. https://doi.org/10.1093/jxb/ert457.

Yordanov, Y. S., Ma, C., Strauss, S. H., & Busov, V. B. (2014). EARLY BUD-BREAK 1 (EBB1) is a regulator of release from seasonal dormancy in poplar trees. *Proceedings of the National Academy of Sciences of the United States of America*, *111*(27), 10001–10006.

Zawaski, C., & Busov, V. B. (2014). Roles of gibberellin catabolism and signaling in growth and physiological response to drought and short-day photoperiods in Populus trees. *PLoS One*, *9*(1), e86217. https://doi.org/10.1371/journal.pone.0086217.

Zhu, Y., Liu, L., Shen, L., & Yu, H. (2016). NaKR1 regulates long-distance movement of FLOWERING LOCUS T in Arabidopsis. *Nature Plants*, *2*(6), 16075. https://doi.org/10.1038/nplants.2016.75.

CHAPTER SEVEN

Digging in wood: New insights in the regulation of wood formation in tree species

Eduardo L.O. Camargo[a,b], Raphaël Ployet[a], Hua Cassan-Wang[a], Fabien Mounet[a], Jacqueline Grima-Pettenati[a,*]

[a]Laboratoire de Recherche en Sciences Végétales, Université de Toulouse, CNRS, UPS, Castanet-Tolosan, France
[b]Laboratório de Genômica e Expressão, Departamento de Genética, Evolução e Bioagentes, Instituto de Biologia, Universidade Estadual de Campinas, Campinas, Brazil
*Corresponding author: e-mail address: grima@lrsv.ups-tlse.fr

This paper is dedicated to the memory of Prof Carl Douglas, pioneer and mentor in this research field who died prematurely.

Contents

Advances in Botanical Research, Volume 89
ISSN 0065-2296
https://doi.org/10.1016/bs.abr.2018.11.007

Abstract

Wood, which represents the most abundant lignocellulosic biomass on earth, fulfils key roles in trees, and is also a raw material for multiple end-uses by mankind. The differentiation of this complex vascular tissue starts with cell division in the vascular cambium and is characterized by a massive deposition of lignified secondary cell walls mainly in fibres and vessels. A transcriptional network underlying this differentiation process ensures a tight regulation of the expression of thousands of genes both at the spatial and temporal levels. Most of our current knowledge of this hierarchical network was extrapolated from studies performed in *Arabidopsis*. Here, we review recent findings on the regulation of wood formation in angiosperm trees species highlighting conserved and distinct mechanisms with *Arabidopsis*. We provide examples shedding light on the central role of auxin and its cross-talk with other hormones at different stages of secondary xylem differentiation. Functional studies of trees' wood-associated transcription factors revealed diversified functions as compared to their *Arabidopsis* orthologs. Sophisticated mechanisms of alternative splicing and cross-regulation between the two distinct groups of top level NAC-domain master regulators were uncovered. These findings underlie the high level of complexity of wood formation in trees and provide a framework for future lines of research in this exiting research field.

1. Introduction

Wood, also called secondary xylem, is a highly specialized vascular tissue characterized by the presence of thick heavily lignified secondary cell walls composed of three main polymers: cellulose, hemicelluloses (e.g. xylan) and lignin. Wood is the most abundant plant biomass on earth and an immense reservoir of fixed carbon for long periods of time. It is used for myriad applications such as construction, pulp and paper making and as a source of renewable energy alternative to fossil fuels. Distinct wood properties are required for the different end-uses of wood as raw material and they are not necessarily the properties which are beneficial for trees fitness and adaptation to environment. Anatomical, chemical and physical properties of wood are determined throughout a complex dynamic process called xylogenesis controlled by both internal signals like hormones and external signals occurring during the tree's life (reviewed in Mauriat et al., 2014).

Wood is produced by the activity of a lateral meristem, the vascular cambium that is responsible for both secondary growth and the perennial life of trees (Déjardin et al., 2010). This particular meristem is composed of two types of stem cells (the fusiform and the ray initials) that divide asymmetrically to maintain the stem cell population and to produce daughter cells.

Cambial cells division rate determines the rate of wood formation. In woody angiosperms, the elongated fusiform initials differentiate into axially oriented woody cells (fibres, vessels, axial parenchyma) ensuring water conduction and mechanical support for the plant body. The nearly isodiametric ray initials give rise to transversely oriented ray parenchyma (contact and isolation ray cells) ensuring transverse conduction and nutrient storage (reviewed in Mellerowicz, Baucher, Sundberg, & Boerjan, 2001; Morris et al., 2016). The differentiation of daughter cells into distinct mature xylem cell-types with specialized functions involves sequential steps including cell expansion, substantial deposition of thick lignified secondary cell walls (SWs), programmed cell death and heartwood maturation (Zhong & Ye, 2015). This temporally and spatially tightly controlled process has driven considerable research attention because of the economic importance of wood.

Most of our current knowledge on the regulation of wood formation was extrapolated from *Arabidopsis thaliana* (Schuetz, Smith, & Ellis, 2013), whose secondary growth is limited to hypocotyl and roots. Despite the fact that this annual plant undergoes true secondary growth (i.e. originating from cambial activity) its secondary xylem lack the parenchyma-like files of cells, the so called "rays", one key feature of angiosperm wood (Chaffey, Cholewa, Regan, & Sundberg, 2002). In contrast to herbaceous annual plants, trees need to synthesize heavily thickened secondary cell walls, in a large and stable quantity to withstand gravity, long-distance conduction as well other environmental factors. Furthermore, wood formation in trees is very complex and involves distinct physiological and anatomic processes that cannot be addressed in the herbaceous model plant (Nieminen, Blomster, Helariutta, & Mähönen, 2015) such as perennial secondary growth, long life-span, cambium dormancy, cambium ageing, juvenile and mature wood, heartwood formation.

In this review, we focus on recent findings on the regulation of wood formation in angiosperm trees species, especially poplar and eucalypts, highlighting conserved and distinct mechanisms with *Arabidopsis*. We first provide illustrative examples of the emerging role of plant hormones in the regulation of wood differentiation (including cell expansion and secondary wall deposition) and on the cross-talk between hormones. We then survey the main transcription factors (TFs) from *Populus* and/or *Eucalyptus* involved in the regulation of secondary cell wall formation and more particularly those that have been recently functionally characterized in trees. In the third part, we describe recent findings on the mechanisms regulating the SW-regulators. We end discussing future outlines and challenges to fill gaps in this exiting research field.

2. Hormonal control of wood formation

It is not our intent here to cover the extensive literature on the hormonal control of cambium activity and cell specification by auxin, cytokinin, gibberellins which were developed in other excellent reviews (Bhalerao & Fischer, 2014; Mauriat et al., 2014; Milhinhos & Miguel, 2013; Nieminen, Robischon, Immanen, & Helariutta, 2012; Smet & De Rybel, 2016; Sorce, Giovannelli, Sebastiani, & Anfodillo, 2013). We will focus on the most recent and significant findings and especially those illustrating the role of auxin and its cross-talk with other hormones in initiating the transcriptional programme of secondary xylem differentiation in trees. Indeed, a recent publication described the spatial distribution of the plant hormones auxin, gibberellin and cytokinin in poplar sections encompassing the phloem, the vascular cambium and the xylem (Immanen et al., 2016). They showed that each hormone exhibits distinct concentration maxima with partially overlapping distribution profiles: auxin peaking in the cambial zone, cytokinins in the developing phloem cells and gibberellin in expanding xylem cells. This suggests that auxin interacts with other hormones to contribute to subsequent cell fate decisions.

2.1 Reading the auxin gradient

In wood-forming tissue, auxin concentrations peak in the cambium and decay rapidly towards the xylem and phloem (for review, see Bhalerao & Fischer, 2014 and references therein, Immanen et al., 2016). The mechanisms underlying the establishment of this gradient and how it could be interpreted have been deeply discussed by Bhalerao and Fischer (2014). The landmark work of Nilsson et al. (2008) showed that auxin-responsive genes in wood-forming tissues of hybrid aspen respond dynamically to changes in cellular auxin levels but the expression patterns of most of them displayed limited correlation with the auxin concentration across this developmental zone. They also demonstrated that the perturbation of auxin signalling by overexpression of stabilized mutated version of the Aux/IAA, *PttIAA3*, in transgenic aspen leads to a reduced cambial cell division while radial extension of the cambial zone increases (Nilsson et al., 2008). This suggests that auxin signalling not only promotes cambium proliferation but also spatially restricts stem cell characteristics within the cambium area (Bhalerao & Fischer, 2014). Notably, the *PttIAA3*-overexpressing trees also produced smaller fibre and vessel cells than wild-type trees supporting a role of auxin not only in regulating the cell divisions that give rise to secondary xylem cells but also in their subsequent expansion and development. A possible interpretation of the auxin gradient proposed by Bhalerao and Fischer (2014) is that high auxin concentrations are a signal for cell division, intermediate levels may promote cell expansion and low levels may be read out as a signal inducing the deposition of secondary cell walls (Bhalerao & Fischer, 2014 and references therein). It is also worth noting that very recently, Johnson et al. (2018) provided evidence that polar auxin transport is important for vessel spatial patterning and size determination and by altering auxin transport, it is possible to shape the basic hydraulic properties of a woody stem.

2.2 The interplay between auxin and the WOX-CLE-PXY complex

Recent work in *Arabidopsis* helped to unravel how auxin can regulate cambial activity. Cambial cell division is promoted by the release of the phloem peptide CLE41 (CLAVATA3/EMBRYO SURROUNDING REGION-RELATED41) which can bind its LRR-RLK receptor PXY (PHLOEM INTERCALATED WITH XYLEM) in the cambium. Receptor signalling then promotes cambial activity through the transcription factor WOX4 (WUSCHEL-related HOMEOBOX4 gene) (Růžička, Ursache, Hejátko, & Helariutta, 2015). WOX4 mediates the auxin-dependent induction of cambium activity and this process is dependent on PXY which is required for a stable auxin-dependent WOX4 activity regulation (Suer, Agusti, Sanchez, Schwarz, & Greb, 2011). Very recently, a breakthrough was achieved by Brackmann et al. (2018) demonstrating the importance of the spatial organization of the auxin signalling in the cambium. They showed the AUXIN RESPONSE FACTORs *ARF3*, *ARF4* and *ARF5/MONOPTEROS* are cambium regulators with different tissue-specificities as well as distinct roles in cambium regulation. ARF3 and ARF4 function redundantly as cambium activators. ARF5 acts by directly attenuating the activity of the *WOX4* gene, specifically in PXY-positive cambium stem cells, in order to foster the transition from stem cells to differentiated vascular cells (Brackmann et al., 2018).

Populus orthologs of *CLE*, *PXY* and *WOX4* have similar expression patterns to those in *Arabidopsis* suggesting that the CLE-PXY-WOX4 signalling pathway is conserved between *Arabidopsis* and woody plants (Ursache, Nieminen, & Helariutta, 2013; Zhang, Nieminen, Serra, & Helariutta, 2014; and references therein). Indeed, Etchells, Mishra,

Kumar, Campbell, and Turner (2015) showed that an aspen receptor kinase PttPXY and its peptide ligand PttCLE41 act to control a multifunctional pathway that regulates both the rate of cambial cell division and woody tissue organization. Whereas ectopic overexpression of *PttPXY* and *PttCLE41* genes in hybrid aspen resulted in vascular tissue abnormalities and poor plant growth, their tissue-specific overexpression generated trees that exhibited a two-fold increase in the rate of wood formation, were taller, and possessed larger leaves compared to the controls. More recently, Kucukoglu, Nilsson, Zheng, Chaabouni, and Nilsson (2017) investigated the role of *WOX4* and *CLE41* orthologs in *Populus* trees. They showed that *PttWOX4* genes (*PttWOX4a/b*) were specifically expressed in the vascular cambium during vegetative growth, but neither after growth cessation nor during dormancy. In transgenic trees with reduced expression levels of *PttWOX4a/b*, both periclinal and anticlinal cambial cell divisions were reduced, causing a dramatic inhibition of secondary growth. This activity involves the positive regulation of *PttWOX4a/b* through *PttCLE41*-related genes likely involving a regulation by auxin. Based on their transcriptional data, the authors concluded that a similar mechanism involving *PttWOX4a/b*, *PttCLE41*-related genes and the gene *PttPXYa* regulates the identity and activity of the vascular cambium in trees (Kucukoglu et al., 2017). The fine spatial organization of signalling events of auxin involving ARF5-dependent attenuation of WOX4 revealed in *Arabidopsis* (Brackmann et al., 2018) needs to be demonstrated in trees.

2.3 The cross-talk between auxin and cytokinins stimulates cambium activity

Cytokinins (CK) signalling is known to be important in the maintenance and proliferation of cambial cells and in cambium cell specification (Bhalerao, Fischer, & Turner, 2016; Milhinhos & Miguel, 2013; and references therein). A reduction in cytokinin level by overexpression of a CK catabolic gene, *Arabidopsis CYTOKININ OXIDASE 2* (*CKX2*), in transgenic poplar leads to a decrease in the number of cambium cells and concomitantly a reduced stem diameter (Nieminen et al., 2008). On the other hand, aspen trees overexpressing the *AtIPT7* gene, encoding one key enzyme in the biosynthesis of major bioactive CKs, displayed stimulated cambial cell division activity resulting in dramatically increased production of the lignocellulosic trunk biomass. The elevation of the CK content led to an increase in cambial auxin concentration, highlighting the interconnected nature of these two hormonal gradients to stimulate cambial activity (Immanen et al., 2016).

2.4 The cross-talk between auxin and gibberellins promotes expansion of cambial derivatives

The role of gibberellin (GA) in wood development in trees is supported by several lines of evidence (Mauriat et al., 2014; Milhinhos & Miguel, 2013; and references therein). For instance, the overexpression of *GIBBERELLIN 20-OXIDASE1* (*GA20ox*), a GA biosynthetic gene, in poplar results in increased growth and xylem fibre length (Eriksson, Israelsson, Olsson, & Moritz, 2000). Overexpression of poplar orthologs of *AtGID1* (*GIBBERELLIN INSENSITIVE DWARF1*) resulted in phenotypes similar to those of the constitutive overexpressors of *GA20ox*, i.e., increases in xylogenesis and cell elongation with the exception of increased fibre length (Mauriat & Moritz, 2009). Noteworthy, the increase in xylogenesis in decapitated aspen trees is stronger when adding both IAA and GA compared to what is observed with adding only one hormone at a time (Björklund, Antti, Uddestrand, Moritz, & Sundberg, 2007). Consistent with the alteration of cambium proliferation observed in poplars affected in GA metabolism, GA stimulates IAA transport and in turn, IAA stimulates GA biosynthesis genes likely through Aux/IAA-ARF signalling elements. Strikingly, both hormones shared a common transcriptomic signature including many transcripts related to cell growth. These findings strongly support that these two hormones play important roles in the post-meristematic expansion of cambial derivatives (Björklund et al., 2007).

The gradual reduction in auxin concentration with increasing distance from the cambium and concurrent increase concentration of active GA may constitute a signal for cells to transition from an expansive phase to one of maturation (Immanen et al., 2016). Building on the hypothesis that the interplay between GA and IAA may act as a signal to initiate fibre differentiation, Johnsson et al. (2018) tested if these hormones were able to regulate the expression of the top level regulators of xylem cell fate and SW deposition, the secondary wall NAC (NAM/ATAF/CUC)-domain transcription factors (see Section 3.1.1 and Table 1). Based on high spatial resolution profiling in *Populus* (Sundell et al., 2017) and comparative phylogeny, they first identified two clades of secondary wall NACs presenting distinct

Table 1 Putative *E. grandis* and *P. trichocarpa* orthologs of *Arabidopsis* NAC domain proteins known to be involved in regulating secondary cell wall biosynthesis.

NAC protein *Arabidopsis*		Putative *Eucalyptus* ortholog (Hussey, Saidi, Hefer, Myburg, & Grima-Pettenati, 2015)	Putative poplar co-orthologs	References
NST	SND1/NST3 (ANAC012)	EgrNAC61	– PtVNS9/PtrWND2A/PtrSND1-B1/PtrNST1-1 (Potri.014G104800) – PtVNS10/PtrWND2B/PtrSND1-B2/PtrNST1-2 (Potri.002G178700)	Ohtani et al. (2011), Zhong, Lee, and Ye (2010a, 2010b), Li et al. (2012), Johnsson et al. (2018)
	NST1 (ANAC043)	EgrNAC49	– PtVNS11/PtrWND1B/PtrSND1-A2/PtrSND1-A2/PtrSND1-2 (Potri.001G448400) – PtVNS12/PtrWND1A/PtrSND1-A1/PtrSND1-1 (Potri.011G153300)	
	NST2 (ANAC066)	—	—	
VND	VND1 (ANAC037) VND2 (ANAC076)	EgrNAC146	– PtVNS01/PtrWND5A/PtrVND6-C1/PtrVND3-1 (Potri.007G014400) – PtVNS02/PtrWND5B/PtrVND6-C2/PtrVND3-2 (Potri.005G116800)	Lin et al. (2013), Ohtani et al. (2011), Zhong et al. (2010a, 2010b), Li et al. (2012), Johnsson et al. (2018), Sundell et al. (2017)
	VND3 (ANAC105)	—	—	
	VND4 (ANAC007) VND5 (ANAC026)	EgrNAC50	– PtVNS03/PtrWND4A/PtrVND6-B2/PtrVND6-3 (Potri.001G120000) – PtVNS04/PtrWND4B/PtrVND6-B1/PtrVND6-4 (Potri.003G113000)	Ohtani et al. (2011), Zhong et al. (2010a, 2010b), Li et al. (2012), Johnsson et al. (2018)

	VND6 (ANAC101)	EgrNAC26	– PtVNS05/PtrWND3A/PtrVND6-A1/PtrVND6-1 (Potri.015G127400) – PtVNS06/PtrWND3B/PtrVND6-A2/PtrVND6-2 (Potri.012G126500)	Ohtani et al. (2011), Zhong et al. (2010a, 2010b), Li et al. (2012), Johnsson et al. (2018)
	VND7 (ANAC030)	EgrNAC75	– PtVNS07/PtrWND6A/PtrVND7-1 (Potri.013G113100) – PtVNS08/PtrWND6B/PtrVND7-2 (Potri.019G083600)	Ohtani et al. (2011), Zhong et al. (2010a, 2010b), Johnsson et al. (2018)
SND	SND2 (ANAC073)	EgrNAC170	– PopNAC154/PtrSND2-2 (Potri.017G016700) – PopNAC156/PtrSND2-1 (Potri.007G135300)	Grant, Fujino, Beers, and Brunner (2010), Johnsson et al. (2018)
	SND3 (ANAC010)	EgrNAC44 EgrNAC45 EgrNAC46 EgrNAC47	– PopNAC157/PtrSND3-1 (Potri.004G049300) – PopNAC105/PtrSND3-2 (Potri.011G058400)	Grant et al. (2010), Johnsson et al. (2018)
		EgrNAC64		
	XND1 (ANAC104)	EgrNAC137 EgrNAC138 EgrNAC139	– PopNAC118 (Potri.003G022800) – PopNAC122 (Potri.005G064100) – PopNAC128 (Potri.001G206900) – PopNAC129 (Potri.007G105000)	Grant et al. (2010), Johnsson et al. (2018)
		EgrNAC152		
	VNI2 (ANAC083)	EgrNAC122	Potri.015G102100	Johnsson et al. (2018)

expression profiles and containing in their promoters, both GA and IAA response elements. Interestingly, they reported that those associated with fibre and SW formation, the so-called "SND/NST", were induced by GA treatment and had a tendency to be repressed by auxin. On the other hand, both GA and auxin could induce vessel-specific "VND" genes. These findings were reflected at the anatomical level. In decapitated poplar stems, exogenous auxin treatment reduced cell wall thickness while GA promotes SW deposition. High concentration of auxin in the cambial and expansion zones represses fibre-specific NAC TFs and allows differentiation of xylem vessels. When auxin concentration decreases and GA levels are high, all wood-associated NACs are induced (Johnsson et al., 2018). This supports a central role of the IAA/GA cross-talk in determining cell fate possibly through the control of SND/NST and VND master regulators and/or direct regulation of SW biosynthesis (Johnson et al., 2018). However, the underlying mechanisms of how auxin and GA coordinate expression of wood-associated NAC TFs remain unclear although the presence of overlapping GA and auxin responses elements in the promoter of some wood-associated NAC genes could provide a way for the suggested IAA/GA cross-talk.

It is also worth noting that deregulation of a SND/NST member can, in turn, affect auxin homeostasis. For instance, transformation of the PtSND2 activator into a strong repressor affects auxin biosynthesis, transport and signalling and, as a result, represses the normal growth and vascular development of transgenic poplar plants (Wang et al., 2014). The promoters of genes involved in auxin biosynthesis, transport and signalling contain SNBE sites suggesting that they could be targets of NAC TFs (Johnson et al., 2018).

2.5 Interaction between auxin and other hormones

Pioneer works on *Zinnia* mesophyll cell cultures demonstrated that brassinosteroids (BRs) are able to regulate xylem differentiation (Yamamoto et al., 2007). In *Arabidopsis* shoots, Ibanes, Fabregas, Chory, and Cano-Delgado (2009) showed that auxin polar transport coupled to BR signalling is required to determine the radial pattern of vascular bundles. Recently, Jin et al. (2017) have overexpressed one poplar ortholog of *AtCYP85A2*, known to encode a bifunctional cytochrome P450 monooxygenase, catalysing a final rate-limiting step in the BR-biosynthetic pathway. Overexpression of *PtCYP85A3* in poplar increased the endogenous BR levels and significantly promoted growth, particularly xylem formation without apparent alteration of SW properties (Jin et al., 2017).

The work of Agusti et al. (2011) revealed a role for strigolactone (SL) signalling in the regulation of secondary growth conserved among species. Exogenous application of artificial SL (GR24) on stems of *Eucalyptus globulus* induced cambium division and *Arabidopsis* SL deficient mutants exhibited less radial growth while presenting high IAA levels and signalling (Agusti et al., 2011). The use of double mutants SL and IAA deficient suggested that SL function predominantly downstream of auxin signalling that positively regulates secondary growth.

2.6 Identifying new Aux/IAAs and ARFs as mediators of auxin signalling in wood formation

Recognition of hormonal signals may require receptors with varying affinities for the respective hormones, which, in the case of auxin, can be mediated through differential expression of Aux/IAAs (Bhalerao & Fischer, 2014). Interestingly, in both poplar and *Eucalyptus*, analyses of xylem gene networks revealed several Aux/IAAs as being tightly correlated with secondary wall TFs (Johnson et al., 2018; R. Ployet et al., unpublished). With exception of *PtIAA3* (Nilsson et al., 2008) whose overexpression reduces the width and the length for both fibre and vessels in transgenic lines in addition to a reduction of cambial cell division, none of the Aux/IAAs has been yet proven to be involved in wood formation. In *Arabidopsis*, ARFs 3 and 4 were recently pointed out as general promoters of cambial activity whereas ARF5 has a more specific role in attenuating the activity of WOX4 to stimulate differentiation (Brackmann et al., 2018). With the objective of identifying auxin signalling mediators involved in wood formation, a genome-wide identification of the ARFs and Aux/IAAs gene families was performed in *Eucalyptus grandis* (Yu et al., 2015, 2014). While the *E. grandis* has one of the largest proportion of tandem duplicated genes (34%; Myburg et al., 2014), these two families which contain 17 ARFs and 24 Aux/IAAs members, respectively, are not affected at all by tandem duplication. They contain much less genes than poplar (35 ARFs, 39 Aux/IAAs) and strikingly even less than *Arabidopsis* (23 ARFs and 29 Aux/IAAs). A combination of comparative phylogenetic analysis and large-scale gene expression profiling enabled the identification of several candidates. Functional characterization of *EgrIAA4* by overexpressing a stabilized version in *Arabidopsis* dramatically impeded plant growth and fertility and induced auxin-insensitive phenotypes. The lignified secondary walls of the interfascicular fibres appeared very late, whereas those of the xylary fibres were virtually undetectable, suggesting that EgrIAA4 may play crucial roles in fibre development and secondary cell wall deposition

(Yu et al., 2015). EgIAA20 and EgIAA9 had a similar inhibiting effects on fibres although the later also triggered prominent vessel formation in secondary xylem (H. Cassan-Wang, unpublished).

2.7 Role of ethylene in xylem cell expansion and maturation

The exogenous application of ethylene stimulates cambium activity but not in the ethylene-insensitive transgenic poplar obtained by overexpressing a dominant negative *Arabidopsis* ethylene receptor *ETR1* (Love et al., 2009). Consistent with these observations, the overexpression of ethylene biosynthesis gene *PttACO1* (ACC oxidase) stimulates cambium proliferation in *Populus*. The authors also showed that ethylene is at the origin of the eccentric cambial activity which gives rise to reaction wood formation in response to leaning.

More recently, Seyfferth et al. (2018) started to elucidate the ethylene signalling pathway in *Populus* trees and identified several putative downstream targets. In silico analysis of the AspWood transcriptome database (http://aspwood.popgenie.org/), which covers all stages of secondary growth in aspen stems, revealed that the ethylene precursor 1-aminocyclopropane-1-carboxylic acid (ACC) is synthesized during xylem expansion and cell maturation. More interestingly, ethylene-mediated transcriptional reprogramming occurs during all stages of secondary growth. The authors also identified new putative regulatory hub genes like *EIN3D* (*ETHYLENE INSENSITIVE 3D*) and 11 *ERFs* (*ETHYLENE RESPONSE FACTORS*) which, for most of them, were connected for the first time to wood formation (Seyfferth et al., 2018).

3. Main actors of the SW regulatory network: NACs and MYBs

The deposition of lignified SWs is a major step in the differentiation of secondary xylem. It is of prime importance for wood for fulfilling its essential roles in trees. It is also determinant for wood end-uses by mankind. In the last two decades, tremendous progress had been made in our understanding of the transcriptional regulation of this strictly coordinated spatiotemporal process thanks to pioneer work using the *Zinnia* and *Arabidopsis* trans-differentiation systems as well as using reverse and forward genetics in the model plant *Arabidopsis* (for reviews, see Demura & Fukuda, 2007; Grima-Pettenati, Soler, Camargo, & Wang, 2012; Hussey, Mizrachi,

Creux, & Myburg, 2013; Nakano, Yamaguchi, Endo, Rejab, & Ohtani, 2015; Schuetz et al., 2013; Umezawa, 2009; Wang & Dixon, 2011; Zhao & Dixon, 2011; Zhong, Lee, et al., 2011; Zhong, McCarthy, Lee, & Ye, 2011; Zhong & Ye, 2009, 2015).

In *Arabidopsis*, a complex hierarchical regulatory network has been proposed to control SW deposition in which the transcription factors NAC (NAM/ATAF/CUC) and MYB (MYeloBlastosis) act as first- and second-level master switches, respectively, to regulate a battery of downstream transcription factors and secondary cell wall biosynthesis genes (Schuetz et al., 2013; Wang & Dixon, 2011). Strong evidence also supports an *Arabidopsis*-like transcriptional cascade in woody angiosperm species (Hussey et al., 2013; Zhong, Lee, et al., 2011; Zhong, McCarthy, et al., 2011; Zhong & Ye, 2015). However, the growing number of functional genomic studies performed in trees highlighted differences in the regulation system for SW thickening between annual herbaceous plants and woody angiosperms. Our intent is to highlight the most significant and recent results in this rapidly evolving area, which illustrate the high complexity of the transcriptional regulation of the SW in trees. We will review recent progress on the functional characterizations of poplar and eucalypts NACs and MYBs and of the emerging mechanisms regulating their activities.

3.1 Wood-associated NAC domain transcription factors

The discovery of the NAC-domain TF genes as the key players regulating the complex transcriptional network leading to wall-thickening cell differentiation represents a milestone in our understanding of the regulation of SW (Kubo et al., 2005).

3.1.1 Top-level NAC master regulators

In *Arabidopsis*, there is clear separation in the expression patterns of the seven VNDs (Vascular-Related NAC Domain: AtVND1 to AtVND7) which are preferentially expressed in xylem vessels and the five NAC genes named either NSTs (NAC Secondary Wall thickening promoting factor) or SNDs (Secondary Wall-Associated NAC Domain: AtNST1, AtNST2 and AtNST3/AtSND1/ANACO12, SND2 and SND3) which are mostly expressed in interfascicular fibres. AtVNDs induce differentiation of vessels whereas NSTs/SNDs regulate SW deposition in fibres (Mitsuda et al., 2007; Mitsuda, Seki, Shinozaki, & Ohme-Takagi, 2005; Yamaguchi, Kubo, Fukuda, & Demura, 2008; Zhong, Demura, & Ye, 2006).

Overexpression of *AtVND6* or *VND7* was shown to induce xylem vessel transdifferentiation both in *Arabidopsis* and in poplar (Kubo et al., 2005; Yamaguchi et al., 2010), suggesting that the molecular mechanism of xylem vessel differentiation is, at least partially, conserved between these two species.

The same poplar genes were named differently depending on the authors (see Table 1) as WNDs (for wood-associated NAC domain transcription factors, Zhong et al., 2010a, 2010b) or VNSs (for VND, NST/SND-, SOMBRERO-related proteins; Ohtani et al., 2011) or PtrSNDs/PtrVNDs (Johnsson et al., 2018; Li et al., 2012). To help the reader finding his way in this confusing nomenclature, we reported in Table 1, the short names of the putative co-orthologs of the *Arabidopsis* genes in poplar and eucalypts, given by different research groups. The poplar orthologs of the *VNDs* and/or *NSTs/SNDs* genes could complement the SW defects of the fibres in the double *nst1/(snd1/nst3/anac012) Arabidopsis* mutant (Zhong & Ye, 2010). However, surprisingly only *PtrWND2B* and *PtrWND6B* were able to induce ectopic deposition of SW when overexpressed in *Arabidopsis* (Zhong et al., 2010a, 2010b). To explain the fact that other VNDs or NST/SNDs members could complement the *nst1/snd1* mutant but are not capable of inducing ectopic expression of SWs, a likely hypothesis is that they need to cooperate with co-factors or other TFs which are only present in cells programmed to be sclerified.

It was initially reported that the fibre- or vessel-specific expression occurring in *Arabidopsis* was not occurring in poplar where all *WNDs/VNS* (both *VND* and *NST/SND*) were expressed in both developing vessels and fibres as well as in xylem ray parenchyma cells. The clear separation of the expression patterns of *VND* and *NST/SND* groups in *Arabidopsis* did not seem to be extensively shared with other plant species including poplar, rice, and maize (Nakano et al., 2015; Ohtani et al., 2011; Zhong, Lee, et al., 2011; Zhong et al., 2010a, 2010b; Zhong, McCarthy, et al., 2011). However, the recent high spatial-resolution RNA sequencing data spanning the secondary phloem, vascular cambium, and wood-forming tissues of *Populus tremula* (Johnsson et al., 2018; Sundell et al., 2017) provided new clues about the expression patterns of the genes of the VNDs and NSTs/SNDs clades which are more complex and subtle than previously thought. Considering cell distance from cambium as a proxy for cell age, Johnsson et al. (2018) showed that *VND6*-orthologs are induced in recently divided cambial xylem initials, while *SND1/NST1* orthologs are expressed in early xylem expansion zone. The expression patterns within the paralogous pairs of

PttSND1, *PttNST1*, and *PttVND6* are highly similar whereas divergent expression profiles were observed within the *PttVND3* and *PttVND7* pairs. Both of the *PttVND7* paralogs are expressed in primary xylem. *PttVND7-1* is induced in the end of the maturation/programmed cell death zone indicating neofunctionalization of this paralog whereas *PttVND7-2* is not detectable in the secondary xylem (Johnsson et al., 2018). These authors further investigated the differences between the PtrSND and PtrVND clades by generating a co-expression network. They showed that *PtrVND6* orthologs together with *PtrVND3-2* formed a cluster connected to the bulk of pectin and xyloglucan biosynthetic genes as well as three primary wall-associated cellulose synthases *PtrCESAs*. *PtrSND1* and *PtrNST1* orthologs were most closely connected to secondary wall-associated *PtrCESAs*.

Using PtrWND2B/6B as tools, Zhong, McCarthy, et al. (2011) have uncovered a suite of up-regulated TFs, many of which were not reported previously. They further showed that PtrWNDs/VNS bind directly to SNBE sites in the promoters of their target genes like *Arabidopsis* SND1 (Zhong, Lee, et al., 2011; Zhong, McCarthy, et al., 2011).

More work is needed to decipher the target genes of the different members of the two clades (VND and NST/SND) since indirect evidence supports the occurrence of a divergence in target genes between clades. Indeed, the identification of co-varying neighbouring genes showed that if there are some common genes shared in the neighbourhoods of members of the same clade, either VND or NST/SND, there is no overlap between members of the two distinct clades (Johnsson et al., 2018). Using *Populus trichocarpa* protoplasts from stem-differentiating xylem, Lin et al. (2013) have started to describe the SND/VND-directed functional hierarchical genetic regulatory network (hGRN) for wood formation. They monitored genome-wide PtrSND1-B1 (PtVNS9/PtrWND2A)-induced gene transactivation responses and used a novel computational method for constructing a hierarchical TF-DNA network involving 76 direct targets, including 8 TFs and 61 enzyme-coding genes previously unidentified as targets (Lin et al., 2013).

3.1.2 Second level NAC proteins involved in wood formation: SND2/SND3

Downstream the top-level NAC regulators, *AtSND2* and *AtSND3*, expressed in SW-associated tissues, are direct target of AtSND1 and function as second level regulators for SW formation (Zhong, Lee, Zhou, McCarthy, & Ye, 2008). Overexpression of *AtSND2* in *Arabidopsis* stems produced a specific increase in xylem fibre wall thickening (Zhong et al., 2008) whereas overexpression of one of the four putative *SND2* co-orthologs in poplar, *PopNAC154*, resulted in a decrease in height and an increase in the proportion of bark to xylem in poplar trees, with no perceptible effect on SW thickness (Grant et al., 2010). This apparent discrepancy between the *SND2* overexpression phenotype in *Arabidopsis* and poplar was initially interpreted as illustrating differences in the regulatory function of *SND2* orthologs between herbaceous and woody plants. However, several lines of evidence suggest that it is more complex than only reflecting species differences. First, overexpression of *AtSND2* in *Eucalyptus* increased the thickness of the SW in fibre cells (Hussey et al., 2011). Second, overexpression of a dominant repressor form of a *SND2* poplar ortholog (PtSND2-SDRX) repressed wood formation and xylem fibres thickness in transgenic poplar lines (Wang et al., 2013). It should be noted that *Eucalyptus* contains one single ortholog of *AtSND2* (*EgrNAC170*, Hussey et al., 2015) whereas poplar has four co-orthologs that may have undergone sub/neofunctionalization. Third, excessive levels of *AtSND2* overexpression have a negative effect on interfascicular fibres SW deposition in *Arabidopsis* (Hussey et al., 2011). The authors postulated that *SND2* overexpression could increase SW deposition within a limited range of overexpression levels, possibly relying on the abundance of additional regulator proteins. The mode of regulation of *SND2* seems complex and requires further investigation.

Recently Johnsson et al. (2018) showed that poplar orthologs of *AtSND2* and *AtSND3* all peak into the xylem maturation zone. While both *PtrSND2* orthologs cluster with the *PtrSND1* and *PtrNST1* orthologs and are closely connected to SW *PtrCESAs*. The *PtrSND3* orthologs sub-cluster distinctly from the others together with primary wall *PtrCESA*s and other cell wall biosynthetic genes. Strikingly, *PtrSND3* orthologs were negatively correlated with the vast majority of the genes in their cluster.

Given the severe growth phenotypes observed in poplar lines either overexpressing *PopNAC154* (Grant et al., 2010) or its dominant repressive form (Wang et al., 2013) and the fact that *PopNAC154* was down-regulated in tension wood (Andersson-Gunnerås et al., 2006; Grant et al., 2010). Jervis et al. (2014) performed a metabolomic study in field-grown poplar overexpressing *PopNAC154*. They observed a substantial increase in arginine content which could be a marker for earlier growth cessation suggesting that *PopNAC154* could also play a role in senescence/dormancy-associated processes. Wang, Li, et al. (2014) and Wang, Tang, et al. (2014) further investigated the phenotypes (reduced growth and xylem formation) caused by dominant repression of *PtSND2* (*PopNAC154*), and found a reduction of auxin biosynthesis, transport and signalling in these transgenic poplar lines.

3.1.3 Other NAC Proteins XND1

XYLEM NAC DOMAIN 1 (XND1) is a negative regulator of SW and programmed cell death in *Arabidopsis* (Zhao, Avci, Grant, Haigler, & Beers, 2008). Its overexpression results in extreme dwarfism associated with the lack of SW synthesis and programmed cell death in xylem vessels. Poplar has four *XND1* co-orthologs *PopNAC118*, *PopNAC122*, *PopNAC128*, *PopNAC129*, each of them exhibiting distinct expression patterns. The overexpression of *PopNAC122* or *PopNAC129* in *Arabidopsis* phenocopied *AtXND1* overexpression in *Arabidopsis* (Grant et al., 2010). When *AtXND1* was overexpressed in poplar vessel diameter and number, plant stature and leaf size were reduced in severely affected lines. The effects of *XND1* overexpression in poplar did not precisely phenocopy those observed in *Arabidopsis* but in both genetic backgrounds overexpression of *XND1* was associated with reduced plant size and inhibition of differentiation of a subset of vascular cell types.

3.2 SW-associated MYB transcription factors

3.2.1 Second-level master regulators

In *Arabidopsis*, *AtMYB46/83* are considered as the unique second-level master regulators since they are the direct targets of top-level VND and NST/SND master switches and are able to activate the promoters of all the three major SW polymers (i.e. cellulose, hemicelluloses and lignins). In poplar, the *AtMYB46/83* co-orthologs, i.e., *PtrMYB2*, *PtrMYB3*, *PtrMYB20*, and *PtrMYB21*, are direct targets of PtWND2B/6B (Zhong, McCarthy, et al., 2011). As their *Arabidopsis* counterparts, they are also capable of activating the biosynthesis pathways of cellulose, xylan, and lignin, leading to ectopic SW deposition when overexpressed in *Arabidopsis* and poplar. Their dominant repression results in a reduction of SW thickening in transgenic poplar wood (McCarthy et al., 2010; Zhong, McCarthy, Haghighat, & Ye, 2013). EgMYB2 from *Eucalyptus* (Goicoechea et al., 2005) is also an ortholog of AMYB46/83 since it is able to activate the entire secondary wall biosynthesis programme when overexpressed and can complement the *Arabidopsis* mutant *atmyb46/83* (Zhong et al., 2010a).

Interestingly, overexpression of *BplMYB46* from *Betula platyphylla* a close ortholog of *PtrMYB2/3/20/21* not only increases lignin deposition, SW thickness and the expression of SW biosynthesis genes but also improves salt and osmotic tolerance in transgenic birch plants (Guo et al., 2017). It should be noted that its overexpression induces overexpression of an ortholog of *AtMYB52* known to confer ABA hypersensitivity and drought tolerance in *Arabidopsis* (Park, Kang, & Kim, 2011). *BplMYB46* is one of the first examples highlighting the cross-talk between the abiotic stress and secondary cell wall biosynthesis pathways in trees (Guo et al., 2017).

One important difference between *Arabidopsis* and poplar is that other targets of PtWND2B/6B than the orthologs of *AtMYB46/83* (*PtrMYB 2, 3, 20* and *21*) are capable of inducing the whole SW transcriptional programme and may function as master switches (Zhong, McCarthy, et al., 2011). Among those, *PtrMYB18* (ortholog of *AtMYB20/43*), *PtrMYB75/92/125/199* (co-orthologs of *AtMYB42/85*), *PtrMYB10/128* (co-orthologs of *AtMYB103*), *PtrMYB74* and *PtrMYB121* can activate the promoter activities of several biosynthesis genes for cellulose, xylan, and lignin (Zhong, Lee, et al., 2011; Zhong, McCarthy, et al., 2011). *PtrMYB74* and *PtrMYB121* were among the 13 targets induced by PtWND2B, for which no ortholog of *Arabidopsis* was previously shown to be involved in the regulation of SW biosynthesis. It is possible that additional master switches have been recruited in poplar to sustain a robust expression of secondary wall biosynthesis genes during wood formation, which requires the deposition of a massive amount of secondary wall components. Another hypothesis is that these genes have specific cell localization and/or have diversified their functions for being involved in responses to stresses. Examples of such a diversification are emerging. *BplMYB46* (Guo et al., 2017) and *PtoMYB170* (Xu et al., 2017, see below) are both at the cross-road between SW formation and response to drought.

Among the poplar wood-associated MYB having diversified their transcriptional regulatory activities from those of their *Arabidopsis* counterparts is *PtrMYB128* (an ortholog of *AtMYB103*). PtrMYB128 is capable to activate the promoters of the biosynthetic genes for all three secondary wall components in transient transactivation assays (Zhong, Lee, et al., 2011; Zhong, McCarthy, et al., 2011). AtMYB103 was first shown to preferentially induce the expression of genes for the biosynthesis of cellulose but not xylan and lignin (Zhong et al., 2008). However, the characterization of two *myb103 Arabidopsis* T-DNA insertion mutants revealed that the main modification at the SW level was a change in lignin monomeric composition (decrease in the S/G ratio) in line with *FERULATE 5 HYDROXYLASE (F5H)* being the main target of AtMYB103 (Öhman et al., 2013).

PtrMYB152 and its paralog *PtrMYB18* are orthologs of the *Arabidopsis* paralogs *AtMYB43* and *AtMYB20*, respectively (Wang, Li, et al., 2014; Wang, Tang, et al., 2014). Both *AtMYB43* and *AtMYB20* were identified as targets of AtSND1 but their functions in SW biosynthesis programme have not been characterized. *PtrMYB18* was identified as up-regulated in response to PtrWND2B, but not *PtrMYB152*. When overexpressed in *Arabidopsis*, *PtrMYB152* increased secondary cell wall thickness, likely caused by increased lignification. Accordingly, elevated expression of genes encoding sets of enzymes in secondary wall biosynthesis was observed (Wang, Li, et al., 2014; Wang, Tang, et al., 2014).

3.2.2 Lignin-specific MYBs

In *Arabidopsis*, three lignin specific MYB TFs have been identified: AtMYB58, AtMYB63 and AtMYB85 (Zhong & Ye, 2009). PtrMYB28 is able to activate specifically lignin biosynthesis genes as its orthologs AtMYB58/MYB63 (Zhong & Ye, 2009). Overexpression of PtoMYB*92*, an ortholog of *MYB85/MYB42* resulted in an increase in SW thickness in stems and ectopic deposition of lignin in leaves. PtoMYB92 specifically activates the expression of lignin biosynthetic genes. Thus, PtoMYB92 is involved in the regulation of SW formation in poplar by controlling the biosynthesis of monolignols as its *Arabidopsis* counterpart (Li et al., 2015).

The respective roles of the two paralogs *PtoMYB170* and *PtoMYB216* (co-orthologs of *AtMYB61*) were examined by Tian et al. (2013) and Xu et al. (2017). Using overexpression in poplar, they first showed that *PtoMYB216* and *PtoMYB170* have conserved functions in regulating lignin biosynthesis. Both TFs specifically activate the expression of the upstream genes in the lignin biosynthetic pathway which results in ectopic deposition of lignin in cells that are normally non-lignified (Tian et al., 2013). However, the CRISPR/Cas9 mutant of *PtoMYB170* displays pendent stem phenotype resulting from reduced lignin deposition and more flexible SW of xylem fibres. This strong phenotype suggests that the functional redundancy between PtoMYB216 and PtoMYB170 is weak because PtoMYB216 cannot replace the function of PtoMYB170 in lignin deposition (Xu et al., 2017). Interestingly, heterologous expression of *PtoMYB170* in transgenic *Arabidopsis* enhanced stomatal closure in the dark and resulted in drought tolerance of the transgenic plants through reduced water loss. *PtoMYB170* was specifically expressed in guard cells of transgenic *Arabidopsis* while *PtoMYB216* was not. These findings illustrate of a neofunctionalization of one of the two paralogs which acquired a new function in enhancing drought tolerance. After BplMYB46 (Guo et al., 2017) PtoMYB170 is the second example of tree SW-TF able to integrate both developmental and stress signals.

3.2.3 Negative regulators

SW formation not only involves transcriptional activators but also entails transcriptional repressors. *Eucalyptus* EgMYB1 represses the expression of secondary wall biosynthesis genes and inhibits secondary wall thickening in fibres when overexpressed in *Arabidopsis* and poplar, suggesting that it is a master transcriptional repressor of secondary wall formation (Legay et al., 2007, 2010). Although, *EgMYB1* is a close ortholog of *AtMYB4*, their respective functions are different since *AtMYB4* has never been reported to regulate SW synthesis but instead it is known to regulate sinapate esters accumulation through its direct target C4H (cinnamate 4-hydroxylase, Jin et al., 2000).

The two *EgMYB1*'s orthologs in *Populus* have been functionally characterized in independent studies. *PdMYB221* was overexpressed in *Arabidopsis* (Tang et al., 2015) and *PtoMYB156* was overexpressed and down-regulated by genome editing in poplar (Yang et al., 2017). Both genes were shown to negatively regulate the secondary wall thicknesses of xylem fibres and the content of cellulose, lignin and hemicelluloses but PtoMYB156 was also shown to repress phenylpropanoid biosynthesis genes, leading to a reduction in the amounts of total phenolic and flavonoid compounds (Yang et al., 2017). Whether PdMYB221 also regulates genes of the phenylpropanoid is still to be determined.

3.3 New NAC and MYB candidates putatively involved in the control of wood formation in trees

The availability of the *E. grandis* genome, the second hardwood forest tree whose genome was sequenced (Myburg et al., 2014), enabled to perform comparative phylogeny studies with woody and nonwoody species. Although the *E. grandis* genome has the very high number of tandem duplicated genes (33%, Myburg et al., 2014), *Eucalyptus* orthologs of the main NAC and R2R3-MYBs controlling secondary cell wall formation and lignin biosynthesis are, in general, not duplicated. For instance, there are single putative *Eucalyptus* orthologs of *Arabidopsis* fibre-associated SND1, SND2 and NST1, vessel-VND6, VND7, MYB46/83 (Hussey et al., 2015; Soler et al., 2015). This contrasts to poplar who underwent recent whole genome duplication (Tuskan et al., 2006) and has in general two duplicated orthologs of one *Arabidopsis* SW main regulatory gene (Tables 1 and 2). In both NAC and MYB families, there are clades exhibiting a high number of tandem duplicated genes. Among those, the "woody-expanded" showing an expansion of members from woody species and the "woody-preferential"

Table 2 MYB transcription factors from angiosperm trees involved in SW formation.

Woody species	TFs	Gene ID	Putative *Arabidopsis* ortholog	Target genes			References
				Lignin	Cellulose	Xylan	
Betula platyphylla	BplMYB46	KP711284	AtMYB46	BplPAL, Bpl4CL, BplCCoAOMT, BplLAC, BplCCR	BplCesA	BplFRA, BplIRX	Guo et al. (2017)
Eucalyptus gunnii	EgMYB1	CAE09058	AtMYB4	EgCCR, EgCAD2, PtrPAL2, Ptr4CL3, PtrCOMT2, PtrHCT1, PtrCCR2, PtrCAD1, PtrC3H3, PtrC4H1, PtrC4H2	PtrCesA1,2,3	PtrFRA8, PtrIRX8	Legay et al. (2007, 2010)
	EgMYB2	AJ576023	AtMYB46	NtPAL, NtC4H, Nt4CL, NtC3H, NtHCT, NtCOMT, NtF5H			Goicoechea et al. (2005)
	EgMYB88	KX470407	NO	PttPAL, Ptt4CL, PttCAld5H			Soler et al. (2017)
Populus deltoides	PdMYB221	Potri.004G174400.1	AtMYB4	PdCOMT2, PdCCR1	PdCesA7/8	PdGT47C	Tang et al. (2015)
Populus trichocarpa	PtrMYB2	Potri.001G258700	AtMYB46/83	At4CL1, AtCCoAOMT, PtrCOMT1, PtrCCoAOMT1	PtrCesA4,8,17	PtrGT43B, PtrGT47e, PtrIRX10	McCarthy et al. (2010), Zhong et al. (2013)
	PtrMYB3	Potri.001G267300	AtMYB46/83	PtrCOMT1, PtrCCoAOMT1	PtrCesA4,8,17	PtrGT43B, PtrGT47e, PtrIRX10	McCarthy et al. (2010), Zhong et al. (2013)

Continued

Table 2 MYB transcription factors from angiosperm trees involved in SW formation.—cont'd

Woody species	TFs	Gene ID	Putative *Arabidopsis* ortholog	Target genes: Lignin	Target genes: Cellulose	Target genes: Xylan	References
	PtrMYB20	Potri.009G061500	AtMYB46/83	PtrCOMT1, PtrCCoAOMT1	PtrCesA4,8,17	PtrGT43B, PtrGT47e, PtrIRX10	McCarthy et al. (2010), Zhong et al. (2013)
	PtrMYB21	Potri.009G053900	AtMYB46/83	PtrCOMT1, PtrCCoAOMT1	PtrCesA4,8,17	PtrGT43B, PtrGT47e, PtrIRX10	McCarthy et al. (2010), Zhong et al. (2013)
	PtrMYB152	Potri.017G130300	AtMYB43/20	At4CL1, AtHCT, AtC3H1, AtCCoAOMT, AtCAD6, PtrPAL4, Pt4CL3, Ptr4CL5, PtrC4H2, PtrCCoAOMT1, PtrCCR2, PtrHCT1	PtrCesA2B,3A	PtrGT43B,D	Li et al. (2014), Wang, Li, et al. (2014), Wang, Tang, et al. (2014)
	PtrMYB28	XM_002307154	AtMYB58/63	At4CL			Zhong and Ye (2009)
Populus tomentosa	PtoMYB92	KP710214	AtMYB42/85	PtoCCOAMT1, PtoCCR2, PtoC3H3		PtoGT8D, PtoGT43B	Li et al. (2015)
	PtoMYB170	KY114929	AtMYB61	PtoCCR2, PtoCCOAOMT1, PtoCOMT, PtoC4H2			Xu et al. (2017)
	PtoMYB216	JQ801749	AtMYB61	PtoPAL4, Pto4CL5, PtoC4H2, PtoC3H3	PtoCesA3A	PtrGT43B,D	Tian et al. (2013)

containing neither *Arabidopsis* nor monocot genes. Combining comparative phylogeny to large scale expression profiling, enabled to point out new NAC and MYB candidate putatively involved in the control of wood formation in trees (Hussey et al., 2015; Soler et al., 2015). Of particular interest are members expressed in vascular cambium and/or differentiating xylem with no ortholog gene in *Arabidopsis*. Among these candidates, *EgMYB88* which belongs to one woody subgroups and is highly and preferentially expressed in vascular cambium was functionally characterized in poplar (Soler et al., 2016). The results suggest that EgMYB88 is an activator of the biosynthesis of some phenylpropanoid-derived secondary metabolites including lignin in cambium and in the first layers of differentiating xylem.

4. Post-transcriptional and post-translational regulations of the SW regulators

4.1 Alternative splicing regulates cell wall thickening

Post-transcriptional regulation plays an important role in modulating VNS/WND activity. A major breakthrough was achieved by Li et al. (2012) who reported the discovery of a stem-differentiating xylem (SDX)-specific alternative SND1 splice variant, PtrSND1-A2^{IR}, that acts as a dominant negative of SND1 transcriptional network genes in poplar. PtrSND1-A2^{IR} which derives from PtrSND1-A2 (also called PtrWND1B/PtrVNS11) lacks DNA binding and transactivation abilities but retains dimerization capability. PtrSND1-A2 and PtrSND1-A2^{IR} can form heterodimers with other PtrSND members. PtrSND1-A2^{IR} which is localized exclusively in cytoplasm is translocated into the nucleus exclusively as a heterodimeric partner with full-size PtrSND1s, where it represses the expression of the *PtrSND1* member genes and their target gene PtrMYB021. This was the first report in plants of the repression of the autoregulation of a TF family by its only splice variant. Zhao, Sun, Xu, Zhang, and Li (2014) further showed that PtrSND1-A2 overexpression enhanced fibre cell wall thickening, while overexpression of PtrWND1B-l (PtrSND1-A2^{IR}) repressed fibre cell wall thickening. Interestingly, its ortholog in *Eucalyptus* (Eugr.E01053) also undergoes a form with a retention of intron 2 whereas such event does not exist in *Arabidopsis*.

Very recently, Lin et al. (2017) reported the discovery of another splice variant, PtrVND6-C1^{IR} family derived from PtrVND6-C1 (PtVNS01/PtrWND5A). It suppresses the protein functions of all PtrVND6 family members. PtrVND6-C1^{IR} also suppresses the expression of all PtrSND1 members, including PtrSND1-A2, demonstrating that PtrVND6-C1^{IR} is the previously unidentified regulator of PtrSND1-A2. Similarly, PtrVND6-C1^{IR} cannot suppress the expression of its cognate transcription factor, PtrVND6-C1. PtrVND6-C1 is suppressed by PtrSND1-A2^{IR}. Both PtrVND6-C1^{IR} and PtrSND1-A2^{IR} cannot suppress their cognate transcription factors but can suppress all members of the other family, indicating that the splice variants from the PtrVND6 and PtrSND1 family may exert reciprocal cross-regulation for complete transcriptional regulation of these two families in wood formation, providing a higher level of regulation to maintain homeostasis for plants to avoid abnormal growth and development.

The importance of alternative splicing (AS) in the regulation of wood formation deserves more attention since 36% of wood-expressed genes were reported to be alternatively spliced in the xylem transcriptome in *P. trichocarpa* (Bao et al., 2013). Xu et al. (2014) found that 28.3% and 20.7% of the highly expressed transcripts were affected by AS events in developing xylem of *Populus* and *Eucalyptus,* respectively. Around 25% of the AS events may cause protein domain modification. About 30% of AS-occurring genes were putative orthologs and 71 conserved AS events were identified in the two species (Xu et al., 2014).

4.2 Protein–protein interactions

Despite the importance of combinatorial control in transcriptional regulation in plants, there are only a handful of examples of the involvement of protein–protein interactions in the network controlling secondary cell wall formation. With the exception of the work on PtrVND and SND/NST homo- or heterodimerization (Li et al., 2012; Shi et al., 2017) most of the studies were performed in *A. thaliana* (Bhargava et al., 2013; Li et al., 2012; Liu & Douglas, 2015; Liu et al., 2014; Yamaguchi et al., 2010). However, recently Soler et al. (2016) found that the *Eucalyptus* transcription factor EgMYB1, which is known to repress lignin biosynthesis, interacts specifically with a linker histone variant, EgH1.3. This interaction enhances the repression of EgMYB1's target genes, strongly limiting the amount of lignin deposited in xylem cell walls. The expression profiles of *EgMYB1* and *EgH1.3* overlap in xylem cells at early stages of their differentiation as well as in mature parenchymatous xylem cells, which have no or only thin lignified secondary cell walls. This suggests that a complex between EgMYB1 and EgH1.3 integrates developmental signals to prevent premature or inappropriate lignification of secondary cell walls, providing a mechanism to fine-tune the differentiation of xylem cells in time and space (Soler et al., 2017).

With the aim of deciphering the *Populus* wood interactome, Petzold et al. (2017) have used yeast-two-hybrid and discovered 165 novel protein–protein interactions and connected networks for 162 distinct genes. Some of these interactions were previously reported but most were novel and relevant to wood formation since they involved proteins known to be involved in SW synthesis (Petzold et al., 2017). This work will provide an excellent framework to further functionally characterize relevant interactions.

5. Concluding remarks

Within the last decade, and particularly within the 5 last years, our understanding of the transcriptional regulation of wood formation in angiosperm trees has substantially progressed. The increasing number of functional characterizations of trees SW-associated TFs revealed that *Arabidopsis* has been instrumental in deciphering conserved mechanisms involved in secondary xylem differentiation. However, substantial differences exist between trees and the model plant and we cannot further rely on simple extrapolations to decipher the transcriptional network underlying wood formation in trees given its higher level of complexity.

A growing number of examples illustrate the diversification of functions of tree wood-associated TFs as compared to their putative orthologs in *Arabidopsis* (Legay et al., 2010; Nakano et al., 2015; Zhong, McCarthy, et al., 2011). Most of the functional studies have focussed on potential orthologs of *Arabidopsis* genes and few investigated the function of TFs absent from the *Arabidopsis* genome (Soler et al., 2015). Those expressed preferentially in vascular cambium or differentiating xylem deserve more attention since they likely play specific roles in trees (Hussey et al., 2015; Soler et al., 2015).

The overexpression strategy has been and still very useful but showed its limits in deciphering the function of wood-associated TFs. An illustrative example is AtMYB103 whose function determined thanks to *Arabidopsis* T-DNA mutants (Öhman et al., 2013) was shown to be different from that deduced from overexpressors (Zhong, Lee, et al., 2011; Zhong, McCarthy, et al., 2011). Implementation of genome editing technologies in poplar (e.g. CRISPR-Cas9) provides a means to engineer strong loss-of-function or null alleles, thereby complementing the lack of mutant in tree species. The powerfulness of this approach was recently highlighted in deciphering the function of two paralogs (PtoMYB126 and PtoMYB170) from *Populus tomentosa* (Xu et al., 2017). As an alternative to poplar transformation which is still a long process as compared to *Arabidopsis*, "Induced somatic sector analysis" (ISSA) is a rapid method implemented both in poplar and *Eucalyptus*, which was successfully used by Hussey et al. (2011) to functionally characterize SND2 in *Eucalyptus*. This method is particularly adapted to study cambium development dynamics in trees (Bossinger & Spokevicius, 2018). Another alternative approach is to use *Eucalyptus* transgenic hairy roots, a versatile and rapid system to dissect the function of genes involved in secondary wall biosynthesis (Plasencia et al., 2016) and in its regulation (Soler et al., 2016).

One exiting outcome from the recent characterization of wood-associated TFs involved in SW is their role in integrating environmental signals leading to increased tolerance to abiotic stresses, (i.e. drought and salinity stresses; Guo et al., 2017; Xu et al., 2017). It is likely that more examples will emerge in the near future as it was shown in *Arabidopsis* that several abiotic and nutritional stresses can co-opt the xylem regulatory network (Taylor-Teeples et al., 2015). Co-optation of a developmental network is a mean to facilitate adaptation to stress. If tolerance to stress is important for all plants because of their sessile lifestyle, it is crucial for long-living trees which have to cope with seasonal variations and climatic changes all along their life. Dissecting the mechanisms and the genes underlying stress tolerance at the wood level is an important challenge to address in the near future given the threat of climate changes on forest trees.

Remarkably, in poplar, alternative splicing (through intron retention) was highlighted as a key mechanism modulating the activities of the wood-associated NAC-domain TFs with reciprocal cross-regulation of the VND and SND clades (Li et al., 2012; Lin et al., 2017; Zhao et al., 2014). Similar splice variants were conserved in *Eucalyptus*, but none was found in *Arabidopsis*. Given the number of alternative splicing events in the wood transcriptome of both *Eucalyptus* and *Populus* (Bao et al., 2013; Xu et al., 2014) and the fact that abiotic stresses modulate landscape of poplar transcriptome via alternative splicing, intron retention and isoform ratio switching (Filichkin et al., 2018), the occurrence of alternative splicing in regulating the activities wood-associated TFs is one important direction to explore in the future. Besides more efforts need to be dedicated to deciphering the wood interactome since not all genes expressed in xylem are regulated at the transcriptional level, and even for those which are regulated at that level, they are often regulated at the post-transcriptional level (Soler et al., 2016). The cellular context in which occur interactions with partners is also an important aspect that needs to be determined.

The large high-resolution profiling of the cambial zone of aspen (Sundell et al., 2017), which is obviously not possible in *Arabidopsis*, has allowed to build co-expression gene networks or system biology approaches to investigate, for instance, the control of wood formation by auxin (Johnsson et al., 2018) and ethylene (Seyfferth et al., 2018). These powerful tools enabled to identify "hubs" and modules which point out hormone signalling candidates, and potential target genes and biological processes under the control of hormones during secondary growth in trees. Using such an approach, it was possible to establish a central role of the IAA/GA cross-talk in determining cell fate possibly through the control of SND/NST and VND master regulators (Johnsson et al., 2018).

The cross-talk between auxin and other hormones seems to be central at all stages of secondary xylem differentiation, i.e., from cambium division to the later stage of SW maturation. This exiting area needs more investigations to identify the genes and the mechanisms integrating this cross-talk, which is likely modulated by environmental signals. Technologies enabling fine spatial resolution will be instrumental in the future to get deeper insights in the regulation of cambial activity and of subsequent xylem differentiation steps.

Finally, new opportunities are created by the development of system genetic approaches and network modelling (Mizrachi & Myburg, 2016) which involves multilevel data integration from a population of genetically diverse individuals and focusses the role of underlying molecular networks in impacting the relationships between traits. Getting a more comprehensive view of the molecular networks underlying wood formation will surely impact tree breeding strategies and genetic engineering approaches to generate trees more suitable for industrial uses and more adapted to environmental changes.

Acknowledgements

The authors thank the University Paul Sabatier Toulouse III, the Centre National pour la Recherche Scientifique and the French Laboratory of Excellence project "TULIP" (ANR-10-LABX-41; ANR-11-IDEX-0002-02) for their financial support. E.L.O.C. acknowledges the PDE-CNPq (202228/2015-0) postdoctoral fellowship.

References

Agusti, J., Herold, S., Schwarz, M., Sanchez, P., Ljung, K., Dun, E. A., et al. (2011). Strigolactone signaling is required for auxin-dependent stimulation of secondary growth in plants. *Proceedings of the National Academy of Sciences of the United States of America*, *108*, 20242–20247.

Andersson-Gunnerås, S., Mellerowicz, E. J., Love, J., Segerman, B., Ohmiya, Y., Coutinho, P. M., et al. (2006). Biosynthesis of cellulose-enriched tension wood in *Populus*: Global analysis of transcripts and metabolites identifies biochemical and developmental regulators in secondary wall biosynthesis. *The Plant Journal*, *45*, 144–165.

Bao, H., Li, E., Mansfield, S. D., Cronk, Q. C. B., El-Kassaby, Y. A., & Douglas, C. J. (2013). The developing xylem transcriptome and genome-wide analysis of alternative splicing in *Populus trichocarpa* (black cottonwood) populations. *BMC Genomics*, *14*(1), 359.

Bhalerao, R. P., & Fischer, U. (2014). Auxin gradients across wood—Instructive or incidental? *Physiologia Plantarum*, *151*, 43–51.

Bhalerao, R. P., Fischer, U., & Turner, S. (2016). Environmental and hormonal control of cambial stem cell dynamics. *Journal of Experimental Botany*, *68*(1), 79–87. https://doi.org/10.1093/jxb/erw466.

Bhargava, A., Ahad, A., Wang, S., Mansfield, S. D., Haughn, G. W., Douglas, C. J., et al. (2013). The interacting MYB75 and KNAT7 transcription factors modulate secondary cell wall deposition both in stems and seed coat in *Arabidopsis*. *Planta*, *237*, 1199–1211.

Björklund, S., Antti, H., Uddestrand, I., Moritz, T., & Sundberg, B. (2007). Cross-talk between gibberellin and auxin in development of *Populus* wood: Gibberellin stimulates polar auxin transport and has a common transcriptome with auxin. *The Plant Journal*, *52*, 499–511.

Bossinger, G., & Spokevicius, A. V. (2018). Sector analysis reveals patterns of cambium differentiation in poplar stems. *Journal of Experimental Botany*, *69*(18), 4339–4348. https://doi.org/10.1093/jxb/ery230.

Brackmann, K., Qi, J., Gebert, M., Jouannet, V., Schlamp, T., Grünwald, K., et al. (2018). Spatial specificity of auxin responses coordinates wood formation. *Nature Communications*, *9*, 875.

Chaffey, N., Cholewa, E., Regan, S., & Sundberg, B. (2002). Secondary xylem development in *Arabidopsis*: A model for wood formation. *Physiologia Plantarum*, *14*, 594–600.

Déjardin, A., Laurans, F., Arnaud, D., Breton, C., Pilate, G., & Leplé, J. C. (2010). Wood formation in Angiosperms. *Comptes Rendus Biologies*, *333*, 325–334.

Demura, T., & Fukuda, H. (2007). Transcriptional regulation in wood formation. *Trends in Plant Science*, *12*(2), 64–70.

Eriksson, M. E., Israelsson, M., Olsson, O., & Moritz, T. (2000). Increased gibberellin biosynthesis in transgenic trees promotes growth, biomass production and xylem fiber length. *Nature Biotechnology*, *18*, 784–788.

Etchells, J. P., Mishra, L. S., Kumar, M., Campbell, L., & Turner, S. R. (2015). Wood formation in trees is increased by manipulating PXY-regulated cell division. *Current Biology*, *25*(8), 1050–1055.

Filichkin, S. A., Hamilton, M., Dharmawardhana, P. D., Singh, S. K., Sullivan, C., Ben-Hur, A., et al. (2018). Abiotic stresses modulate landscape of poplar transcriptome via alternative splicing, differential intron retention, and isoform ratio switching. *Frontiers in Plant Science*, *9*, 1–17.

Goicoechea, M., Lacombe, E., Legay, S., Mihaljevic, S., Rech, P., Jauneau, A., et al. (2005). EgMYB2, a new transcriptional activator from *Eucalyptus xylem*, regulates secondary cell wall formation and lignin biosynthesis. *The Plant Journal*, *43*, 553–567.

Grant, E. H., Fujino, T., Beers, E. P., & Brunner, A. M. (2010). Characterization of NAC domain transcription factors implicated in control of vascular cell differentiation in *Arabidopsis* and *Populus*. *Planta*, *232*, 337–352.

Grima-Pettenati, J., Soler, M., Camargo, E. L. O., & Wang, H. (2012). Transcriptional regulation of the lignin biosynthetic pathway revisited: New players and insights. *Advances in Botanical Research*, *61*, 173–218. https://doi.org/10.1016/B978-0-12-416023-1.00006-9.

Guo, H., Wang, Y., Wang, L., Hu, P., Wang, Y., Jia, Y., et al. (2017). Expression of the MYB transcription factor gene BplMYB46 affects abiotic stress tolerance and secondary cell wall deposition in *Betula platyphylla*. *Plant Biotechnology Journal*, *15*, 107–121.

Hussey, S. G., Mizrachi, E., Creux, N. M., & Myburg, A. A. (2013). Navigating the transcriptional roadmap regulating plant secondary cell wall deposition. *Frontiers in Plant Science, 4,* 1–21.

Hussey, S. G., Mizrachi, E., Spokevicius, A. V., Bossinger, G., Berger, D. K., & Myburg, A. A. (2011). SND2, a NAC transcription factor gene, regulates genes involved in secondary cell wall development in *Arabidopsis* fibres and increases fibre cell area in *Eucalyptus. BMC Plant Biology, 11,* 173.

Hussey, S. G., Saidi, M. N., Hefer, C. A., Myburg, A. A., & Grima-Pettenati, J. (2015). Structural, evolutionary and functional analysis of the NAC domain protein family in *Eucalyptus. The New Phytologist, 206,* 1337–1350.

Ibanes, M., Fabregas, N., Chory, J., & Cano-Delgado, A. I. (2009). Brassinosteroid signaling and auxin transport are required to establish the periodic pattern of *Arabidopsis* shoot vascular bundles. *Proceedings of the National Academy of Sciences of the United States of America, 106,* 13630–13635.

Immanen, J., Nieminen, K., Smolander, O. P., Kojima, M., Alonso Serra, J., Koskinen, P., et al. (2016). Cytokinin and auxin display distinct but interconnected distribution and signaling profiles to stimulate cambial activity. *Current Biology, 26,* 1990–1997.

Jervis, J., Hildreth, S. B., Sheng, X., Beers, E. P., Brunner, A. M., & Helm, R. F. (2014). A metabolomic assessment of NAC154 transcription factor overexpression in field grown poplar stem wood. *Phytochemistry, 115,* 112–120.

Jin, H., Cominelli, E., Bailey, P., Parr, A., Mehrtens, F., Jones, J., et al. (2000). Transcriptional repression by AtMYB4 controls production of UV-protecting sunscreens in *Arabidopsis. The EMBO Journal, 19,* 6150–6161.

Jin, Y. L., Tang, R. J., Wang, H. H., Jiang, C. M., Bao, Y., Yang, Y., et al. (2017). Overexpression of *Populus trichocarpa CYP85A3* promotes growth and biomass production in transgenic trees. *Plant Biotechnology Journal, 15,* 1309–1321.

Johnson, D., Eckart, P., Alsamadisi, N., Noble, H., Martin, C., & Spicer, R. (2018). Polar auxin transport is implicated in vessel differentiation and spatial patterning during secondary growth in *Populus. American Journal of Botany, 105,* 186–196.

Johnsson, C., Jin, X., Xue, W., Dubreuil, C., Lezhneva, L., & Fischer, U. (2018). The plant hormone auxin directs timing of xylem development by inhibition of secondary cell wall deposition through repression of secondary wall NAC-domain transcription factors. *Physiologia Plantarum,* 1–24. https://doi.org/10.1111/ppl.12766. (28 May 2018).

Kubo, M., Udagawa, M., Nishikubo, N., Horiguchi, G., Yamaguchi, M., Ito, J., et al. (2005). Transcription switches for protoxylem and metaxylem vessel formation. *Genes & Development, 19,* 1855–1860.

Kucukoglu, M., Nilsson, J., Zheng, B., Chaabouni, S., & Nilsson, O. (2017). WUSCHEL-RELATED HOMEOBOX4 (WOX4)-like genes regulate cambial cell division activity and secondary growth in *Populus* trees. *The New Phytologist, 215,* 642–657.

Legay, S., Lacombe, E., Goicoechea, M., Brière, C., Séguin, A., Mackay, J., et al. (2007). Molecular characterization of EgMYB1, a putative transcriptional repressor of the lignin biosynthetic pathway. *Plant Science, 173,* 542–549.

Legay, S., Sivadon, P., Blervacq, A. S., Pavy, N., Baghdady, A., Tremblay, L., et al. (2010). EgMYB1, an R2R3 MYB transcription factor from *Eucalyptus* negatively regulates secondary cell wall formation in *Arabidopsis* and poplar. *The New Phytologist, 188,* 774–786.

Li, Q., Lin, Y.-C., Sun, Y.-H., Song, J., Chen, H., Zhang, X.-H., et al. (2012). Splice variant of the SND1 transcription factor is a dominant negative of SND1 members and their regulation in *Populus trichocarpa. Proceedings of the National Academy of Sciences of the United States of America, 109*(36), 14699–14704.

Li, C., Wang, X., Lu, W., Liu, R., Tian, Q., Sun, Y., et al. (2014). A poplar R2R3-MYB transcription factor, PtrMYB152, is involved in regulation of lignin biosynthesis during secondary cell wall formation. *Plant Cell, Tissue and Organ Culture, 119*(3), 553–563. https://doi.org/10.1007/s11240-014-0555-8.

Li, C., Wang, X., Ran, L., Tian, Q., Fan, D., & Luo, K. (2015). PtoMYB92 is a transcriptional activator of the lignin biosynthetic pathway during secondary cell wall formation in *Populus tomentosa. Plant & Cell Physiology, 56*, 2436–2446.

Lin, Y.-C. J., Chen, H., Li, Q., Li, W., Wang, J. P., Shi, R., et al. (2017). Reciprocal cross-regulation of VND and SND multigene TF families for wood formation in Populus trichocarpa. *Proceedings of the National Academy of Sciences USA, 114*(45), E9722–E9729. https://doi.org/10.1073/pnas.1714422114.

Lin, Y. C., Li, W., Sun, Y. H., Kumari, S., Wei, H., Li, Q., et al. (2013). SND1 transcription factor-directed quantitative functional hierarchical genetic regulatory network in wood formation in *Populus trichocarpa. The Plant Cell, 25*, 4324–4341.

Liu, Y., & Douglas, C. J. (2015). A role for OVATE FAMILY PROTEIN1 (OFP1) and OFP4 in a BLH6-KNAT7 multi-protein complex regulating secondary cell wall formation in *Arabidopsis thaliana. Plant Signaling & Behavior, 10*(7), e1033126.

Liu, Y., You, S., Taylor-Teeples, M., Li, W. L., Schuetz, M., Brady, S. M., et al. (2014). BEL1-LIKE HOMEODOMAIN6 and KNOTTED ARABIDOPSIS THALIANA7 interact and regulate secondary cell wall formation via repression of REVOLUTA. *The Plant Cell, 26*, 4843–4861.

Love, J., Bjorklund, S., Vahala, J., Hertzberg, M., Kangasjarvi, J., & Sundberg, B. (2009). Ethylene is an endogenous stimulator of cell division in the cambial meristem of *Populus. Proceedings of the National Academy of Sciences of the United States of America, 106*, 5984–5989.

Mauriat, M., Le Provost, G., Rozenberg, P., Delzon, S., Breda, N., Clair, B., et al. (2014). Wood formation in trees. In K. G. Ramawat, J. M. Merillon, & M. R. Ahuja (Eds.), *Tree biotechnology* (p. 656). CRC Press.

Mauriat, M., & Moritz, T. (2009). Analyses of GA20ox- and GID1-over-expressing aspen suggest that gibberellins play two distinct roles in wood formation. *The Plant Journal, 58*, 989–1003.

McCarthy, R. L., Zhong, R., Fowler, S., Lyskowski, D., Piyasena, H., Carleton, K., et al. (2010). The poplar MYB transcription factors, PtrMYB3 and PtrMYB20, are involved in the regulation of secondary wall biosynthesis. *Plant & Cell Physiology, 51*, 1084–1090.

Mellerowicz, E. J., Baucher, M., Sundberg, B., & Boerjan, W. (2001). Unravelling cell wall formation in the woody dicot stem. *Plant Molecular Biology, 47*, 239–274.

Milhinhos, A., & Miguel, C. M. (2013). Hormone interactions in xylem development: A matter of signals. *Plant Cell Reports, 32*, 867–883.

Mitsuda, N., Iwase, A., Yamamoto, H., Yoshida, M., Seki, M., Shinozaki, K., et al. (2007). NAC transcription factors, NST1 and NST3, are key regulators of the formation of secondary walls in woody tissues of *Arabidopsis. The Plant Cell, 19*, 270–280.

Mitsuda, N., Seki, M., Shinozaki, K., & Ohme-Takagi, M. (2005). The NAC transcription factors NST1 and NST2 of *Arabidopsis* regulate secondary wall thickenings and are required for anther dehiscence. *The Plant Cell, 17*, 2993–3006.

Mizrachi, E., & Myburg, A. A. (2016). Systems genetics of wood formation. *Current Opinion in Plant Biology, 30*, 94–100.

Morris, H., Plavcová, L., Cvecko, P., Fichtler, E., Gillingham, M. A. F., Martínez-Cabrera, H. I., et al. (2016). A global analysis of parenchyma tissue fractions in secondary xylem of seed plants. *The New Phytologist, 209*, 1553–1565.

Myburg, A. A., Grattapaglia, D., Tuskan, G. A., Hellsten, U., Hayes, R. D., Grimwood, J., et al. (2014). The genome of *Eucalyptus grandis. Nature, 510*, 356–362.

Nakano, Y., Yamaguchi, M., Endo, H., Rejab, N. A., & Ohtani, M. (2015). NAC-MYB-based transcriptional regulation of secondary cell wall biosynthesis in land plants. *Frontiers in Plant Science, 6*, 288.

Nieminen, K., Blomster, T., Helariutta, Y., & Mähönen, A. P. (2015). Vascular cambium development. *The Arabidopsis Book, 13*, e0177.

Nieminen, K., Immanen, J., Laxell, M., Kauppinen, L., Tarkowski, P., Dolezal, K., et al. (2008). Cytokinin signaling regulates cambial development in poplar. *Proceedings of the National Academy of Sciences of the United States of America, 105*(50), 20032–20037. https://doi.org/10.1073/pnas.0805617106.

Nieminen, K., Robischon, M., Immanen, J., & Helariutta, Y. (2012). Towards optimizing wood development in bioenergy trees. *The New Phytologist, 194*, 46–53.

Nilsson, J., Karlberg, A., Antti, H., Lopez-Vernaza, M., Mellerowicz, E., Perrot-Rechenmann, C., et al. (2008). Dissecting the molecular basis of the regulation of wood formation by auxin in hybrid aspen. *The Plant Cell, 20*, 843–855.

Öhman, D., Demedts, B., Kumar, M., Gerber, L., Gorzsás, A., Goeminne, G., et al. (2013). MYB103 is required for FERULATE-5-HYDROXYLASE expression and syringyl lignin biosynthesis in *Arabidopsis* stems. *The Plant Journal, 73*(1), 63–76.

Ohtani, M., Nishikubo, N., Xu, B., Yamaguchi, M., Mitsuda, N., Goué, N., et al. (2011). A NAC domain protein family contributing to the regulation of wood formation in poplar. *The Plant Journal, 67*, 499–512.

Park, M. Y., Kang, J.-Y., & Kim, S. Y. (2011). Overexpression of AtMYB52 confers ABA hypersensitivity and drought tolerance. *Molecules and Cells, 31*, 447–454.

Petzold, H. E., Rigoulot, S. B., Zhao, C., Chanda, B., Sheng, X., Zhao, M., et al. (2017). Identification of new protein–protein and protein–DNA interactions linked with wood formation in *Populus trichocarpa*. *Tree Physiology, 38*, 362–377.

Plasencia, A., Soler, M., Dupas, A., Ladouce, N., Silva-Martins, G., Martinez, Y., et al. (2016). *Eucalyptus* hairy roots, a fast, efficient and versatile tool to explore function and expression of genes involved in wood formation. *Plant Biotechnology Journal, 14*(6), 1381–1393.

Růžička, K., Ursache, R., Hejátko, J., & Helariutta, Y. (2015). Xylem development—From the cradle to the grave. *The New Phytologist, 207*, 519–535.

Schuetz, M., Smith, R., & Ellis, B. (2013). Xylem tissue specification, patterning, and differentiation mechanisms. *Journal of Experimental Botany, 64*, 11–31.

Seyfferth, C., Wessels, B., Jokipii-Lukkari, S., Sundberg, B., Delhomme, N., Felten, J., et al. (2018). Ethylene-related gene expression networks in wood formation. *Frontiers in Plant Science, 9*, 272.

Shi, R., Wang, J. P., Lin, Y. C., Li, Q., Sun, Y. H., Chen, H., et al. (2017). Tissue and cell-type co-expression networks of transcription factors and wood component genes in *Populus trichocarpa*. *Planta, 245*, 927–938.

Smet, W., & De Rybel, B. (2016). Genetic and hormonal control of vascular tissue proliferation. *Current Opinion in Plant Biology, 29*, 50–56.

Soler, M., Camargo, E. L. O., Carocha, V., Cassan-Wang, H., San Clemente, H., Savelli, B., et al. (2015). The *Eucalyptus grandis* R2R3-MYB transcription factor family: Evidence for woody growth-related evolution and function. *The New Phytologist, 206*, 1364–1377.

Soler, M., Plasencia, A., Larbat, R., Pouzet, C., Jauneau, A., Rivas, S., et al. (2017). The Eucalyptus linker histone variant EgH1.3 cooperates with the transcription factor EgMYB1 to control lignin biosynthesis during wood formation. *New Phytologist, 213*(1), 287–299. https://doi.org/10.1111/nph.14129.

Soler, M., Plasencia, A., Lepikson-Neto, J., Camargo, E. L. O., Dupas, A., Ladouce, N., et al. (2016). The woody-preferential gene EgMYB88 regulates the biosynthesis of phenylpropanoid-derived compounds in wood. *Frontiers in Plant Science, 7*, 1422.

Sorce, C., Giovannelli, A., Sebastiani, L., & Anfodillo, T. (2013). Hormonal signals involved in the regulation of cambial activity, xylogenesis and vessel patterning in trees. *Plant Cell Reports, 32*, 885–898.

Suer, S., Agusti, J., Sanchez, P., Schwarz, M., & Greb, T. (2011). WOX4 imparts auxin responsiveness to cambium cells in *Arabidopsis*. *Plant Cell, 23*, 3247–3259.

Sundell, D., Street, N. R., Kumar, M., Mellerowicz, E. J., Kucukoglu, M., Johnsson, C., et al. (2017). AspWood: High-spatial-resolution transcriptome profiles reveal uncharacterized modularity of wood formation in *Populus tremula*. *Plant Cell*, *29*, 1585–1604.

Tang, X., Zhuang, Y., Qi, G., Wang, D., Liu, H., Wang, K., et al. (2015). Poplar PdMYB221 is involved in the direct and indirect regulation of secondary wall biosynthesis during wood formation. *Scientific Reports*, *5*, 12240.

Taylor-Teeples, M., Lin, L., de Lucas, M., Turco, G., Toal, T. W., Gaudinier, A., et al. (2015). An *Arabidopsis* gene regulatory network for secondary cell wall synthesis. *Nature*, *517*, 571–575.

Tian, Q., Wang, X., Li, C., Lu, W., Yang, L., Jiang, Y., et al. (2013). Functional characterization of the poplar R2R3-MYB transcription factor PtoMYB216 involved in the regulation of lignin biosynthesis during wood formation. *PLoS One*, *8*(10), e76369.

Tuskan, G. A., Difazio, S., Jansson, S., Bohlmann, J., Grigoriev, I., Hellsten, U., et al. (2006). The genome of black cottonwood, *Populus trichocarpa* (Torr. & Gray). *Science*, *313*, 1596–1604.

Umezawa, T. (2009). The cinnamate/monolignol pathway. *Phytochemistry Reviews*, *9*(1), 1–17. https://doi.org/10.1007/s11101-009-9155-3.

Ursache, R., Nieminen, K., & Helariutta, Y. (2013). Genetic and hormonal regulation of cambial development. *Physiologia Plantarum*, *147*, 36–45.

Wang, H.-Z., & Dixon, R. A. (2011). On-off switches for secondary cell wall biosynthesis. *Molecular Plant*, *5*(2), 297–303.

Wang, S., Li, E., Porth, I., Chen, J., Mansfield, S. D., & Douglas, C. J. (2014). Regulation of secondary cell wall biosynthesis by poplar R2R3 MYB transcription factor PtrMYB152 in *Arabidopsis*. *Scientific Reports*, *4*, 5054.

Wang, H. H., Tang, R. J., Liu, H., Chen, H. Y., Liu, J. Y., Jiang, X. N., et al. (2013). Chimeric repressor of PtSND2 severely affects wood formation in transgenic *Populus*. *Tree Physiology*, *33*, 878–886.

Wang, H., Tang, R., Wang, C., Qi, Q., Gai, Y., Jiang, X., et al. (2014). Functional repression of PtSND2 represses growth and development by disturbing auxin biosynthesis, transport and signaling in transgenic poplar. *Tree Physiology*, *35*, 95–105.

Xu, C., Fu, X., Liu, R., Guo, L., Ran, L., Li, C., et al. (2017). PtoMYB170 positively regulates lignin deposition during wood formation in poplar and confers drought tolerance in transgenic *Arabidopsis*. *Tree Physiology*, *37*, 1713–1726.

Xu, P., Kong, Y., Song, D., Huang, C., Li, X., & Li, L. (2014). Conservation and functional influence of alternative splicing in wood formation of *Populus* and *Eucalyptus*. *BMC Genomics*, *15*, 1–12.

Yamaguchi, M., Goué, N., Igarashi, H., Ohtani, M., Nakano, Y., Mortimer, J. C., et al. (2010). VASCULAR-RELATED NAC-DOMAIN6 and VASCULAR-RELATED NAC-DOMAIN7 effectively induce transdifferentiation into xylem vessel elements under control of an induction system. *Plant Physiology*, *153*, 906–914.

Yamaguchi, M., Kubo, M., Fukuda, H., & Demura, T. (2008). Vascular-related NAC-DOMAIN7 is involved in the differentiation of all types of xylem vessels in *Arabidopsis* roots and shoots. *The Plant Journal*, *55*, 652–664.

Yamamoto, R., Fujioka, S., Iwamoto, K., Demura, T., Takatsuto, S., Yoshida, S., et al. (2007). Co-regulation of brassinosteroid biosynthesis-related genes during xylem cell differentiation. *Plant & Cell Physiology*, *48*, 74–83.

Yang, L., Zhao, X., Ran, L., Li, C., Fan, D., & Luo, K. (2017). PtoMYB156 is involved in negative regulation of phenylpropanoid metabolism and secondary cell wall biosynthesis during wood formation in poplar. *Scientific Reports*, *7*, 1–14.

Yu, H., Soler, M., Clemente, H. S., Mila, I., Paiva, J. A. P., Myburg, A. A., et al. (2015). Comprehensive genome-wide analysis of the Aux/IAA gene family in *Eucalyptus*: Evidence for the role of EgrIAA4 in wood formation. *Plant & Cell Physiology*, *56*, 700–714.

Yu, H., Soler, M., Mila, I., Clemente, H. S., Savelli, B., Dunand, C., et al. (2014). Genome-wide characterization and expression profiling of the AUXIN RESPONSE FACTOR (ARF) gene family in *Eucalyptus grandis*. *PLoS One, 9*, 13.
Zhang, J., Nieminen, K., Serra, J. A. A., & Helariutta, Y. (2014). The formation of wood and its control. *Current Opinion in Plant Biology, 17*, 56–63.
Zhao, C., Avci, U., Grant, E. H., Haigler, C. H., & Beers, E. P. (2008). XND1, a member of the NAC domain family in *Arabidopsis thaliana*, negatively regulates lignocellulose synthesis and programmed cell death in xylem. *The Plant Journal, 53*, 425–436.
Zhao, Q., & Dixon, R. A. (2011). Transcriptional networks for lignin biosynthesis: More complex than we thought? *Trends in Plant Science, 16*(4), 227–233. https://doi.org/10.1016/j.tplants.2010.12.005.
Zhao, Y., Sun, J., Xu, P., Zhang, R., & Li, L. (2014). Intron-mediated alternative splicing of WOOD-ASSOCIATED NAC TRANSCRIPTION FACTOR1B regulates cell wall thickening during fiber development in Populus species. *Plant Physiology, 164*(2), 765–776. https://doi.org/10.1104/pp.113.231134.
Zhong, R., Demura, T., & Ye, Z. H. (2006). SND1, a NAC domain transcription factor, is a key regulator of secondary wall synthesis in fibers of *Arabidopsis*. *The Plant Cell, 18*, 3158–3170.
Zhong, R., Lee, C., McCarthy, R. L., Reeves, C. K., Jones, E. G., & Ye, Z. H. (2011). Transcriptional activation of secondary wall biosynthesis by rice and maize NAC and MYB transcription factors. *Plant & Cell Physiology, 52*, 1856–1871.
Zhong, R., Lee, C., & Ye, Z. H. (2010a). Functional characterization of poplar wood-associated NAC domain transcription factors. *Plant Physiology, 152*, 1044–1055.
Zhong, R., Lee, C., & Ye, Z.-H. (2010b). Evolutionary conservation of the transcriptional network regulating secondary cell wall biosynthesis. *Trends in Plant Science, 15*, 625–632.
Zhong, R., Lee, C., Zhou, J., McCarthy, R. L., & Ye, Z. H. (2008). A battery of transcription factors involved in the regulation of secondary cell wall biosynthesis in *Arabidopsis*. *The Plant Cell, 20*, 2763–2782.
Zhong, R., McCarthy, R. L., Haghighat, M., & Ye, Z. H. (2013). The poplar MYB master switches bind to the SMRE site and activate the secondary wall biosynthetic program during wood formation. *PLoS One, 8*, 18–21.
Zhong, R., McCarthy, R. L., Lee, C., & Ye, Z. H. (2011). Dissection of the transcriptional program regulating secondary wall biosynthesis during wood formation in poplar. *Plant Physiology, 157*, 1452–1468.
Zhong, R., & Ye, Z. H. (2009). Transcriptional regulation of lignin biosynthesis. *Plant Signaling & Behavior, 4*, 1028–1034.
Zhong, R., & Ye, Z. H. (2010). The poplar PtrWNDs are transcriptional activators of secondary cell wall biosynthesis. *Plant Signaling & Behavior, 5*, 469–472.
Zhong, R., & Ye, Z.-H. (2015). Secondary cell walls: Biosynthesis, patterned deposition and transcriptional regulation. *Plant & Cell Physiology, 56*, 195–214.

Further reading

Matsumoto-Kitano, M., Kusumoto, T., Tarkowski, P., Kinoshita-Tsujimura, K., Vaclavikova, K., Miyawaki, K., et al. (2008). Cytokinins are central regulators of cambial activity. *Proceedings of the National Academy of Sciences of the United States of America, 105*, 20027–20031.

CHAPTER EIGHT

Cell wall pectins in tree growth and woody biomass utilization

Scott A. Harding[a,b,c,*], **Chung-Jui Tsai**[a,b,c]
[a]Warnell School of Forestry and Natural Resources, University of Georgia, Athens, GA, United States
[b]Department of Genetics, University of Georgia, Athens, GA, United States
[c]Department of Plant Biology, University of Georgia, Athens, GA, United States
*Corresponding author: e-mail address: sharding@uga.edu

Contents

Abstract

Pectins are networks of dynamically and variably modified, galacturonan-rich polymers that occur predominantly in the primary cell wall and the middle lamella between cells. Despite their low abundance in lignocellulosic feedstocks, pectins can have outsized, and often negative effects on post-harvest processing. There is compelling evidence that pectins interfere with bio-based saccharification. Transgenic strategies for improving woody biomass utilization by pectin manipulation have therefore been explored. While much research has focused on the apoplastic, *in muro* steps of pectin modification during cell wall remodelling, recent biotechnological successes suggest potential for substantial and informative gains by targeting biosynthetic steps in the Golgi apparatus. At the same time, it must be recognized that pectins are essential for plant fitness. They comprise a substantial fraction of the primary cell wall, which, during rapid growth and differentiation, constitutes most of the cell wall. Pectin involvement in plant

Advances in Botanical Research, Volume 89
ISSN 0065-2296
https://doi.org/10.1016/bs.abr.2018.11.008

defence, stomatal dynamics, embolism control, root nutrient uptake and trace metal sequestration will also need to be taken into account to maximize the potential for durable gains. Early reports suggest that there is potential for bottom line success, not only for saccharification efficiency, but for biomass yields as well. Longer-term field testing across more varied environments, with more detailed phenotyping, will provide valuable insights for the future of pectin manipulation in woody biomass improvement.

1. Introduction

Much of the general perception about pectins probably derives from the changes they undergo during the later stages of fruit and vegetable softening (Marin-Rodriguez, Orchard, & Seymour, 2002; Tucker et al., 2017). However, pectins undergo impactful change throughout cell wall development in all organs of all plants. Ultimately, in the case of woody species, pectins have a noticeable presence in the tissues that give rise to biomass used as lignocellulosic feedstocks. Although pectins are much less abundant on a dry mass basis than the main load-bearing polymers cellulose and lignin, their importance to plant form and fitness has led to great biotechnological interest (Xiao & Anderson, 2013). From multivariate statistical analysis of growth and saccharification trade-offs for a wide range of transgenic poplar (*Populus tremula* × *Populus tremuloides*) lines, the case was recently made that minor sugar components associated with the pectin fraction should be viewed as important new markers for tree biomass improvement (Escamez et al., 2017). In this chapter, we will touch upon some of the ways in which pectins hold promise for improving post-harvest utilization of woody plant biomass. The involvement of pectins in reproductive processes will not be addressed.

1.1 Pectin evolution and cellular diversification

It is generally considered that pectins comprise four distinct structural types in the cell walls of higher plants (Harholt, Suttangkakul, & Scheller, 2010; Mohnen, 2008). Homogalacturonan (HG) and rhamnogalacturonan I (RGI) can make up as much as 90% of that pectin (Mohnen, 2008). HG is an unbranched polymer of α-1,4-linked D-galacturonic acid residues while RGI is made up of a heavily galactosylated/arabinosylated (branched) backbone of disaccharide (α-1,4-D-galacturonic acid-α-1,2-L-rhamnose) repeats. The less abundant components are xylogalacturonan which comprises an HG backbone decorated with monosaccharides, primarily of xylose, and rhamnogalacturonan II (RGII). In evolutionary terms, 12 of

16 pectin-related gene families predate the moss *Physcomitrella patens* (McCarthy, Der, Honaas, dePamphilis, & Anderson, 2014). HG, the most abundant of the cell wall pectins, also predates the mosses, but the absence of several gene family representatives in *P. patens* suggests a relatively simple pectin matrix, perhaps one that lacks RGI, xyloglucans and RGII, in these non-vascular plants. RGII exhibits a complex, though highly conserved structure, and is clearly limited to vascular tracheophytes with upright growth habits (Matsunaga et al., 2004). Mutations that affect RGII structure and ability to form borate-mediated crosslinks result in severe growth abnormalities (O'Neill, Eberhard, Albersheim, & Darvill, 2001). Normal regulation of cell wall lignification also appears to depend on RGII (Voxeur et al., 2017).

As pectins have become more structurally complex during higher plant diversification, new roles for the HG backbone have probably evolved. From phylogenetic studies, it seems likely that *in muro* modifications to the HG backbone, notably changes in its methylester status, are under more elaborate regulation in vascular taxa, including *Populus*, than in non-vascular taxa (McCarthy et al., 2014; Wang et al., 2013). For example, the number of pectin methylesterase genes is similar in vascular and non-vascular taxa, but the ratio of regulatory pectin methylesterase inhibitor (PMEI) genes to pectin methylesterase (PME) genes is 1:1 in the ancestral angiosperm *Amborella*, 2:1 in *Populus* and *Arabidopsis*, and 1:20 or less in the mosses (Wang et al., 2013). HG methyl status exhibits spatiotemporal variation consistent with roles in coordinating cell wall plasticity, rigidity, and cell-cell adhesion during the process of expansion growth and differentiation in higher plants (Senechal, Wattier, Rusterucci, & Pelloux, 2014). In poplars, and presumably other woody perennials, pectin methylester status also changes seasonally (Catesson, 1994). Not surprisingly then, PME and PMEI have drawn interest along with several other HG pectin modifying enzymes for biomass trait manipulation as will be described later in this chapter.

1.2 Pectin accrual and function during cell wall development

A plant cell wall begins its formation as an intracellular pectin-rich plate within a raft of microtubules and microfilaments known as the phragmoplast (Daher & Braybrook, 2015). As the cell plate grows towards the walls of the mother cell, it surrounds itself with a plasma membrane, effectively becoming an apoplastic compartment supplied by the phragmoplast (Dhonukshe et al., 2006). Pectin oligomers are synthesized in the Golgi apparatus and delivered to the apoplast by vesicle trafficking (Mohnen, 2008). An array of methyl-, acetyl-, and glycosyl-transferases orchestrate biosynthesis and sorting of the pectin oligomers within the various Golgi lumen compartments in accordance with still-evolving models (Atmodjo, Hao, & Mohnen, 2013). Deposition of pectins at the cell plate begins earlier than that of cellulose, but little is known about how deposition of those polymers is orchestrated *in planta* (Samuels, Giddings, & Staehelin, 1995). *In vitro* reconstitution experiments in which cellulose synthesized by bacteria is secreted into varied concentrations and size fractions of pectin oligomers support the idea that the pectin milieu initially affects the extensibility, and ultimately 'finalizes' the porosity of the resulting cellulose matrix (Chanliaud & Gidley, 1999; Lopez-Sanchez et al., 2016).

To carry out its function, the Golgi apparatus ultimately depends on the availability and uptake of sugar nucleotides from the cytosol (Seifert, 2004). Various dehydrogenases, epimerases, mutases and decarboxylases operate in a cytosolic network of feedback-regulated pathways to supply the Golgi apparatus with the required donor sugars (Caffall & Mohnen, 2009; Harholt et al., 2010). A simplified depiction of the carbohydrate interconversions and their compartmentalization for pectin biosynthesis is presented in Fig. 1. The HG pectin backbone is made up of galacturonic acid which is an epimer of glucuronic acid derived from UDP-glucose via UDP-glucose dehydrogenase. Biosynthesis of cellulose and glucuronic acid compete for cytosolic UDP-glucose, with the enzyme UDP-glucose-dehydrogenase having a key role in mediating that competition (Klinghammer & Tenhaken, 2007). Despite the many steps in between, correlations with SNPs, gene expression, and enzyme activity link UDP-glucose dehydrogenase to key biomass attributes in various species of *Populus* and *Salix* (Johansson, Sterky, Amini, Lundeberg, & Kleczkowski, 2002; Liu et al., 2015; Serapiglia, Cameron, Stipanovic, & Smart, 2012; Tian, Du, Li, & Zhang, 2014). This clearly implicates an element of metabolic control over the distribution of photosynthate between different cell wall polymers. However, it is the Golgi apparatus steps immediately upstream of pectin oligomer formation, and the apoplastic steps during cell wall remodelling that most clearly relate to pectin homeostasis. Together, those steps have been the focus of most pectin-based strategies to improve tree biomass utilization (see discussion further below).

Pectins exert some of their most well-defined controls over plant growth and morphogenesis during rapid expansion of the primary cell wall. In meristematic zones, cells or suites of cells need to maintain adhesion while being

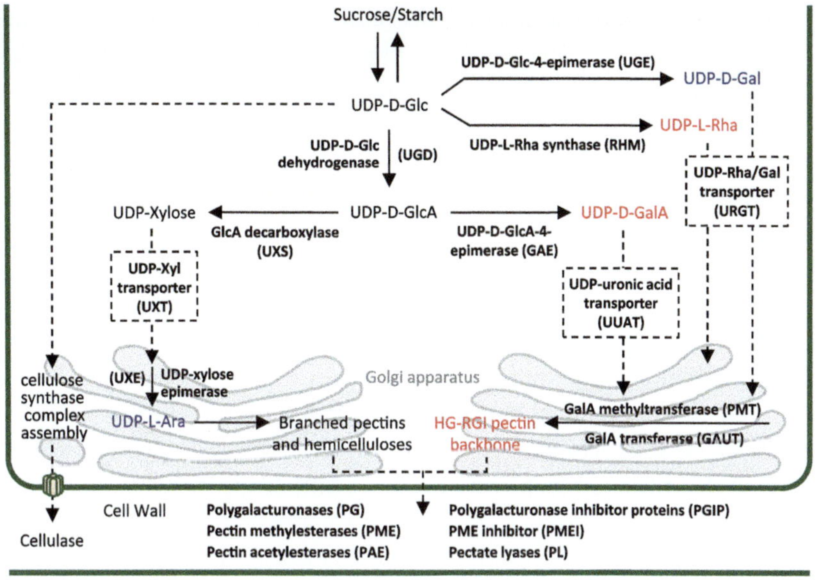

Fig. 1 Illustrative flowchart highlights metabolic origins of carbohydrates that make up the pectin backbone constituents HG and RGI (red), and the most abundant of the other pectin sugars, arabinose (Ara) and galactose (Gal) (blue). Transport into the Golgi apparatus is mediated by UDP-xylose transporters (UXT) (Ebert et al., 2015); UDP-rhamnose/galactose transporters (URGT) (Rautengarten et al., 2014); and UDP-uronic acid transporters (UUAT) (Saez-Aguayo et al., 2017). Most of the pathway steps have been previously summarized (Caffall & Mohnen, 2009), although the role of UDP-xylose synthase (UXS) has recently been updated (Kuang et al., 2016). Pectin remodelling takes place in the cell wall (bottom) and is mediated by the enzymes listed. Chemical abbreviations: arabinose (Ara); galactose (Gal); glucose (Glc); rhamnose (Rha); galacturonic acid (GalA); and glucuronic acid (GlcA).

allowed to expand and divide differentially. Pectin biosynthesis mutants exhibit loss of intercellular adhesion and abnormal growth (Bouton et al., 2002; Iwai, Masaoka, Ishii, & Satoh, 2002). Also critical is the preservation of cell-cell adhesion for plasmodesmatal integrity and uninterrupted intercellular trafficking of metabolites and signals (Singh, Eland, Harholt, Scheller, & Marchant, 2005). In concert with expansion and differentiation, polygalacturonases (PG), PG inhibitor proteins (PGIP), pectate lyases (PL) and PL-like, PME, PMEI, and pectin acetylesterases (PAE) all act in the apoplastic space to modulate pectin network plasticity and orchestrate integration with other cell wall carbohydrates. This involves establishing a variety of covalent and non-covalent linkages, in some cases mediated by extensins and arabinogalactan proteins AGPs (Cannon et al., 2008).

After cell expansion slows, the bulk of the original pectin matrix remains in the middle lamella between cells where the first phase of lignin polymerization then begins (Donaldson, 2001; Terashima, Fukushima, Sano, & Takabe, 1988; Wi, Singh, Lee, & Kim, 2005). *In vitro* lignification studies have shown that the presence of pectins has a governing effect on the dispersion of lignin and the proportion of alkyl-aryl-ether bonds in the cellulose matrix (Lairez et al., 2005; Touzel, Chabbert, Monties, Debeire, & Cathala, 2003). One study provided *in vivo* support for the importance of pectins to lignification in *Arabidopsis* by showing that *PG* overexpression affected the timing of lignin accrual (Xiao et al., 2017). In asking whether pectins might also have direct roles in secondary cell wall formation or structure, it should be noted that pectins are found in, or are linked to, the secondary cell wall as well as the cellulose-rich tension wood (TW) fibres of poplar (Guedes et al., 2017; Xiao & Anderson, 2013). From radial expression gradients of aspen wood, one reliable marker of pectin biosynthesis, the Clade A galacturonic acid transferase (*GAUT1*), exhibits strong expression from cambial division well into the phase of secondary cell wall accrual and lignification (Sundell et al., 2017). A rather novel set of studies found that pectins were affected in transgenic *P. tremula* × *alba* that overexpressed tubulin genes (Harding et al., 2018; Swamy et al., 2015). Although pectin levels were only affected in leaves, pectin integration with cell wall polymers was clearly altered in stem xylem (Swamy et al., 2015). These suggest that the relevance of pectin deposition and remodelling is not strictly limited to the expansion of differentiating xylem fibre and vessel cells, but also extends towards the determination of secondary cell wall properties.

The nature of the associations between pectins and other cell wall polymers continues to be explored at the evolutionary and biochemical levels, and to yield major revisions in our understanding of cell wall organization and function (Anderson, 2016; Harholt et al., 2010; Saffer, Carpita, & Irish, 2017; Tan et al., 2013; Wang & Hong, 2016). Until recently, pectins were regarded to provide a semi-autonomous soft matrix in which interactions between cellulose and hemicellulose or lignin determined the main load bearing characteristics and expansion of the cell wall (Carpita & Gibeaut, 1993; Cosgrove, 2001; Hongo, Sato, Yokoyama, & Nishitani, 2012). It is now widely regarded, based on evidence of pectin cross-linkages with cellulose microfibrils, that pectins constitute part of a single covalently linked network that defines the main load-bearing properties of the cell wall (Braybrook & Jonsson, 2016; Chebli & Geitmann, 2017; Hofte, 2015; Tan et al., 2013). A hypothetical framework that accords well with this was recently offered, in which HG remodelling is incorporated into the acid-growth theory of cell wall expansion (Hocq, Pelloux, & Lefebvre, 2017). pH homeostasis in the apoplastic space can therefore be considered to have a more HG-centric effect than previously thought on acid-modulated cell wall expansion.

1.3 Pectins in post-harvest biomass utilization

Pectins account for as much as 35% of primary cell wall mass, but the percentage is much smaller in thick walled vascular cells that comprise lignocellulosic feedstocks. After cellulose and xylans, the most abundant structural carbohydrate in sapwood and heartwood of several poplar and aspen hybrids is pectin-derived galacturonic acid at 2–3% dry mass (Willfor, Sundberg, Pranovich, & Holmbom, 2005). Clearly there is substantial dilution of tissue pectins after secondary wall thickening is complete. Nevertheless, several findings point towards the outsized and direct influence of pectins on post-harvest utilization of woody biomass, and to opportunities for biotechnological exploitation of pectin-cell wall interactions. From the perspective of plant biomass utilization, pectins have a detrimental effect on paper production (Thornton, Ekman, Holmbom, & Eckerman, 1993; Willfor et al., 2005). It is also generally the case that pectins interfere with biomass saccharification (Chung et al., 2014). Gene expression of certain pectin modifying enzymes has recently been found to correlate even more strongly than lignin content with reduced saccharification potential of poplar woody biomass (Wildhagen et al., 2018). An association study of *Eucalyptus pilularis* populations provided evidence linking solid wood characteristics, such as shrinkage and collapse of drying timber, with two isoforms of PME (Sexton et al., 2012). Pectins also cross link with cell wall extensions which directly hinder digestion of pre-treated hybrid aspen (*P. alba* × *tremuloides*) biomass (Fleming, Decker, & Bedinger, 2016). In the following sections, it will become evident that transgenic strategies for improving woody biomass utilization by pectin manipulation, especially its down-regulation, bear promise.

2. Potential targets for pectin manipulation

2.1 Pectin-degrading enzymes

Two decades ago, powerful methodologies for gene discovery using developing wood as a model system began to shed light on promising targets for future study (Hertzberg et al., 2001). Among the most strongly up-regulated genes in xylem-forming tissues of fast-growing transgenic hybrid aspen (*P. tremula* × *tremuloides*) was a *PL* gene (Israelsson et al., 2003). There are 30 PL-like genes in the *P. trichocarpa* genome assembly v3 (Bai et al., 2017). The PLs, like other pectin-hydrolysing enzymes, operate in the cell wall apoplast (Fig. 1), where they cleave the α-(1,4)-glycosidic bond between unmethylated galacturonic acid units of HG (Marin-Rodriguez et al., 2002). A number of years subsequent to the discovery of Israelsson et al. (2003), overexpression of *PL1–27* in transgenic hybrid aspen was successfully used to decrease pectin levels and improve growth and saccharification (Biswal et al., 2014). Indeed, the capacity to degrade pectins is quite important for cell expansion during plant growth, and for microbial saccharification of plant biomass after harvest (Chung et al., 2014). Interestingly, transgenic overexpression of several individual *PL*-like genes reduced xylem development and growth in *P. tremula* × *alba* (Bai et al., 2017). The conflicting outcomes could reflect differences in the degree or timing of pectin hydrolysis that may inform future studies.

A second class of apoplastic HG pectin-degrading enzymes is the PG. The large *PG* gene family and expression of some of its members in wood forming and other tissues have been described for *Populus* (Yang, Liu, Wang, & Zeng, 2013). PGs (EC3.2.1.15) and PLs (EC4.2.2.2) degrade the HG backbone by hydrolytic and β-eliminative cleavage, respectively, and exhibit much different pH optima and calcium ion sensitivities (Pagel & Heitefuss, 1990). Nevertheless, only PL has been targeted for transgenic manipulation in *Populus* (Bai et al., 2017; Biswal et al., 2014). Up-regulated PG activity in transgenic rice had a mildly negative effect on growth but decreased drought tolerance (Liu et al., 2014). PG activity is inhibited *in vivo* by PGIP which have drawn a great deal of interest due to their importance in pathogen resistance (Kalunke et al., 2015). There is one report that two PGIPs might be involved in the resistance to *Marssonina brunnea* in *P. deltoides* (Cheng, Cao, Pan, Wang, & Huang, 2008). Constitutive overexpression of *PGIP* in tobacco increased resistance to pathogen attack by altering cell wall structure (Nguema-Ona et al., 2013). From the perspective of wood formation and post-harvest processing, PG manipulation perhaps in combination with PL and PGIP would seem to be of further exploratory interest.

2.2 Pectin esterases

PME and PAE also operate in the apoplastic space of the cell wall pectin network (Fig. 1). PME has been the most extensively studied of the two esterases (Wolf & Greiner, 2012). Pectin polymers are exported from the Golgi network to the apoplast in a highly methylesterified state (O'Neill, Albersheim, & Darvill, 1990). During cell expansion, and in accordance with diurnal rhythms, developmental stage and environmental cues, apoplastic PMEs remove methyl groups from the HG backbone as has been described for *Populus* and other taxa (Huve et al., 2007; Micheli, Sundberg, Goldberg, & Richard, 2000; Willats et al., 2001). This exposes carboxyl groups which can engage in charge-charge repulsion or calcium bridged cross-links. The spatial pattern of HG demethylation influences pectin porosity and swelling as well as the frequency of calcium-mediated cross linkages that lend to cell wall stiffening for support and defence (Micheli, 2001; Navazio et al., 2002; Pelloux, Rusterucci, & Mellerowicz, 2007). Despite these important regulatory roles, PME activity can be manipulated without lethal effects. The effect of PME manipulation on woody biomass saccharification has not yet been explored, even though hybrid aspen transgenic lines with reduced *PME* expression and altered fibre characteristics are available (Siedlecka et al., 2008). PME manipulation to improve wood fibre characteristics showed promise in preliminary observations (Siedlecka et al., 2008), but negative side effects, including an increased propensity for the formation of undesirable tyloses, were later discovered (Lesniewska et al., 2017). Tylose blockages would presumably restrict growth and impede infiltration of enzyme cocktails for post-harvest biomass saccharification as well.

An alternative approach tested the idea of overexpressing PMEI in *Arabidopsis* to improve growth as well as post-harvest biomass saccharification (Lionetti et al., 2010). However, when that approach was applied to in rice (*Oryza sativa*), growth was compromised (Nguyen, Jeong, Jeon, Kim, & Lee, 2017). Clearly, strategies which deploy PME manipulation to improve wood utilization can be expected to face special technical challenges. For one thing, numerous *PME* and PMEI genes are present in the *Populus* genome (Wang et al., 2013). In addition, *PME* and PMEI expression can change in response to many biotic and abiotic cues, including drought, wounding and pathogen attack (An et al., 2008; Harding et al., 2018; Senechal et al., 2014; Xiong, Yang, Fu, & Tao, 2015). Evidently, precise control of HG demethylation by both PME and PMEI is important for normal growth and defence (An et al., 2008; Willats et al., 2001). The potential for pleiotropic effects by altering the control of HG methylester status certainly exists as discovered in the case of aspen (Lesniewska et al., 2017).

PAE, like PME, has been a target of manipulation in *Populus* and other taxa (de Souza, Hull, Gille, & Pauly, 2014; Gou et al., 2012). Galacturonic acid residues in the HG and RGI backbones can be sporadically *O*-acetylated at the C2 and C3 positions by an unknown mechanism (Gille & Pauly, 2012). Unlike pectin methylation, pectin acetylation is thought to stiffen cell walls (Orfila et al., 2012). Calcium binding by PME-demethylated pectin can be sterically influenced by acetylation of the HG backbone (Ralet, Crepeau, Buchholt, & Thibault, 2003). This raises the possibility for complex interactions between HG methylation, acetylation and apoplastic calcium that might complicate biotechnological outcomes. There is evidence from *Arabidopsis* mutants that reduced PAE activity has a strong negative effect on stem growth, but surprisingly, also a potentially beneficial effect on saccharification despite the high degree of acetylation (de Souza et al., 2014). Of note, it is generally considered that acetyl groups inhibit biomass digestion (Gille & Pauly, 2012). Suppression of *PAE* in transgenic *Populus* lines has been found to promote digestibility, but perhaps surprisingly in light of the enhanced growth reported in de Souza et al. (2014), overexpression of the poplar *PAE* gene in tobacco (*Nicotiana tabacum*) reduced growth (Gou et al., 2012).

2.3 Pectin biosynthetic enzymes

From the above-mentioned advances, it is evident that a number of bioengineering strategies for pectin manipulation have focused on the apoplastic steps where modification and remodelling take place (Fig. 1). However, recent successes suggest potential for substantial and informative gains by targeting the biosynthetic steps. HG biosynthesis depends in large part on GAUTs encoded by a 20-member gene family in *Populus* (Caffall, Pattathil, Phillips, Hahn, & Mohnen, 2009). GAUT proteins are associated with the Golgi complex and certain GAUTs transfer galacturonic acid from UDP-galacturonic acid onto an HG acceptor (Sterling et al., 2006) (Fig. 1). Deployment of strategies that target pectin biosynthesis directly via GAUT has led to increased biomass yields and conversion in *Populus* (Biswal et al., 2015). GAUT12, one of several galacturonosyl transferases expressed in wood forming tissues of *Populus*, was targeted for down-regulation in that work. GAUT12 is a clade C member that is involved with pectin as well as xylan biosynthesis (Caffall & Mohnen, 2009). Recently, transgenic *Populus* lines were generated in which a Clade A GAUT4, more specifically involved in HG pectin biosynthesis, was down-regulated (Biswal et al., 2018). The yield enhancements and digestibility gains were greater than those reported earlier with GAUT12-down-regulation and were more directly attributable to reduced HG pectin content. Based on studies with

PME, PAE, GAUT, and *PL*, the potential for economically significant gains from pectin manipulation seems to be there despite known interactions with signalling pathways that coordinate both growth and defence (Brutus, Sicilia, Macone, Cervone, & De Lorenzo, 2010; Wolf et al., 2014).

Other pectin biosynthetic steps include methylation by pectin methyl transferases (PMT), and tautomeric interconversions of sugars, as well as their trafficking into the Golgi compartments (Harholt et al., 2010). While PME operates in the apoplast as mentioned, various Golgi-associated PMT or putative PMT proteins catalyse methylation of pectin/xyloglucuronan oligo intermediates during biosynthesis (Caffall & Mohnen, 2009; Kim, Held, Zemelis, Wilkerson, & Brandizzi, 2015; Qu et al., 2016; Urbanowicz et al., 2012). These PMTs are distinct from other methyltransferases, those that facilitate monolignol biosynthesis, for example, and they are highly specific for their carbohydrate acceptors (Urbanowicz et al., 2012). Recent findings from several putative PMT mutants (see below) should yield insights about the coordination of apoplastic PME and Golgi-localized PMT activities in the regulation of pectin methylester status.

In leaves of *Arabidopsis*, cell wall pectin methylester status was increased by overexpressing *PMTs* encoded by *COTTON GOLGI RELATED* (Kim et al., 2015; Weraduwage et al., 2016). PMT activity increased disproportionately more than the methyl status of cell wall uronic acid residues (Kim et al., 2015), which suggests that apoplastic PME activity may have exerted a degree of compensatory demethylation. Nevertheless, clear effects of PMT overexpression on leaf size and intercellular air spaces were reported (Weraduwage et al., 2016). Elsewhere, it was found that *Arabidopsis* overexpressing *INCREASED CAMBIAL ACTIVITY*, a putative pectin methyltransferase, exhibited increased PME activity and increased cell division activity in the cambium (Kim et al., 2016). Cell wall composition was not analysed, but cytokinin and gibberellic acid (GA) levels increased. The authors concluded that methanol, due presumably to compensatory PME activity, stimulated hormone production and plant growth (Kim et al., 2016). To date, there are no reports on *PMT* expression or manipulation in *Populus*. In light of the earlier reports on increased expression of pectin modifying genes in high-GA hybrid aspen lines with accelerated growth (Israelsson et al., 2003), the interplay between pectin biogenesis, apoplastic demethylation and growth warrants continued study. Findings utilizing wood-forming perennials as models may broaden the impact of the recent studies with herbaceous annuals if seasonal effects on pectin methyl status, as reported for aspen (Catesson, 1994), are also incorporated.

3. Roles for pectins in poplar growth and fitness

3.1 Tension wood

The wood of trees is not uniform in structure or composition, and this is partly due to how trees deal with mechanical forces that accompany tilting, swaying or changes in load distribution due to crown growth. The entire process during which cell wall biogenesis is reprogrammed to enable stems and branches to counteract perceived changes in their gravitational field has been recently reviewed (Groover, 2016). Reprogramming leads to the formation of reaction wood, a component of which in hardwood species is tension wood (TW) with enhanced contractile properties that strengthen the wood or lead to corrective growth in juvenile parts of the stem. Interest in TW formation has grown since recognition that enzymatic saccharification for biofuels production correlated positively with TW content of *Salix* feedstocks (Brereton et al., 2012). A high resolution proteomic survey of developing TW in hybrid aspen revealed late up-regulation of three pathways for the biosynthesis of RGI pectin sugars, as well as the likely involvement of a putative xylan xylosyltransferase for the biosynthesis of the RGII pectin sugar apiose (Bygdell et al., 2017). Xylan levels decreased in the TW-forming cells which led the authors to conclude that the increased activity of the pathways described was most likely for pectin rather than hemicellulose biosynthesis. UDP-glucose dehydrogenase was also up-regulated, consistent with the diversion of sugar nucleotides towards pectin biosynthesis (Bygdell et al., 2017). An independent report reinforces the notion that RGI pectin is in fact one of the key determinants of TW tensile strength in *Populus* (Guedes et al., 2017). One simplistic but plausible assessment of the TW studies together with that of Biswal et al. (2018) is that simultaneous manipulation to reduce HG levels and increase RG pectin abundance represents an avenue for improved production and saccharification of poplar biomass.

3.2 Stem hydraulic properties

One side effect of engineering trees with reduced pectin accrual or altered pectin properties might be a change in their susceptibility to cavitation. The possibility seems real because most reports indicate that pectins are present during some, if not all stages of pit membrane development

(Hafren & Westermark, 2001; Herbette et al., 2015; Kim & Daniel, 2013; van Doorn, Hiemstra, & Fanourakis, 2011). Pit membranes limit the formation of embolisms and comprise 50–90% of hydraulic conductivity in angiosperm tree species (Choat, Brodie, Cobb, Zwieniecki, & Holbrook, 2006). Specifically, it is pit membrane thickness and pore size that determine cavitation vulnerability (Venturas, Sperry, & Hacke, 2017). Infusion treatments with enzymatic hydrolysis and TEM showed that cellulose and pectins contributed importantly to pit membrane structure and cavitation control in beech (*Fagus sylvatica*) and *P. tremula* (Dusotoit-Coucaud, Brunel, Tixier, Cochard, & Herbette, 2014). Based on the *in vitro* model studies of pectin-cellulose matrix formation (Chanliaud & Gidley, 1999; Lopez-Sanchez et al., 2016), one role of pectins in developing pit membranes might be to influence cellulose matrix porosity which would then be preserved even in the event of developmental hydrolysis of those pectins, as is thought to occur when xylem matures. Impaired pit membrane function and xylem vessel cavitation can occur due to mechanical strains from growth or abiotic stresses, including drought and repeated freeze-thaw cycles (Venturas et al., 2017). Due partly to their rapid growth, poplars are especially prone to drought-induced cavitation (Fichot, Brignolas, Cochard, & Ceulemans, 2015). Embolisms caused by freeze-thaw cycles also bring to mind the point that the tree hydraulic system functions across a wide temperature range. Whether pit membranes adjust in response to temperature remains unreported, but the degree of pectin methylesterification changes seasonally in the cambium of *Populus* (Catesson, 1994). Pectin modification by PME at cooler temperatures appears to be an important leaf adaptive trait (Solecka, Zebrowski, & Kacperska, 2008). Based on the conclusions of those authors, changes in pit membrane porosity due to pectin manipulation could potentially affect freezing point depression as well as propagation of ice crystals in xylem vessels of trees.

3.3 Leaf stomatal behaviour

It has been recognized for some time that pectin structure, specifically the presence of arabinan-rich RGI pectin side chains, contributes to the modulation of stomatal opening and closing (Jones, Milne, Ashford, & McQueen-Mason, 2003). Later, it was shown that PME function has an important role in setting the dynamic range for stomatal opening (Amsbury et al., 2016). More recently, in experiments with tubulin-overexpressing *Populus* lines that also exhibited striking increases in leaf cell wall pectin levels, stomatal opening and closing were slowed (Harding et al., 2018). The increased methylesterification of pectins in the transgenic poplar lines and the increased ratio of HG to RGI were both consistent with the mechanistic arguments made in Jones et al. (2003) and Amsbury et al. (2016) that the pectin changes contributed to reduced guard cell wall flexibility (Harding et al., 2018).

3.4 Root-rhizosphere interactions

Since one of the attractions of trees as biomass crops is their tolerance for marginal planting sites, breeding or other strategies intended to genetically improve biomass potential may include the targeting of root traits that can facilitate good performance under limiting conditions. One of the most limiting soil nutrients for agriculture worldwide is phosphorus (P). Roots therefore have evolved multiple strategies for acquiring, storing and remobilizing soil P, and it now appears that storage and remobilization of some of that P directly involve cell wall pectins (Zhu et al., 2015; Zhu, Zhu, Zhao, Zheng, & Shen, 2016). Those authors reported that under P-deficiency, ethylene-induced increases in root pectin level correlated with remobilization of cell wall-adsorbed P in rice (*O. sativa*). The specific molecular events surrounding either the ethylene-promoted pectin increase or the P release in those roots were not described, but were found to be ABA-sensitive (Yu, Zhu, Li, Kronzucker, & Shi, 2016). Although a mechanism for solubilizing certain P-immobilizing salts by pectin carboxyl groups was proposed, pectin methylesterification was not reported (Zhu et al., 2016). It may be enlightening that in ripening fruit, ethylene promotes PME activity which in turn exposes carboxyl groups, facilitates access by other hydrolytic enzymes to the pectin matrix, and promotes expansion of middle lamella and tissue softening (Tucker et al., 2017).

Although pectin methylesterification was not evaluated in the rice studies, changes in root pectin methylesterification are important for root sequestration of heavy metals commonly found in contaminated environments (Krzeslowska et al., 2016). Harmful trace metals are bound in the pectin fraction of root tip cells and then further sequestered by various other mechanisms, including callose layering (Krzeslowska, 2011). Changes in pectin methylesterification affect the availability of charged carboxyl groups which then are free to interact with heavy metals in the uptake stream. Cadmium, lead and, in acidic soils, aluminium (Al^{3+}) can all outcompete calcium which is normally coordinated with these binding sites, to become sequestered in the root cell walls. In a study of several genotypes of *Populus*, Al^{3+} resistance was correlated with the partitioning of Al^{3+} into the apoplastic fraction rather than the symplastic fraction of the roots (Smith, Naik, & Cumming, 2011). Al^{3+} stress increased root pectin levels in all of the poplar lines; however, the resistant lines retained Al^{3+} in the cell wall fraction. As in the rice studies of Zhu et al. (2016), neither PME activity nor the degree of pectin methylesterification was reported. Elsewhere, transcript level changes for nine genes associated with pectin biosynthesis and modification after Al^{3+} treatment were reported in a *Populus* microarray study (Grisel et al., 2010). Four of those affected *PME* and PMEI expression, but both up- and down-regulation occurred implying a complex protective mechanism. The authors suggested that forest soils are often acidic, and that trees have evolved Al^{3+} tolerance or avoidance mechanisms which their findings now shed light upon. One lesson to be drawn from these examples is that bioengineering approaches for cell wall pectin manipulation for biomass traits will need to consider the ramifications for root functions.

4. Conclusions

To date, only a handful of reports demonstrate potential for manipulating pectins to improve woody plant biomass production and utilization. Pectins have largely negative effects on post-harvest utilization, but they are also varied and dynamic structures essential for plant productivity. Roles in stomatal behaviour, embolism control, root nutrient uptake and sequestration of metals will need to be accommodated when considering biotechnological means to reduce pectin accrual or alter its chemical properties for woody biomass utilization. Initial studies suggest that there is potential for improved bioprocessing efficiency, but longer-term field testing across more varied environments, with more detailed phenotyping, will provide valuable insights for the future of this approach.

Acknowledgements

Research in our lab has been supported in part by the Plant Feedstock Genomics for Bioenergy Program (grant no. DE-SC0008470) and The Center for Bioenergy Innovation (a Department of Energy's Bioenergy Research Center), both under the Office of Biological and Environmental Research in the DOE Office of Science; and by the Georgia Research Alliance-Hank Haynes Forest Biotechnology Endowment.

References

Amsbury, S., Hunt, L., Elhaddad, N., Baillie, A., Lundgren, M., Verhertbruggen, Y., et al. (2016). Stomatal function requires pectin de-methyl-esterification of the guard cell wall. *Current Biology*, *26*, 2899–2906.

An, S. H., Sohn, K. H., Choi, H. W., Hwang, I. S., Lee, S. C., & Hwang, B. K. (2008). Pepper pectin methylesterase inhibitor protein CaPMEI1 is required for antifungal activity, basal disease resistance and abiotic stress tolerance. *Planta*, *228*, 61–78.

Anderson, C. T. (2016). We be jammin': An update on pectin biosynthesis, trafficking and dynamics. *Journal of Experimental Botany*, *67*, 495–502.

Atmodjo, M. A., Hao, Z. Y., & Mohnen, D. (2013). Evolving views of pectin biosynthesis. *Annual Review of Plant Biology*, *64*, 747–779.

Bai, Y., Wu, D., Liu, F., Li, Y. Y., Chen, P., Lu, M. Z., et al. (2017). Characterization and functional analysis of the poplar pectate lyase-like gene PtPL1-18 reveal its role in the development of vascular tissues. *Frontiers in Plant Science*, *8*, 17.

Biswal, A. K., Atmodjo, M. A., Li, M., Baxter, H. L., Yoo, C. G., Pu, Y. Q., et al. (2018). Sugar release and growth of biofuel crops are improved by downregulation of pectin biosynthesis. *Nature Biotechnology*, *36*, 249–257.

Biswal, A. K., Hao, Z. Y., Pattathil, S., Yang, X. H., Winkeler, K., Collins, C., et al. (2015). Downregulation of GAUT12 in *Populus deltoides* by RNA silencing results in reduced recalcitrance, increased growth and reduced xylan and pectin in a woody biofuel feedstock. *Biotechnology for Biofuels*, *8*, 41.

Biswal, A. K., Soeno, K., Gandla, M. L., Immerzeel, P., Pattathil, S., Lucenius, J., et al. (2014). Aspen pectate lyase PtxtPL1-27 mobilizes matrix polysaccharides from woody tissues and improves saccharification yield. *Biotechnology for Biofuels*, *7*, 11.

Bouton, S., Leboeuf, E., Mouille, G., Leydecker, M. T., Talbotec, J., Granier, F., et al. (2002). QUASIMODO1 encodes a putative membrane-bound glycosyltransferase required for normal pectin synthesis and cell adhesion in *Arabidopsis*. *Plant Cell*, *14*, 2577–2590.

Braybrook, S. A., & Jonsson, H. (2016). Shifting foundations: The mechanical cell wall and development. *Current Opinion in Plant Biology*, *29*, 115–120.

Brereton, N. J. B., Ray, M. J., Shield, I., Martin, P., Karp, A., & Murphy, R. J. (2012). Reaction wood—A key cause of variation in cell wall recalcitrance in willow. *Biotechnology for Biofuels*, *5*, 83.

Brutus, A., Sicilia, F., Macone, A., Cervone, F., & De Lorenzo, G. (2010). A domain swap approach reveals a role of the plant wall-associated kinase 1 (WAK1) as a receptor of oligogalacturonides. *Proceedings of the National Academy of Sciences of the United States of America*, *107*, 9452–9457.

Bygdell, J., Srivastava, V., Obudulu, O., Srivastava, M. K., Nilsson, R., Sundberg, B., et al. (2017). Protein expression in tension wood formation monitored at high tissue resolution in *Populus*. *Journal of Experimental Botany*, *68*, 3405–3417.

Caffall, K. H., & Mohnen, D. (2009). The structure, function, and biosynthesis of plant cell wall pectic polysaccharides. *Carbohydrate Research*, *344*, 1879–1900.

Caffall, K. H., Pattathil, S., Phillips, S. E., Hahn, M. G., & Mohnen, D. (2009). *Arabidopsis thaliana* T-DNA mutants implicate GAUT genes in the biosynthesis of pectin and xylan in cell walls and seed testa. *Molecular Plant*, *2*, 1000–1014.

Cannon, M. C., Terneus, K., Hall, Q., Tan, L., Wang, Y. M., Wegenhart, B. L., et al. (2008). Self-assembly of the plant cell wall requires an extensin scaffold. *Proceedings of the National Academy of Sciences of the United States of America*, *105*, 2226–2231.

Carpita, N. C., & Gibeaut, D. M. (1993). Structural models of primary-cell walls in flowering plants: Consistency of molecular-structure with the physical-properties of the walls during growth. *The Plant Journal*, *3*, 1–30.

Catesson, A. M. (1994). Cambial ultrastructure and biochemistry: Changes in relation to vascular tissue differentiation and the seasonal cycle. *International Journal of Plant Sciences*, *155*, 251–261.

Chanliaud, E., & Gidley, M. J. (1999). *In vitro* synthesis and properties of pectin/*Acetobacter xylinus* cellulose composites. *The Plant Journal*, *20*, 25–35.

Chebli, Y., & Geitmann, A. (2017). Cellular growth in plants requires regulation of cell wall biochemistry. *Current Opinion in Cell Biology*, *44*, 28–35.

Cheng, Q., Cao, Y. Z., Pan, H. X., Wang, M. X., & Huang, M. R. (2008). Isolation and characterization of two genes encoding polygalacturonase-inhibiting protein from *Populus deltoides*. *Journal of Genetics and Genomics*, *35*, 631–638.

Choat, B., Brodie, T. W., Cobb, A. R., Zwieniecki, M. A., & Holbrook, N. M. (2006). Direct measurements of intervessel pit membrane hydraulic resistance in two angiosperm tree species. *American Journal of Botany*, *93*, 993–1000.

Chung, D., Pattathil, S., Biswal, A. K., Hahn, M. G., Mohnen, D., & Westpheling, J. (2014). Deletion of a gene cluster encoding pectin degrading enzymes in *Caldicellulosiruptor bescii* reveals an important role for pectin in plant biomass recalcitrance. *Biotechnology for Biofuels*, 7, 147.

Cosgrove, D. J. (2001). Wall structure and wall loosening. A look backwards and forwards. *Plant Physiology*, *125*, 131–134.

Daher, F. B., & Braybrook, S. A. (2015). How to let go: Pectin and plant cell adhesion. *Frontiers in Plant Science*, *6*, 523.

de Souza, A., Hull, P. A., Gille, S., & Pauly, M. (2014). Identification and functional characterization of the distinct plant pectin esterases PAE8 and PAE9 and their deletion mutants. *Planta*, *240*, 1123–1138.

Dhonukshe, P., Baluska, F., Schlicht, M., Hlavacka, A., Samaj, J., Friml, J., et al. (2006). Endocytosis of cell surface material mediates cell plate formation during plant cytokinesis. *Developmental Cell*, *10*, 137–150.

Donaldson, L. A. (2001). Lignification and lignin topochemistry—An ultrastructural view. *Phytochemistry*, *57*, 859–873.

Dusotoit-Coucaud, A., Brunel, N., Tixier, A., Cochard, H., & Herbette, S. (2014). Hydrolase treatments help unravel the function of intervessel pits in xylem hydraulics. *Physiologia Plantarum*, *150*, 388–396.

Ebert, B., Rautengarten, C., Guo, X. Y., Xiong, G. Y., Stonebloom, S., Smith-Moritz, A. M., et al. (2015). Identification and characterization of a golgi-localized UDP-xylose transporter family from *Arabidopsis*. *Plant Cell*, *27*, 1218–1227.

Escamez, S., Gandla, M. L., Derba-Maceluch, M., Lundqvist, S. O., Mellerowicz, E. J., Jonsson, L. J., et al. (2017). A collection of genetically engineered *Populus* trees reveals wood biomass traits that predict glucose yield from enzymatic hydrolysis. *Scientific Reports*, 7, 15798.

Fichot, R., Brignolas, F., Cochard, H., & Ceulemans, R. (2015). Vulnerability to drought-induced cavitation in poplars: Synthesis and future opportunities. *Plant, Cell & Environment*, *38*, 1233–1251.

Fleming, M. B., Decker, S. R., & Bedinger, P. A. (2016). Investigating the role of extensin proteins in poplar biomass recalcitrance. *BioResources*, *11*, 4727–4744.

Gille, S., & Pauly, M. (2012). O-Acetylation of plant cell wall polysaccharides. *Frontiers in Plant Science*, *3*, 12.

Gou, J. Y., Miller, L. M., Hou, G. C., Yu, X. H., Chen, X. Y., & Liu, C. J. (2012). Acetylesterase-mediated deacetylation of pectin impairs cell elongation, pollen germination, and plant reproduction. *Plant Cell*, *24*, 50–65.

Grisel, N., Zoller, S., Kunzli-Gontarczyk, M., Lampart, T., Munsterkotter, M., Brunner, I., et al. (2010). Transcriptome responses to aluminum stress in roots of aspen (*Populus tremula*). *BMC Plant Biology*, *10*, 185.

Groover, A. (2016). Gravitropisms and reaction woods of forest trees—Evolution, functions and mechanisms. *New Phytologist, 211*, 790–802.

Guedes, F. T. P., Laurans, F., Quemener, B., Assor, C., Laine-Prade, V., Boizot, N., et al. (2017). Non-cellulosic polysaccharide distribution during G-layer formation in poplar tension wood fibers: Abundance of rhamnogalacturonan I and arabinogalactan proteins but no evidence of xyloglucan. *Planta, 246*, 857–878.

Hafren, J., & Westermark, U. (2001). Distribution of acidic and esterified polygalacturonans in sapwood of spruce, birch and aspen. *Nordic Pulp & Paper Research Journal, 16*, 284–290.

Harding, S. A., Hu, H., Nyamdari, B., Xue, L. J., Naran, R., & Tsai, C. J. (2018). Tubulins, rhythms and cell walls in poplar leaves: It's all in the timing. *Tree Physiology, 38*, 397–408.

Harholt, J., Suttangkakul, A., & Scheller, H. V. (2010). Biosynthesis of pectin. *Plant Physiology, 153*, 384–395.

Herbette, S., Bouchet, B., Brunel, N., Bonnin, E., Cochard, H., & Guillon, F. (2015). Immunolabelling of intervessel pits for polysaccharides and lignin helps in understanding their hydraulic properties in *Populus tremula* x *alba*. *Annals of Botany, 115*, 187–199.

Hertzberg, M., Aspeborg, H., Schrader, J., Andersson, A., Erlandsson, R., Blomqvist, K., et al. (2001). A transcriptional roadmap to wood formation. *Proceedings of the National Academy of Sciences of the United States of America, 98*, 14732–14737.

Hocq, L., Pelloux, J., & Lefebvre, V. (2017). Connecting homogalacturonan-type pectin remodeling to acid growth. *Trends in Plant Science, 22*, 20–29.

Hofte, H. (2015). The yin and yang of cell wall integrity control: Brassinosteroid and FERONIA signaling. *Plant and Cell Physiology, 56*, 224–231.

Hongo, S., Sato, K., Yokoyama, R., & Nishitani, K. (2012). Demethylesterification of the primary wall by PECTIN METHYLESTERASE35 provides mechanical support to the *Arabidopsis* stem. *Plant Cell, 24*, 2624–2634.

Huve, K., Christ, M. M., Kleist, E., Uerlings, R., Niinemets, U., Walter, A., et al. (2007). Simultaneous growth and emission measurements demonstrate an interactive control of methanol release by leaf expansion and stomata. *Journal of Experimental Botany, 58*, 1783–1793.

Israelsson, M., Eriksson, M. E., Hertzberg, M., Aspeborg, H., Nilsson, P., & Moritz, T. (2003). Changes in gene expression in the wood-forming tissue of transgenic hybrid aspen with increased secondary growth. *Plant Molecular Biology, 52*, 893–903.

Iwai, H., Masaoka, N., Ishii, T., & Satoh, S. (2002). A pectin glucuronyltransferase gene is essential for intercellular attachment in the plant meristem. *Proceedings of the National Academy of Sciences of the United States of America, 99*, 16319–16324.

Johansson, H., Sterky, F., Amini, B., Lundeberg, J., & Kleczkowski, L. A. (2002). Molecular cloning and characterization of a cDNA encoding poplar UDP-glucose dehydrogenase, a key gene of hemicellulose/pectin formation. *Biochimica et Biophysica Acta, 1576*, 53–58.

Jones, L., Milne, J. L., Ashford, D., & McQueen-Mason, S. J. (2003). Cell wall arabinan is essential for guard cell function. *Proceedings of the National Academy of Sciences of the United States of America, 100*, 11783–11788.

Kalunke, R. M., Tundo, S., Benedetti, M., Cervone, F., De Lorenzo, G., & D'Ovidio, R. (2015). An update on polygalacturonase-inhibiting protein (PGIP), a leucine-rich repeat protein that protects crop plants against pathogens. *Frontiers in Plant Science, 6*, 17.

Kim, J. S., & Daniel, G. (2013). Developmental localization of homogalacturonan and xyloglucan epitopes in pit membranes varies between pit types in two poplar species. *IAWA Journal, 34*, 245–262.

Kim, S. J., Held, M. A., Zemelis, S., Wilkerson, C., & Brandizzi, F. (2015). CGR2 and CGR3 have critical overlapping roles in pectin methylesterification and plant growth in *Arabidopsis thaliana*. *Plant Journal, 82*, 208–220.

Kim, H., Kojima, M., Choi, D., Park, S., Matsui, M., Sakakibara, H., et al. (2016). Overexpression of INCREASED CAMBIAL ACTIVITY, a putative methyltransferase, increases cambial activity and plant growth. *Journal of Integrative Plant Biology, 58*, 874–889.

Klinghammer, M., & Tenhaken, R. (2007). Genome-wide analysis of the UDP-glucose dehydrogenase gene family in *Arabidopsis*, a key enzyme for matrix polysaccharides in cell walls. *Journal of Experimental Botany*, *58*, 3609–3621.

Krzeslowska, M. (2011). The cell wall in plant cell response to trace metals: Polysaccharide remodeling and its role in defense strategy. *Acta Physiologiae Plantarum*, *33*, 35–51.

Krzeslowska, M., Rabeda, I., Basinska, A., Lewandowski, M., Mellerowicz, E. J., Napieralska, A., et al. (2016). Pectinous cell wall thickenings formation—A common defense strategy of plants to cope with Pb. *Environmental Pollution*, *214*, 354–361.

Kuang, B. Q., Zhao, X. H., Zhou, C., Zeng, W., Ren, J. L., Ebert, B., et al. (2016). Role of UDP-glucuronic acid decarboxylase in xylan biosynthesis in *Arabidopsis*. *Molecular Plant*, *9*, 1119–1131.

Lairez, D., Cathala, B., Monties, B., Bedos-Belval, F., Duran, H., & Gorrichon, L. (2005). Aggregation during coniferyl alcohol polymerization in pectin solution: A biomimetic approach of the first steps of lignification. *Biomacromolecules*, *6*, 763–774.

Lesniewska, J., Ohman, D., Krzeslowska, M., Kushwah, S., Barciszewska-Pacak, M., Kleczkowski, L. A., et al. (2017). Defense responses in aspen with altered pectin methylesterase activity reveal the hormonal inducers of tyloses. *Plant Physiology*, *173*, 1409–1419.

Lionetti, V., Francocci, F., Ferrari, S., Volpi, C., Bellincampi, D., Galletti, R., et al. (2010). Engineering the cell wall by reducing de-methyl-esterified homogalacturonan improves saccharification of plant tissues for bioconversion. *Proceedings of the National Academy of Sciences of the United States of America*, *107*, 616–621.

Liu, J. W., Hai, G. H., Wang, C., Cao, S. Q., Xu, W. J., Jia, Z. G., et al. (2015). Comparative proteomic analysis of *Populus trichocarpa* early stem from primary to secondary growth. *Journal of Proteomics*, *126*, 94–108.

Liu, H. H., Ma, Y., Chen, N., Guo, S. Y., Liu, H. L., Guo, X. Y., et al. (2014). Overexpression of stress-inducible OsBURP16, the beta subunit of polygalacturonase 1, decreases pectin content and cell adhesion and increases abiotic stress sensitivity in rice. *Plant, Cell & Environment*, *37*, 1144–1158.

Lopez-Sanchez, P., Martinez-Sanz, M., Bonilla, M. R., Wang, D. J., Walsh, C. T., Gilbert, E. P., et al. (2016). Pectin impacts cellulose fibre architecture and hydrogel mechanics in the absence of calcium. *Carbohydrate Polymers*, *153*, 236–245.

Marin-Rodriguez, M. C., Orchard, J., & Seymour, G. B. (2002). Pectate lyases, cell wall degradation and fruit softening. *Journal of Experimental Botany*, *53*, 2115–2119.

Matsunaga, T., Ishii, T., Matsumoto, S., Higuchi, M., Darvill, A., Albersheim, P., et al. (2004). Occurrence of the primary cell wall polysaccharide rhamnogalacturonan II in pteridophytes, lycophytes, and bryophytes. Implications for the evolution of vascular plants. *Plant Physiology*, *134*, 339–351.

McCarthy, T. W., Der, J. P., Honaas, L. A., dePamphilis, C. W., & Anderson, C. T. (2014). Phylogenetic analysis of pectin-related gene families in *Physcomitrella patens* and nine other plant species yields evolutionary insights into cell walls. *BMC Plant Biology*, *14*, 79.

Micheli, F. (2001). Pectin methylesterases: Cell wall enzymes with important roles in plant physiology. *Trends in Plant Science*, *6*, 414–419.

Micheli, F., Sundberg, B., Goldberg, R., & Richard, L. (2000). Radial distribution pattern of pectin methylesterases across the cambial region of hybrid aspen at activity and dormancy. *Plant Physiology*, *124*, 191–199.

Mohnen, D. (2008). Pectin structure and biosynthesis. *Current Opinion in Plant Biology*, *11*, 266–277.

Navazio, L., Moscatiello, R., Bellincampi, D., Baldan, B., Meggio, F., Brini, M., et al. (2002). The role of calcium in oligogalacturonide-activated signalling in soybean cells. *Planta*, *215*, 596–605.

Nguema-Ona, E., Moore, J. P., Fagerstrom, A. D., Fangel, J. U., Willats, W. G. T., Hugo, A., et al. (2013). Overexpression of the grapevine PGIP1 in tobacco results in compositional changes in the leaf arabinoxyloglucan network in the absence of fungal infection. *BMC Plant Biology, 13*, 46.

Nguyen, H. P., Jeong, H. Y., Jeon, S. H., Kim, D., & Lee, C. (2017). Rice pectin methylesterase inhibitor28 (OsPMEI28) encodes a functional PMEI and its overexpression results in a dwarf phenotype through increased pectin methylesterification levels. *Journal of Plant Physiology, 208*, 17–25.

O'Neill, M., Albersheim, P., & Darvill, A. (1990). The pectic polysaccharides of primary cell walls. In P. M. Dey (Ed.), *Methods in plant biochemistry: Vol. 2* (pp. 415–441): Academic Press.

O'Neill, M. A., Eberhard, S., Albersheim, P., & Darvill, A. G. (2001). Requirement of borate cross-linking of cell wall rhamnogalacturonan II for *Arabidopsis* growth. *Science, 294*, 846–849.

Orfila, C., Dal Degan, F., Jorgensen, B., Scheller, H. V., Ray, P. M., & Ulvskov, P. (2012). Expression of mung bean pectin acetyl esterase in potato tubers: Effect on acetylation of cell wall polymers and tuber mechanical properties. *Planta, 236*, 185–196.

Pagel, W., & Heitefuss, R. (1990). Enzyme activities in soft rot pathogenesis of potato tubers: Effects of calcium, pH, and degree of pectin esterification on the activities of polygalacturonase and pectate lyase. *Physiological and Molecular Plant Pathology, 37*, 9–25.

Pelloux, J., Rusterucci, C., & Mellerowicz, E. J. (2007). New insights into pectin methylesterase structure and function. *Trends in Plant Science, 12*, 267–277.

Qu, L. H., Wu, C. Y., Zhang, F., Wu, Y. Y., Fang, C. Y., Jin, C., et al. (2016). Rice putative methyltransferase gene OsTSD2 is required for root development involving pectin modification. *Journal of Experimental Botany, 67*, 5349–5362.

Ralet, M. C., Crepeau, M. J., Buchholt, H. C., & Thibault, J. F. (2003). Polyelectrolyte behaviour and calcium binding properties of sugar beet pectins differing in their degrees of methylation and acetylation. *Biochemical Engineering Journal, 16*, 191–201.

Rautengarten, C., Ebert, B., Moreno, I., Temple, H., Herter, T., Link, B., et al. (2014). The golgi-localized bifunctional UDP-rhamnose/UDP-galactose transporter family of *Arabidopsis. Proceedings of the National Academy of Sciences of the United States of America, 111*, 11563–11568.

Saez-Aguayo, S., Rautengarten, C., Temple, H., Sanhueza, D., Ejsmentewicz, T., Sandoval-Ibanez, O., et al. (2017). UUAT1 is a golgi-localized UDP-uronic acid transporter that modulates the polysaccharide composition of *Arabidopsis* seed mucilage. *Plant Cell, 29*, 129–143.

Saffer, A. M., Carpita, N. C., & Irish, V. F. (2017). Rhamnose-containing cell wall polymers suppress helical plant growth independently of microtubule orientation. *Current Biology, 27*, 2248–2259.

Samuels, A. L., Giddings, T. H., & Staehelin, L. A. (1995). Cytokinesis in tobacco BY-2 and root tip cells: A new model of cell plate formation in higher plants. *Journal of Cell Biology, 130*, 1345–1357.

Seifert, G. J. (2004). Nucleotide sugar interconversions and cell wall biosynthesis: How to bring the inside to the outside. *Current Opinion in Plant Biology, 7*, 277–284.

Senechal, F., Wattier, C., Rusterucci, C., & Pelloux, J. (2014). Homogalacturonan-modifying enzymes: Structure, expression, and roles in plants. *Journal of Experimental Botany, 65*, 5125–5160.

Serapiglia, M. J., Cameron, K. D., Stipanovic, A. J., & Smart, L. B. (2012). Correlations of expression of cell wall biosynthesis genes with variation in biomass composition in shrub willow (*Salix* spp.) biomass crops. *Tree Genetics & Genomes, 8*, 775–788.

Sexton, T. R., Henry, R. J., Harwood, C. E., Thomas, D. S., McManus, L. J., Raymond, C., et al. (2012). Pectin methylesterase genes influence solid wood properties of *Eucalyptus pilularis. Plant Physiology, 158*, 531–541.

Siedlecka, A., Wiklund, S., Peronne, M. A., Micheli, F., Lesniewska, J., Sethson, I., et al. (2008). Pectin methyl esterase inhibits intrusive and symplastic cell growth in developing wood cells of *Populus*. *Plant Physiology*, *146*, 554–565.

Singh, S. K., Eland, C., Harholt, J., Scheller, H. V., & Marchant, A. (2005). Cell adhesion in *Arabidopsis thaliana* is mediated by ECTOPICALLY PARTING CELLS 1—A glycosyltransferase (GT64) related to the animal exostosins. *Plant Journal*, *43*, 384–397.

Smith, E., Naik, D., & Cumming, J. R. (2011). Genotypic variation in aluminum resistance, cellular aluminum fractions, callose and pectin formation and organic acid accumulation in roots of *Populus* hybrids. *Environmental and Experimental Botany*, *72*, 182–193.

Solecka, D., Zebrowski, J., & Kacperska, A. (2008). Are pectins involved in cold acclimation and de-acclimation of winter oil-seed rape plants? *Annals of Botany*, *101*, 521–530.

Sterling, J. D., Atmodjo, M. A., Inwood, S. E., Kolli, V. S. K., Quigley, H. F., Hahn, M. G., et al. (2006). Functional identification of an *Arabidopsis* pectin biosynthetic homogalacturonan galacturonosyltransferase. *Proceedings of the National Academy of Sciences of the United States of America*, *103*, 5236–5241.

Sundell, D., Street, N. R., Kumar, M., Mellerowicz, E. J., Kucukoglu, M., Johnsson, C., et al. (2017). AspWood: High-spatial-resolution transcriptome profiles reveal uncharacterized modularity of wood formation in *Populus tremula*. *Plant Cell*, *29*, 1585–1604.

Swamy, P. S., Hu, H., Pattathil, S., Maloney, V. J., Xiao, H., Xue, L. J., et al. (2015). Tubulin perturbation leads to unexpected cell wall modifications and affects stomatal behaviour in Populus. *Journal of Experimental Botany*, *66*(20), 6507–6518.

Tan, L., Eberhard, S., Pattathil, S., Warder, C., Glushka, J., Yuan, C. H., et al. (2013). An *Arabidopsis* cell wall proteoglycan consists of pectin and arabinoxylan covalently linked to an arabinogalactan protein. *Plant Cell*, *25*, 270–287.

Terashima, N., Fukushima, K., Sano, Y., & Takabe, K. (1988). Heterogeneity in formation of lignin. 10. Visualization of lignification process in differentiating xylem of pine by microautoradiography. *Holzforschung*, *42*, 347–350.

Thornton, J., Ekman, R., Holmbom, B., & Eckerman, C. (1993). Release of potential anionic trash in peroxide bleaching of mechanical pulp. *Paperi ja Puu—Paper and Timber*, *75*, 426–431.

Tian, J. X., Du, Q. Z., Li, B. L., & Zhang, D. Q. (2014). Single nucleotide polymorphisms in the 5' UTR of UDP-glucose dehydrogenase (PtUGDH) associate with wood properties in *Populus tomentosa*. *Tree Genetics & Genomes*, *10*, 339–354.

Touzel, J. P., Chabbert, B., Monties, B., Debeire, P., & Cathala, B. (2003). Synthesis and characterization of dehydrogenation polymers in Gluconacetobacter xylinus cellulose and cellulose/pectin composite. *Journal of Agricultural and Food Chemistry*, *51*, 981–986.

Tucker, G., Yin, X. R., Zhang, A. D., Wang, M. M., Zhu, Q. G., Liu, X. F., et al. (2017). Ethylene and fruit softening. *Food Quality and Safety*, *1*, 253–267.

Urbanowicz, B. R., Pena, M. J., Ratnaparkhe, S., Avci, U., Backe, J., Steet, H. F., et al. (2012). 4-O-Methylation of glucuronic acid in *Arabidopsis* glucuronoxylan is catalyzed by a domain of unknown function family 579 protein. *Proceedings of the National Academy of Sciences of the United States of America*, *109*, 14253–14258.

van Doorn, W. G., Hiemstra, T., & Fanourakis, D. (2011). Hydrogel regulation of xylem water flow: An alternative hypothesis. *Plant Physiology*, *157*, 1642–1649.

Venturas, M. D., Sperry, J. S., & Hacke, U. G. (2017). Plant xylem hydraulics: What we understand, current research, and future challenges. *Journal of Integrative Plant Biology*, *59*, 356–389.

Voxeur, A., Soubigou-Taconnat, L., Legee, F., Sakai, K., Antelme, S., Durand-Tardif, M., et al. (2017). Altered lignification in mur1-1 a mutant deficient in GDP-L-fucose synthesis with reduced RG-II cross linking. *PLoS One*, *12*, e0184820.

Wang, T., & Hong, M. (2016). Solid-state NMR investigations of cellulose structure and interactions with matrix polysaccharides in plant primary cell walls. *Journal of Experimental Botany*, *67*, 503–514.

Wang, M. J., Yuan, D. J., Gao, W. H., Li, Y., Tan, J. F., & Zhang, X. L. (2013). A comparative genome analysis of PME and PMEI families reveals the evolution of pectin metabolism in plant cell walls. *PLoS One*, *8*, e72082.

Weraduwage, S. M., Kim, S. J., Renna, L., Anozie, F. C., Sharkey, T. D., & Brandizzi, F. (2016). Pectin methylesterification impacts the relationship between photosynthesis and plant growth. *Plant Physiology*, *171*, 833–848.

Wi, S. G., Singh, A. P., Lee, K. H., & Kim, Y. S. (2005). The pattern of distribution of pectin, peroxidase and lignin in the middle lamella of secondary xylem fibres in alfalfa (*Medicago sativa*). *Annals of Botany*, *95*, 863–868.

Wildhagen, H., Paul, S., Allwright, M., Smith, H. K., Malinowska, M., Schnabel, S. K., et al. (2018). Genes and gene clusters related to genotype and drought-induced variation in saccharification potential, lignin content and wood anatomical traits in *Populus nigra*. *Tree Physiology*, *38*, 320–339.

Willats, W. G. T., Orfila, C., Limberg, G., Buchholt, H. C., van Alebeek, G., Voragen, A. G. J., et al. (2001). Modulation of the degree and pattern of methyl-esterification of pectic homogalacturonan in plant cell walls: Implications for pectin methyl esterase action, matrix properties, and cell adhesion. *Journal of Biological Chemistry*, *276*, 19404–19413.

Willfor, S., Sundberg, A., Pranovich, A., & Holmbom, B. (2005). Polysaccharides in some industrially important hardwood species. *Wood Science and Technology*, *39*, 601–617.

Wolf, S., & Greiner, S. (2012). Growth control by cell wall pectins. *Protoplasma*, *249*, 169–175.

Wolf, S., van der Does, D., Ladwig, F., Sticht, C., Kolbeck, A., Schurholz, A. K., et al. (2014). A receptor-like protein mediates the response to pectin modification by activating brassinosteroid signaling. *Proceedings of the National Academy of Sciences of the United States of America*, *111*, 15261–15266.

Xiao, C. W., & Anderson, C. T. (2013). Roles of pectin in biomass yield and processing for biofuels. *Frontiers in Plant Science*, *4*, 67.

Xiao, C. W., Barnes, W. J., Zamil, M. S., Yi, H., Puri, V. M., & Anderson, C. T. (2017). Activation tagging of *Arabidopsis* POLYGALACTURONASE INVOLVED IN EXPANSION2 promotes hypocotyl elongation, leaf expansion, stem lignification, mechanical stiffening, and lodging. *Plant Journal*, *89*, 1159–1173.

Xiong, J., Yang, Y. J., Fu, G. F., & Tao, L. X. (2015). Novel roles of hydrogen peroxide (H_2O_2) in regulating pectin synthesis and demethylesterification in the cell wall of rice (*Oryza sativa*) root tips. *New Phytologist*, *206*, 118–126.

Yang, Z. L., Liu, H. J., Wang, X. R., & Zeng, Q. Y. (2013). Molecular evolution and expression divergence of the *Populus* polygalacturonase supergene family shed light on the evolution of increasingly complex organs in plants. *New Phytologist*, *197*, 1353–1365.

Yu, F. W., Zhu, X. F., Li, G. J., Kronzucker, H. J., & Shi, W. M. (2016). The chloroplast protease AMOS1/EGY1 affects phosphate homeostasis under phosphate stress. *Plant Physiology*, *172*, 1200–1208.

Zhu, X. F., Wang, Z. W., Wan, J. X., Sun, Y., Wu, Y. R., Li, G. X., et al. (2015). Pectin enhances rice (*Oryza sativa*) root phosphorus remobilization. *Journal of Experimental Botany*, *66*, 1017–1024.

Zhu, X. F., Zhu, C. Q., Zhao, X. S., Zheng, S. J., & Shen, R. F. (2016). Ethylene is involved in root phosphorus remobilization in rice (*Oryza sativa*) by regulating cell-wall pectin and enhancing phosphate translocation to shoots. *Annals of Botany*, *118*, 645–653.

CHAPTER NINE

Applications of transgenic approaches in developing an understanding of drought tolerance in poplar

Edward G. Kirby*, Juan Jesús Molina-Rueda
Department of Biological Sciences, Rutgers University, Newark, NJ, United States
*Corresponding author: e-mail address: ekirby@rutgers.edu

Contents

Abstract

This paper will present an overview of genetic approaches to develop an understanding of interacting roles of reduced nitrogen in conferring drought tolerance in poplar. Included in the discussion are: (1) the critical role that drought tolerant phenotypes of forest trees will play in a worldwide environment faced with the challenges of climate change; (2) a summary of transgenic technologies as applied to poplar with assessment of recent transgenic approaches into biological mechanisms for drought tolerance; and (3) the relationship between nitrogen assimilation and drought tolerance in poplar. Inherent in this discussion is assessment of the complexity and pleiotropic nature of drought tolerant phenotypes, including physiological, biochemical, morphological, and genetic components (Allwright & Taylor, 2016; Borland et al., 2015; Harfouche, Meilan, & Altman, 2014; Polle & Chen, 2015). Central to this discussion are the

Advances in Botanical Research, Volume 89
ISSN 0065-2296
https://doi.org/10.1016/bs.abr.2018.11.009

considerable advances in poplar biology and genome analysis resulting in defining poplar as a model forest species (Bradshaw, Ceulemans, Davis, & Stettler, 2000; Cronk, 2005; Taylor, 2002). Focus is not only on the application of GE technologies for expression of specific transgenes to produce new phenotypes, but also on assessment of specific phenotypes created through GE approaches that can allow identification of specific genes and gene expression networks correlated with the induced phenotype that can be incorporated into advanced breeding approaches.

1. The need for drought tolerant energy crops, including poplar

The current worldwide environment is facing significant challenges directly associated with increased release of heat-trapping greenhouse gases. These changes have not only resulted in areas with significant increases in the occurrences of drought, but predictions of long term increase in global mean temperatures (IPCC, 2014; Walsh et al., 2014). Moreover, if the trend continues, increases in global temperatures will result in changing precipitation patterns reflected in drought in certain regions and increased precipitation in others.

Increasing episodes of drought have led to decreased agricultural and forest productivity in expanding regions worldwide (Blum, 1996; Bogardi et al., 2012; Boisvenue & Running, 2006). Although irrigation can provide relief from drought conditions, these practices are costly and can result in increased salt content and erosion of agricultural soils (Aragüés et al., 2011; Pagliai, Vignozzi, & Pellegrini, 2004). Establishment of new varieties of agricultural and energy crops characterized by enhanced drought tolerance are essential to address the challenges of climate change. This is of increasing importance for biofuel crops being deployed to address the alternative energy demands, including poplar (Ye et al., 2013).

2. Transgenic technologies for conferring drought tolerance in forest species

2.1 *Agrobacterium* transformations

Instrumental to genetic improvement of energy crops are applications of evolving biotechnology. These approaches pose significant opportunities for improvement of important agronomic traits of forest species including enhanced biomass production and drought and salt tolerance (for reviews:

Dinneny et al., 2008; Dubouzet, Strabala, & Wagner, 2013; Polle, Janz, Teichmann, & Lipka, 2013; Wang, Cai, Xu, Wang, & Dai, 2016; Wang, Liu, et al., 2016; Yadav, Ayre, & Bush, 2015). Both forward and reverse genetic approaches focus on regulation of expression of specific genes and introduction and expression of transgenes conferring desired phenotypes. These approaches routinely employ *Agrobacterium*-mediated transformations.

Various poplar species and poplar hybrids have been genetically engineered using *Agrobacterium tumefaciens* vectors, ranging from the initial reports (Leple, Brasileiro, Michel, et al., 1992; Parsons, Sinkar, Stettler, Nester, & Gordon, 1986) to current approaches. Poplar transformation protocols and protocols for construction of T-DNA plasmid vectors for poplar transformations have been extensively reviewed (Busov et al., 2005; Cseke, Cseke, & Podila, 2007; Maheshwari & Kovalchuk, 2016; Movahedi, Zhang, Amirian, & Zhuge, 2014; Yevtushenko & Misra, 2010). *Agrobacterium*-mediated transformations in poplar are typically carried out using various ex vitro and in vitro explants, including stem internodes, hypocotyls, cell suspension cultures, petioles, stem segments, and leaves. New approaches for effective construction of T-DNA transformation vectors for delivering transgenes to poplar have recently been reported. These involve Gateway-compatible binary vectors based on bacterial artificial chromosomes which allow the insertion of single-copy transgenes with high efficiency (Tark-Dame et al., 2018).

2.2 RNA interference (RNAi)

RNA interference (RNAi) approaches have also been employed (Katoch & Thakur, 2013). RNAi is a sequence-specific gene silencing mechanism, triggered by the introduction of dsRNA leading to mRNA degradation and regulation of targeted gene expression. Several reports describe the use of RNAi in regulating gene expression in poplar (Meyer et al., 2004). Included are the impact of endo-β-1,4-glucanases on cell wall development (Glass, Barkwill, Unda, & Mansfield, 2015), down regulation of plasma membrane intrinsic proteins (PIPs) (Bi et al., 2015), the roles of several genes in conferring drought resistance and biomass production including a phytoalexin PAD4, legion simulation disease 1 (LDS1), and enhanced disease susceptibility (EDS1) (Szechyńska-Hebda, Czarnocka, Hebda, & Karpiński, 2016), and down-regulation of KORRIGAN-like endo-β-1,4-glucanase (Kalluri et al., 2016). Li et al. have shown stable transgene suppression via RNAi in field trials of transgenics (Li, Brunner, Meilan, & Strauss, 2009). Recently RNAi has been used in poplar to reduce expression of a tonoplast sucrose efflux transporter, *PtaSUT4* (Xue, Frost, Tsai, & Harding, 2016). These RNAi plants exhibit reduced shoot growth compared to the wild type (WT) under sustained mild drought leading the authors to conclude that steepened a shoot-root sugar gradient in RNAi plants increases sensitivity of root tips to decreasing soil water availability.

2.3 CRISPR/Cas9 technology

RNA-guided genome editing using the type II clustered regularly interspaced short palindromic repeats (CRISPR)-associated protein (Cas) offers significant opportunities for editing plant genomes (Samanta, Dey, & Gayen, 2016). Several reports focus on genome editing in poplar (Demirci, Zhang, & Unver, 2018; Tsai & Xue, 2015). Reports from poplar include target editing of genomic sites of the phytoene desaturase gene 8 (*PtoPDS*) (Fan et al., 2015). Zhou, Jacobs, Xue, Harding, and Tsai (2015) have explored CRISPR-induced mutations in poplar to reveal 4-coumarate: CoA ligase (4-CL) specificities and redundancy. Recently Elorriaga, Klocko, Ma, and Strauss (2018) have targeted the poplar homologue of the floral meristem identity gene, *LEAFY* (*LFY*), and two poplar orthologues of the floral organ identity gene *AGAMOUS* (*AG*) and shown variation in mutation spectra achieved using the CRISPR/Cas9 approach. The CRISPR/Cas9 gene editing and evolving technologies are certain to become of increasing importance in regulating gene expression in crop and biomass species.

Caution has been recommended in deploying genetically engineered trees, either in specific field trials or in unconfined commercial forests without significant environmental risk assessment (Häggman et al., 2013; Strauss et al., 2017). Areas of concern focus on the impacts of genetically engineered perennial species on ecological biodiversity and societal concerns and support for implementation of these technologies.

3. What transgenic technologies have revealed about drought tolerance in poplar

A number of recent reports describe efforts to enhance drought tolerance in poplar using GE technology. These reports reveal underlying biological factors and genetic contributions to this important trait. The reports summarized below focus on three major areas that contribute to this important phenotype in poplar: enhanced ability of drought resistant phenotypes to reduce the effects of drought-induced reactive oxygen species, ABA responses on stomatal behaviour and tissue patterning related to reduced leaf transpiration, and drought tolerance enhanced by increased nitrogen assimilation.

In *Populus trichocarpa*, a number of pyrabactin resistance-like ABA receptors have been identified. Two of these receptors (PtPYRLs) positively regulate drought tolerance (Yu et al., 2016). The physiological role of these ABA receptors has recently been studied (Yu et al., 2017). Constitutive overexpression of *PtPRYL1* and *PtPRYL5* in transgenic poplars increased drought tolerance through ABA signalling leading to increases in stomatal closure and accompanying decreases in leaf water loss during drought. Overexpression lines also exhibited improved ROS scavenging and activation of antioxidant enzymes during drought. Interestingly, Wang, Cai, et al. (2016) and Wang, Liu, et al. (2016) recently reported that overexpression of *PdEPF1*, a member of the EPIDERMAL PATTERNING FACTOR gene family isolated from a fast growing clone of *Populus nigra*, resulted in increased water deficit tolerance and water use efficiency in *Paulownia tomentosa*. Physiological and morphological analysis showed that transgenic lines exhibited 30% reduction in transpiration rates and 28% reduction in abaxial stomate density.

Cellular signalling plays key roles in stress responses. Included are components of the MAP kinase signalling cascade. Overexpression of *PtMKK4* in transgenic poplar significantly enhanced drought tolerance, in part by lowering levels of ROS and increasing activities of antioxidant enzymes including POD and SOD (Wang, Shen, Xu, & Guo, 2014; Wang, Su, et al., 2014). Zhang et al. (2017) reported that overexpression of *AtNDPK2*, a nucleotide diphosphate kinases associated with H_2O_2-mediated mitogen-activated protein kinase signalling in *Arabidopsis*, resulted in lower levels of ROS and enhanced growth and drought tolerance in poplar.

Transgenic poplars expressing enhanced nitrogen assimilation as conferred by ectopic expression of pine cytosolic glutamine synthetase (*PsGS1a*) exhibit interesting pleiotropic phenotypes, including drought tolerance (Molina-Rueda & Kirby, 2015). Investigation of the mechanisms accounting for drought tolerance in GS poplars indicates that overexpression of the heterologous pine *GS1a* enhanced sustained photosynthetic electron transport capacity during severe stomatal limitation. The data also suggest that ectopic expression of cytosolic GS increases photorespiratory activity, and that this serves as a protective sink for electrons from photosynthetic reaction centres (El-Khatib, Hamerlynck, Gallardo, & Kirby, 2004).

In silico characterization of the SOD gene family of *P. trichocarpa* and assessment of transcript levels for the SOD gene family in various tissues of GS transgenic and wild type poplars subjected to drought treatments have shown that drought tolerant GS poplars have altered patterns of SOD expression when compared with the wild type under drought conditions. Specific SODs appear strongly correlated with the drought tolerant phenotype (Molina-Rueda, Tsai, & Kirby, 2013). Moreover, *PtGS1a* overexpression also results in alterations in drought-related metabolites including higher levels of GABA, hydroxyproline, and putrescine (Molina-Rueda & Kirby, 2015).

4. Nitrogen utilization in trees and the central role of glutamine synthetase

The availability of inorganic nitrogen in soils is often a limiting factor in plant growth and development. Most tree species grow in areas where nitrate and/or ammonium are the major forms of nitrogen available in soil (Wolt & Wolt, 1994).

One main difference between herbaceous and woody plants is the manner by which soil nitrate is assimilated. In many tree species the majority of nitrate taken up from soil is reduced in roots to ammonium by the action of both nitrate reductase (NR) and nitrite reductase (NiR). NR and NiR require cytosolic NAD(P)H and plastidic NAD(P)H (or reduced ferredoxin), respectively, for their reactions (for review, see Gallardo, Fu, Jing, Kirby, & Cánovas, 2003; Kirby, Gallardo, Man, & El-Khatib, 2006; Wang, Shen, et al., 2014; Wang, Su, et al., 2014; Xu, Fan, & Miller, 2012). Ammonium ions produced by NR and NiR are incorporated into the pool of organic nitrogen-containing compounds through reactions catalysed by glutamine synthetase (GS) and glutamate synthase (GOGAT) in the GS/GOGAT cycle (Fig. 1). GS plays the central role in the complex matrix of plant nitrogen metabolism since the products of glutamine synthetase, glutamate and glutamine are precursors of all organic nitrogen compounds in plants (Häusler, Ludewig, & Krueger, 2014).

In angiosperms GS is a holoenzyme composed of eight identical polypeptide subunits and exists as two major isoenzymes, GS1 and GS2, located in different subcellular compartments: GS1 located in the cytoplasm and GS2 localized in chloroplasts. The two major GS isoenzymes display non-overlapping roles. GS1, the cytosolic isoform, is the predominant enzyme in roots and non-photosynthetic tissues. The biological role of

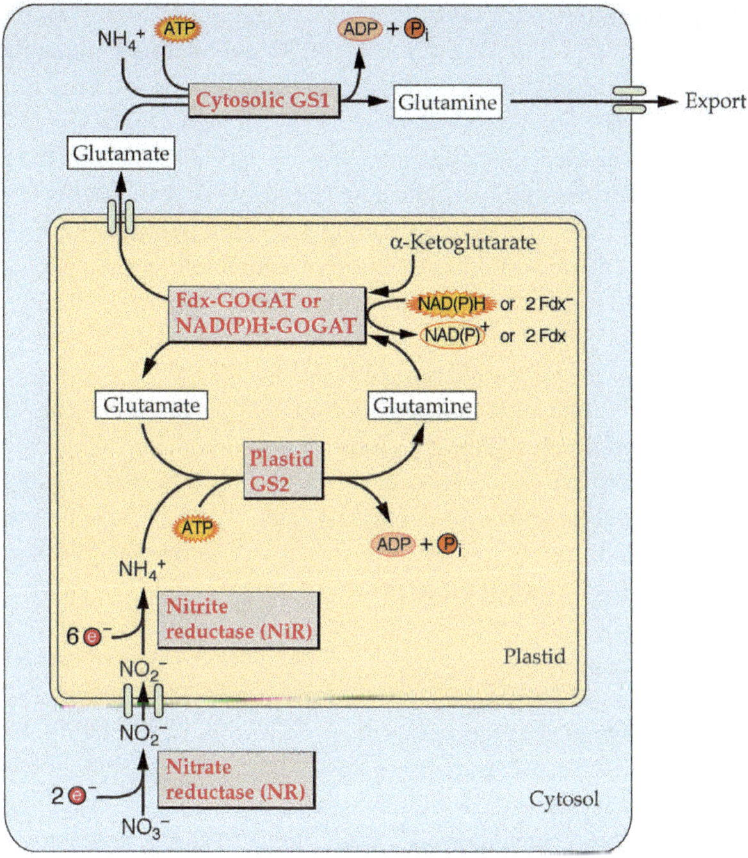

Fig. 1 A simplified diagram of the pathway for assimilation of nitrogen in angiosperms. The isoenzymes of glutamine synthetase (GS) are present on both plastids (GS2) and the cytoplasm (GS1). Fdx, ferredoxin. *Buchanan, B. B., Gruissem, W., & Jones, R. L. (2000).* Biochemistry and molecular biology of plants. *Rockville, Maryland, USA: American Society of Plant Biologists, by permission.*

GS1 is still not well-defined, but is involved in the primary assimilation of ammonium from the soil and in recycling of ammonium released through metabolic processes other than photorespiration (i.e. phenylpropanoid biosynthesis, protein metabolism including storage proteins, and chlorophyll breakdown). Since GS1 in leaves is expressed specifically in the vascular bundles, a role for GS1 in transport of glutamine to other plant organs has been proposed (Bernard & Habash, 2009; Betti et al., 2012).

In photosynthetic tissues of angiosperms, plastidic GS2 is responsible for the assimilation of ammonium derived from nitrate reduction and photorespiration (Ireland & Lea, 1999). GS2 is expressed during plastid differentiation (Gálvez, Gallardo, & Cánovas, 1990) and under stress conditions, including water stress (Bauer et al., 1997) and pathogen attack (Perez-Garcia, Canovas, Gallardo, Hirel, & Devicente, 1995). In leaves, GS2 works with plastidic ferredoxin-dependent glutamate synthase (Fd-GOGAT) in the production of glutamate from 2-oxoglutarate (Fig. 1). GS1 works with plastidic NADH-GOGAT to produce glutamate in non-photosynthetic cells, especially in nitrogen fixing nodules in the roots of leguminous species. GS1 comprises a small gene family of 2–5 members. In contrast, GS2, Fd-GOGAT, and NADH-GOGAT gene families consist of only 1–2 genes (Lam, Coschigano, Oliveira, Melo-Oliveira, & Coruzzi, 1996).

The GS gene family in *P. trichocarpa* consists of eight different genes exhibiting all structural and regulatory elements consistent with their roles as functional genes (Castro-Rodríguez et al., 2011). Poplar GS family members are organized in four groups of duplicated genes, three of which code for cytosolic GS isoforms (GS1) and one which codes for the chloroplastic GS isoform (GS2) (Castro-Rodríguez et al., 2011).

Based on molecular, biochemical, and cytological studies, there is currently no evidence for the presence of a plastidic GS2 in conifers (García-Gutiérrez, Cánovas, & Ávila, 2018; Perez-Garcia et al., 1995). However, two GS1 isoenzymes are reported in pine (GS1a and GS1b) with different patterns of expression and distinct molecular and kinetic properties (Avila, Suárez, Gómez-Maldonado, & Cánovas, 2001; De la Torre et al., 2002). While GS1b exhibits an expression pattern similar to angiosperm GS1, pine GS1a expression is associated with chloroplast development, suggesting that its physiological role may be similar to that of GS2 in angiosperms (Gallardo et al., 2003).

4.1 Nitrogen use efficiency (NUE)

The need to develop crop and energy plants with enhanced ability to take up and assimilate soil nutrients has been a focus of research efforts. NUE in plants may be divided into three main physiological components: (1) nitrogen uptake and transport; (2) nitrogen assimilation; (3) remobilization of assimilated nitrogen (Hirel, Le Gouis, Ney, & Gallais, 2007). Approaches to improve NUE in crop plants have focused on both genetic and transgenic approaches to address these components (Han, Okamoto, Beatty, Rothstein, & Good, 2015; Li, Hu, & Chu, 2017; McAllister, Beatty, & Good, 2012).

Transgenic crop plants characterized by ectopic expression of NUE transgenes may display enhanced phenotypes. NUE transgenes that have been studied include high affinity nitrogen transporters, nitrate and nitrite reductase, amino transferases (AspAT, GdhA, ASN1, AlaAT), cytosolic and plastidic glutamine synthetase (GS1 and GS2), glutamate synthase (Fd-GOGAT and NADH-GOGAT), and regulatory/transcription factors (ANR1, Dof1, GLB1) (for reviews, see Andrews, Lea, Raven, & Lindsey, 2004; Coque & Gallais, 2006; Fageria & Baligar, 2005; Good, Shrawat, & Muench, 2004; Hirel et al., 2007).

Transgenic approaches have produced mixed results, some showing enhancements in N transport, N assimilation, and plant yield; others showing minimal effects on targeted phenotypes. Interestingly, several reports have described engineering increased nitrogen use efficiency in canola and rice using a barley alanine aminotransferase (*AlaAT*) driven by root-specific promoters (Good et al., 2007; Shrawat, Carroll, DePauw, Taylor, & Good, 2008). Results demonstrate success in enhancing NUE by manipulating a downstream step in nitrogen metabolism. However, analysis of the rice transcriptome in roots and shoots of AlaAT transformants revealed little impact of the transgene on expression profiles (Beatty, Shrawat, Carroll, Zhu, & Good, 2009). Although several genes induced in roots and shoots were identified, this study did not assess expression patterns in transgenics under conditions of limited nitrogen availability. Because multiple genetic components contribute to NUE, it is generally concluded that improvement of NUE must take both genetic (breeding) strategies and transgenic approaches (Kant, Bi, & Rothstein, 2011; McAllister et al., 2012; Thomsen, Eriksson, Møller, & Schjoerring, 2014). Although the need to improve NUE in forest trees must be addressed for economic, bioenergy, and environmental reasons (Man, Boriel, El-Khatib, & Kirby, 2005), little research has focused on this fundamental aspect of forest yield.

Hybrid poplar plants (*Populus tremula* × *P. alba*, INRA 717-1-B4) expressing ectopically pine glutamine synthetase (*GS1a*) display a number of significant pleiotropic phenotypes. Included are enhanced growth (Gallardo et al., 1999; Zhong et al., 2004) enhanced nitrogen use efficiency (NUE) (Man et al., 2005), enhanced resistance to phosphinothricin herbicides (Pascual, Jing, Kirby, Cánovas, & Gallardo, 2008) altered auxin homeostasis (Man, Pollmann, Weiler, & Kirby, 2011), alterations in wood chemistry and fibre characteristics (Coleman, Cánovas, Man, Kirby, & Mansfield, 2012), and drought tolerance (El-Khatib et al., 2004; Molina-Rueda & Kirby, 2015; Molina-Rueda et al., 2013). Ectopic expression of pine GS1a in transgenic poplars results in higher GS activities and glutamine levels than in wild type plants (Gallardo et al., 1999; Man et al., 2005; Molina-Rueda & Kirby, 2015). Furthermore, expression of the pine GS in poplar results in higher levels of metabolically related amino acids, including glutamate and γ-amino butyric acid (GABA), amino acids that can act as signalling molecules in plants (Bouché & Fromm, 2004; Forde & Lea, 2007; Häusler et al., 2014). GABA metabolism in plants is induced in response to biotic and abiotic stresses (Kinnersley & Turano, 2000), including salinity (Zhang, Zhang, Du, Chen, & Tang, 2011) and drought (Bor et al., 2009). Additionally, GABA has been shown to regulate expression of hormone-related genes (Kathiresan, Tung, Chinnappa, & Reid, 1997; Lancien & Roberts, 2006; Shi et al., 2010).

5. The relationship between nitrogen assimilation and drought tolerance in hybrid poplar

Recognizing the importance of developing an understanding of drought tolerance, we have investigated mechanisms for enhanced drought tolerance observed in GS poplar in ecophysiological studies (El-Khatib et al., 2004; Molina-Rueda & Kirby, 2015; Molina-Rueda et al., 2013; Zhong et al., 2004). In comparison to leaves of WT controls, GS poplars maintain expression of fundamental enzymes (GS, GOGAT, and Rubisco) under water stress and during recovery. Overexpression of GS also sustained photosynthetic electron transport capacity under severe stomatal limitations (El-Khatib et al., 2004). This suggests that ectopic expression of GS increases photorespiratory activity, thus serving as an effective protective energy sink to light-harvesting capacity. Moreover, overexpression of GS1 in transgenic poplars enhances antioxidant defence systems resulting in increased resistance to photooxidation during drought (Molina-Rueda & Kirby, 2015; Molina-Rueda et al., 2013).

Comparative transcriptome analysis was undertaken to establish patterns of differential gene expression in drought tolerant GS transgenic poplars and the wild type (WT) control. The data allow construction of co-expression networks for both GS and WT poplars revealing interesting changes associated with expression of the GS transgene (Molina-Rueda, Du, Wang, Tsai, & Kirby, in preparation). Summarized below, these studies reveal new hypotheses about mechanisms for drought tolerance in poplar and add to the understanding of drought tolerance mechanisms, an increasingly important and complex agronomic trait (Civelek & Lusis, 2014; McCune & Schimenti, 2012; Porth et al., 2013).

To undertake these studies, RNA was extracted from four tissue types (young leaves, mature leaves, stems, and roots) of GS and WT poplars grown in greenhouse studies under well-watered, drought, and drought recovery conditions. RNA was purified, amplified, and transcriptome analysis undertaken utilizing a custom Agilent microarray (GE 4x44K, AMADID # 021876). The SLIM multiple hypothesis testing algorithm (Wang, Tuominen, & Tsai, 2011) revealed a total of 6217 genes whose expression was significantly altered (minimum of twofold differences) between the GS transgenic and the wild type in at least one tissue and condition (Fig. 2). In every tissue and condition, differentially expressed (DE) genes segregated into two groups: genes with higher transcript levels in the GS transgenic than in the WT (up-regulated genes), and genes with lower transcript level in the GS than in the WT (down-regulated genes). Additionally, genes up- or down-regulated in more than one growth condition were identified (overlaps in diagrams of Fig. 2).

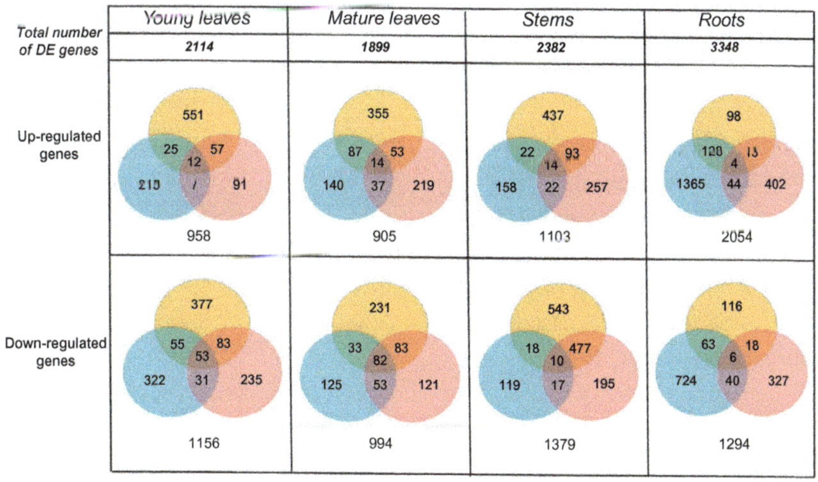

Fig. 2 Venn diagram representation of numbers of differentially expressed (DE) genes with higher (up regulated) and lower (down regulated) transcript levels (minimum of twofold differences; SLIM-2X; Wang et al., 2011) in GS transgenic compared to WT tissues under well-watered (blue), drought (orange), and recovery (pink) conditions. A total of 6217 DE genes were identified. Samples for microarray analysis were processed as described in (Molina-Rueda et al., 2013). A custom microarray GE 4x44K, AMADID # 021876 (Agilent, Santa Clara CA, USA) was used for transcript analysis.

Functional category analysis including AgriGO (Du, Zhou, Ling, Zhang, & Su, 2010), and MapMan (v3.5.1) (Thimm et al., 2004) using the *P. trichocarpa* map (JGIv1.1) of these 6217 genes by tissue and condition revealed interesting assignments. Significantly enriched functional categories of DE genes were classified according to the closest *Arabidopsis* orthologues under hierarchical and functional categories (BINs). Functional categories differentially expressed among tissues and conditions included stress, signalling, hormone, RNA regulation, cell wall, secondary metabolism, photosynthesis, and lipid. Stress related BINs presented the lowest associated *P*-values in the four tissues, indicating that stress categories were significantly up-regulated during drought (Molina-Rueda et al., in preparation)

Comparative gene network analysis was the undertaken to dissect further the transcriptional responses of GS and WT poplars to drought. Weighted gene correlation networks were separately constructed for GS and WT plants. While the two networks had largely similar distribution patterns and edge numbers (connectivities), some striking differences were observed among the most highly connected edges (edges associated with the greatest number of nodes). As shown in Fig. 3, the top 40,000 edges ranked by topological overlap matrix (TOM values) (connection strengths) were consistently associated with greater numbers of genes (nodes) in the GS network than in the WT network (Fig. 3, left axis). However, the corresponding TOM values were consistently higher in the WT network than in GS network (Fig. 3, right axis). This suggests that nodes (genes) in the GS network were associated with extended neighbourhoods, but individual nodes are not as densely connected in the GS network when compared to the WT network.

Functional categories in which the percentage of differentially connected genes is significantly different in the GS and WT networks with regard to the remaining functional categories were selected (MapMan). The following categories contained genes more highly connected (at least twofold) in the GS network vs the WT network "hormone metabolism", "regulation of transcription" and "calcium signalling". The "stress" and "protein" categories showed the opposite trend, i.e., less connected in the GS network vs the WT network.

Considering that "stress" and "hormone" related genes showed the highest differences in connectivity between the two networks, these two categories were chosen for cluster analysis (Ramoni, Sebastiani, & Kohane, 2002). A cluster containing 31 genes under the "stress" BIN was identified and approximately one-half of the genes in this cluster fell under the "stress.

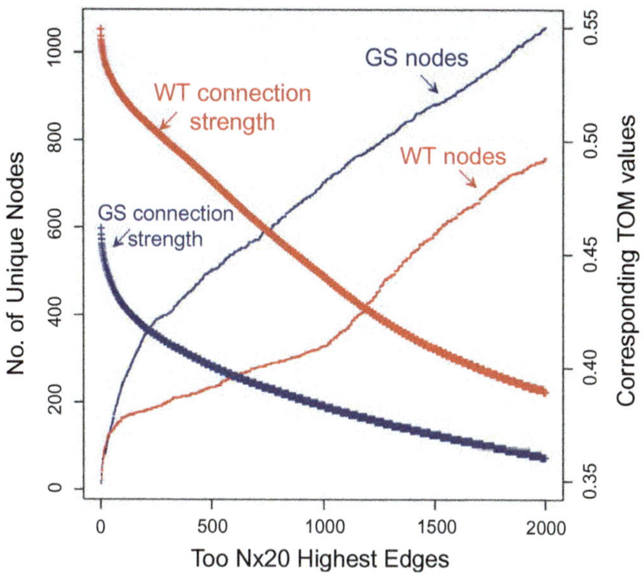

Fig. 3 Behaviour of GS and wild type networks. GS transgenic and wild type transcript profiles were analysed separately to construct two weighted co-expression networks, and topological overlap matrix (TOM) analysis was used as a connectivity measure for network construction (Zhang & Horvath, 2005). Among the strongest edges ranked by the topological overlap matrix (TOM) connectivity measure, the numbers of nodes in the GS network (blue up-line) are greater than in the WT network (red up-line), and that the overall connectivity in WT network (red down-line) is greater than in GS network (blue down-line).

abiotic.heat" subcategory. Transcript levels for genes in this cluster were significantly higher in GS than in WT conditions, especially under drought conditions in aerial tissues.

Among all the DE genes selected, only three genes showed significantly higher transcript levels in GS than in WT in all tissues under drought. Of these three genes, one encodes a chloroplastic iron superoxide dismutase (*PtFSD2.1*) and the second encodes a plasma membrane ion transporter involved in providing iron to SODs. We previously analysed the SOD gene family in poplar and reported that expression patterns of these and other SOD-related genes is enhanced in all tissues of GS poplars during drought (Molina-Rueda et al., 2013).

The third gene showing higher transcript levels in GS than in WT is highly similar to an *Arabidopsis* orthologue which encodes galactinol synthase 1 (*AtGolS1*). Interestingly, stress-inducible galactinol synthase in chickpea

(*CaGolS*) is implicated in heat and oxidative stress tolerance by reducing excessive ROS accumulation (Salvi, Kamble, & Majee, 2017).

A strongly supported cluster included many of the JA-related genes which showed clear down-regulation in aerial tissues under all three growth conditions. Eight jasmonate-related genes were identified in this cluster, including the transcription factor MYC2, which differentially modulates diverse jasmonate-dependent functions in *Arabidopsis* (Dombrecht et al., 2007).

Biochemical and molecular factors contributing to drought tolerance in GS poplars are summarized in Fig. 4. Overexpression of pine GS leads to higher GS activity, lower levels of soluble inorganic nitrogen (nitrate and ammonium), and higher levels of organic nitrogen compounds, including glutamine, glutamate and GABA (Man et al., 2005; Molina-Rueda & Kirby, 2015).

Both GABA and PII have been proposed as carbon-nitrogen sensors (Chellamuthu et al., 2014; Shelp, Bozzo, et al., 2012; Shelp, Mullen, & Waller, 2012). Higher PII transcript levels were detected in GS poplars. PII has been connected to sensing of nitrogen status in plants (Gent & Forde, 2017). GABA is produced by glutamate decarboxylase (GAD), an enzyme whose activity is regulated by calcium and pH. Transcript levels of GAD are higher in GS than in WT poplars (Molina-Rueda & Kirby, 2015). Moreover, GAD participates in the GABA shunt, a pathway that has been related to ROS detoxification (Zhang et al., 2011). GABA can act also as osmolyte contributing to drought tolerance and as signalling molecule (Shelp, Mullen, et al., 2012).

Enhanced NUE in GS poplars leads to higher nitrogen status and alterations in C/N balance, which may increase carbon demand leading to enhanced photosynthesis (El-Khatib et al., 2004; Man et al., 2011). However, photosynthesis is a source of reactive oxygen species (ROS) including ion superoxide (Foyer & Shigeoka, 2011). Superoxide dismutases (SODs) constitute a first line of defence against ROS (Alscher, Erturk, & Heath, 2002). Several genes encoding SODs and related chaperones were up-regulated in GS poplars during drought (Molina-Rueda et al., 2013). We observed coordinated down-regulation of the plastidic copper SODs (PtCDS2s) and up-regulation of the plastidic iron SODs (PtFSDs), at the mRNA as well as activity levels in drought-stressed GS transgenics. This is consistent with preferential allocation of Cu cofactors to plastocyanin to sustain high rates of photosynthesis in the GS transgenics under drought. The model is further supported by down-regulation of several chloroplastic

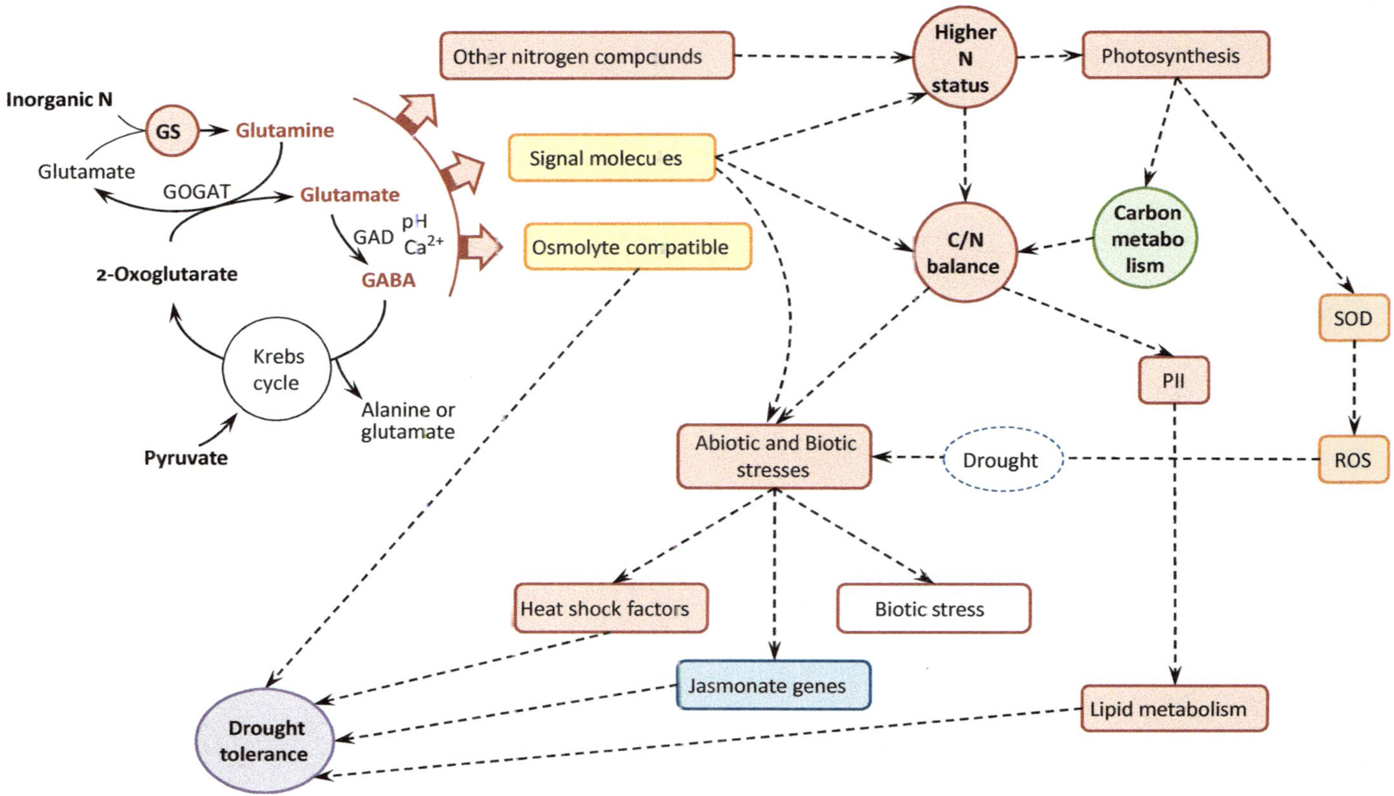

Fig. 4 Schematic representation of the principal elements contributing to enhanced nitrogen assimilation and drought tolerance in GS poplars and their interactions. Overexpression of pine *GS1a* leads to higher GS activity, lower cellular levels of inorganic nitrogen content and higher organic nitrogen compounds, including glutamine, glutamate and GABA. Enhanced nitrogen use efficiency in GS leads to a higher nitrogen status and alterations in C/N balance, which increase carbon demand leading to enhanced photosynthesis. Photosynthesis is a source of ROS. Alterations in SOD expression and activities during drought in GS plants could contribute ROS detoxification. Both GABA and PII has been proposed as carbon-nitrogen sensors.

polyphenol oxidases (PPOs), another class of Cu protein, in the GS poplar during drought conditions. In summary, these results suggest that alterations in N metabolism in GS transgenics cause differential regulation of genes involved in ROS protection under drought conditions leading to drought tolerance observed in the transgenics. Cu homeostasis and antioxidant regulation in response to altered N metabolism in the GS poplars need to be further investigated.

6. Conclusions

An overview of the applications of genetic engineering approaches to develop an understanding of interacting roles of reduced nitrogen in conferring drought tolerance in poplar is presented. Overexpression of glutamine synthetase (GS) causes alterations to nitrogen metabolism including higher production of free amino acids including glutamine, glutamate, and GABA. In addition to contributing to enhanced growth, these metabolites can act as signals driving changes in expression of stress-related genes with relevant roles in drought tolerance. The transcriptomic approach has revealed significant differences in gene expression patterns in GS transgenic poplars compared to the wild type control. Included are genes with functional annotations under photosynthesis, ROS, and heat stresses, lipid metabolism, and jasmonate-related categories. Preliminary co-expression network analysis indicates that GS overexpression results in significant network rewiring, especially in local networks, associated to these genes (Molina-Rueda et al., in preparation). These studies also shed light on compensatory regulation between chloroplastic CSDs and FSDs, and suggest that a copper-mediated mechanism represents a common response to oxidative stress. Further studies into the interactions between nitrogen metabolism and expression of stress-related genes are indicated. This approach will contribute to development of strategies for developing new genotypes of drought tolerant energy crops, including poplar, and provide further insights into the complex interacting factors related to the acquisition of drought tolerance, a trait of significant and increasing importance in our changing environment.

Acknowledgements

We gratefully acknowledge the efforts of the laboratory of Dr. Chung Jui Tsai at the University of Georgia for patient collaboration on the poplar transcriptome analysis and expression network analysis.

References

Allwright, M. R., & Taylor, G. (2016). Molecular breeding for improved second generation bioenergy crops. *Trends in Plant Science*, *21*(1), 43–54. https://doi.org/10.1016/j.tplants.2015.10.002.

Alscher, R. G., Erturk, N., & Heath, L. S. (2002). Role of superoxide dismutases (SODs) in controlling oxidative stress in plants. *Journal of Experimental Botany*, *53*(372), 1331–1341.

Andrews, M., Lea, P. J., Raven, J. A., & Lindsey, K. (2004). Can genetic manipulation of plant nitrogen assimilation enzymes result in increased crop yield and greater N-use efficiency? An assessment. *Annals of Applied Biology*, *145*(1), 25–40. https://doi.org/10.1111/j.1744-7348.2004.tb00356.x.

Aragüés, R., Urdanoz, V., Çetin, M., Kirda, C., Daghari, H., Ltifi, W., et al. (2011). Soil salinity related to physical soil characteristics and irrigation management in four Mediterranean irrigation districts. *Agricultural Water Management*, *98*(6), 959–966. https://doi.org/10.1016/j.agwat.2011.01.004.

Avila, C., Suárez, M. F., Gómez-Maldonado, J., & Cánovas, F. M. (2001). Spatial and temporal expression of two cytosolic glutamine synthetase genes in Scots pine: Functional implications on nitrogen metabolism during early stages of conifer development. *Plant Journal*, *25*(1), 93–102. https://doi.org/10.1046/j.1365-313X.2001.00938.x.

Bauer, D., Biehler, K., Fock, H., Carrayol, E., Hirel, B., Migge, A., et al. (1997). A role for cytosolic glutamine synthetase in the remobilization of leaf nitrogen during water stress in tomato. *Physiologia Plantarum*, *99*(2), 241–248. https://doi.org/10.1034/j.1399-3054.1997.990205.x.

Beatty, P. H., Shrawat, A. K., Carroll, R. T., Zhu, T., & Good, A. G. (2009). Transcriptome analysis of nitrogen-efficient rice over-expressing alanine aminotransferase. *Plant Biotechnology Journal*, 7(6), 562–576. https://doi.org/10.1111/j.1467-7652.2009.00424.x.

Bernard, S. M., & Habash, D. Z. (2009). The importance of cytosolic glutamine synthetase in nitrogen\assimilation and recycling. *The New Phytologist*, *182*(3), 608–620.

Betti, M., García-Calderón, M., Pérez-Delgado, C. M., Credali, A., Estivill, G., Galván, F., et al. (2012). Glutamine synthetase in legumes: Recent advances in enzyme structure and functional genomics. *International Journal of Molecular Sciences*, *13*(7), 7994–8024. https://doi.org/10.3390/ijms13077994.

Bi, Z., Merl-Pham, J., Uehlein, N., Zimmer, I., Mühlhans, S., Aichler, M., et al. (2015). RNAi-mediated downregulation of poplar plasma membrane intrinsic proteins (PIPs) changes plasma membrane proteome composition and affects leaf physiology. *Journal of Proteomics*, *128*, 321–332. https://doi.org/10.1016/j.jprot.2015.07.029.

Blum, A. (1996). Crop responses to drought and the interpretation of adaptation. *Plant Growth Regulation*, *20*(2), 135–148.

Bogardi, J. J., Dudgeon, D., Lawford, R., Flinkerbusch, E., Meyn, A., Pahl-Wostl, C., et al. (2012). Water security for a planet under pressure: Interconnected challenges of a changing world call for sustainable solutions. *Current Opinion in Environmental Sustainability*, *4*(1), 35–43. https://doi.org/10.1016/j.cosust.2011.12.002.

Boisvenue, C., & Running, S. W. (2006). Impacts of climate change on natural forest productivity—Evidence since the middle of the 20th century. *Global Change Biology*, *12*(5), 862–882. https://doi.org/10.1111/j.1365-2486.2006.01134.x.

Bor, M., Seckin, B., Ozgur, R., Yılmaz, O., Ozdemir, F., & Turkan, I. (2009). Comparative effects of drought, salt, heavy metal and heat stresses on gamma-aminobutryric acid levels of sesame (Sesamum indicum L.). *Acta Physiologiae Plantarum*, *31*(3), 655–659. https://doi.org/10.1007/s11738-008-0255-2.

Borland, A. M., Wullschleger, S. D., Weston, D. J., Hartwell, J., Tuskan, G. A., Yang, X., et al. (2015). Climate-resilient agroforestry: Physiological responses to climate change and engineering of crassulacean acid metabolism (CAM) as a mitigation strategy. *Plant, Cell and Environment*, *38*(9), 1833–1849. https://doi.org/10.1111/pce.12479.

Bouché, N., & Fromm, H. (2004). GABA in plants: Just a metabolite? *Trends in Plant Science, 9*(3), 110–115. https://doi.org/10.1016/j.tplants.2004.01.006.

Bradshaw, H. D., Ceulemans, R., Davis, J., & Stettler, R. (2000). Emerging model systems in plant biology: Poplar (Populus) as a model forest tree. *Journal of Plant Growth Regulation, 19*(3), 306–313. https://doi.org/10.1007/s003440000030.

Busov, V. B., Brunner, A. M., Meilan, R., Filichkin, S., Ganio, L., Gandhi, S., et al. (2005). Genetic transformation: A powerful tool for dissection of adaptive traits in trees. *New Phytologist, 167*(1), 9–18. https://doi.org/10.1111/j.1469-8137.2005.01412.x.

Castro-Rodríguez, V., García-Gutiérrez, A., Canales, J., Avila, C., Kirby, E. G., & Cánovas, F. M. (2011). The glutamine synthetase gene family in Populus. *BMC Plant Biology, 11*, 119. https://doi.org/10.1186/1471-2229-11-119.

Chellamuthu, V.-R., Ermilova, E., Lapina, T., Lüddecke, J., Minaeva, E., Herrmann, C., et al. (2014). A widespread glutamine-sensing mechanism in the plant kingdom. *Cell, 159*(5), 1188–1199. https://doi.org/10.1016/j.cell.2014.10.015.

Civelek, M., & Lusis, A. J. (2014). Systems genetics approaches to understand complex traits. *Nature Reviews. Genetics, 15*(1), 34–48. https://doi.org/10.1038/nrg3575.

Coleman, H. D., Cánovas, F. M., Man, H., Kirby, E. G., & Mansfield, S. D. (2012). Enhanced expression of glutamine synthetase (GS1a) confers altered fibre and wood chemistry in field grown hybrid poplar (Populus tremula X alba) (717-1B4). *Plant Biotechnology Journal, 10*(7), 883–889. https://doi.org/10.1111/j.1467-7652.2012.00714.x.

Coque, M., & Gallais, A. (2006). Genomic regions involved in response to grain yield selection at high and low nitrogen fertilization in maize. *Theoretical and Applied Genetics, 112*(7), 1205–1220. https://doi.org/10.1007/s00122-006-0222-5.

Cronk, Q. C. B. (2005). Plant eco-devo: The potential of poplar as a model organism. *New Phytologist, 166*(1), 39–48. https://doi.org/10.1111/j.1469-8137.2005.01369.x.

Cseke, L. J., Cseke, S. B., & Podila, G. K. (2007). High efficiency poplar transformation. *Plant Cell Reports, 26*(9), 1529–1538. https://doi.org/10.1007/s00299-007-0365-0.

De la Torre, F., García-Gutiérrez, A., Crespillo, R., Cantón, F. R., Ávila, C., & Cánovas, F. M. (2002). Functional expression of two pine glutamine synthetase genes in bacteria reveals that they encode cytosolic holoenzymes with different molecular and catalytic properties. *Plant and Cell Physiology, 43*(7), 802–809. https://doi.org/10.1093/pcp/pcf094.

Demirci, Y., Zhang, B., & Unver, T. (2018). CRISPR/Cas9: An RNA-guided highly precise synthetic tool for plant genome editing. *Journal of Cellular Physiology, 233*(3), 1844–1859. https://doi.org/10.1002/jcp.25970.

Dinneny, J. R., Long, T. A., Wang, J. Y., Jung, J. W., Mace, D., Pointer, S., et al. (2008). Cell identity mediates the response of Arabidopsis roots to abiotic stress. *Science (New York, N.Y.), 320*(5878), 942–945. https://doi.org/10.1126/science.1153795.

Dombrecht, B., Xue, G. P., Sprague, S. J., Kirkegaard, J. A., Ross, J. J., Reid, J. B., et al. (2007). MYC2 differentially modulates diverse jasmonate-dependent functions in Arabidopsis. *The Plant Cell, 19*(7), 2225–2245. https://doi.org/10.1105/tpc.106.048017.

Du, Z., Zhou, X., Ling, Y., Zhang, Z., & Su, Z. (2010). agriGO: A GO analysis toolkit for the agricultural community. *Nucleic Acids Research, 38*(Web Server issue), W64–W70. https://doi.org/10.1093/nar/gkq310.

Dubouzet, J. G., Strabala, T. J., & Wagner, A. (2013). Potential transgenic routes to increase tree biomass. *Plant Science, 212*, 72–101. https://doi.org/10.1016/j.plantsci.2013.08.006.

El-Khatib, R. T., Hamerlynck, E. P., Gallardo, F., & Kirby, E. G. (2004). Transgenic poplar characterized by ectopic expression of a pine cytosolic glutamine synthetase gene exhibits enhanced tolerance to water stress. *Tree Physiology, 24*(7), 729–736. https://doi.org/10.1093/treephys/24.7.729.

Elorriaga, E., Klocko, A. L., Ma, C., & Strauss, S. H. (2018). Variation in mutation spectra among CRISPR/Cas9 mutagenized poplars. *Frontiers in Plant Science, 9*(May), 594. https://doi.org/10.3389/fpls.2018.00594.

Fageria, N. K., & Baligar, V. (2005). Enhancing nitrogen use efficiency in crop plants. *Advances in Agronomy*, *88*, 97–185. https://doi.org/10.1016/S0065-2113(05)88004-6.

Fan, D., Liu, T., Li, C., Jiao, B., Li, S., Hou, Y., et al. (2015). Efficient CRISPR/Cas9-mediated targeted mutagenesis in Populus in the first generation. *Scientific Reports*, *5*, 1–7. https://doi.org/10.1038/srep12217.

Forde, B. G., & Lea, P. J. (2007). Glutamate in plants: Metabolism, regulation, and signalling. *Journal of Experimental Botany*, *58*(9), 2339–2358. https://doi.org/10.1093/jxb/erm121.

Foyer, C. H., & Shigeoka, S. (2011). Understanding oxidative stress and antioxidant functions to enhance photosynthesis. *Plant Physiology*, *155*(1), 93–100. https://doi.org/10.1104/pp.110.166181.

Gallardo, F., Fu, J., Canton, F., Garcia-Gutierrez, A., Canovas, F., & Kirby, E. (1999). Expression of a conifer glutamine synthetase gene in transgenic poplar. *Planta*, *210*(1), 19–26. Retrieved from http://www.ncbi.nlm.nih.gov/pubmed/10592028.

Gallardo, F., Fu, J., Jing, Z. P., Kirby, E. G., & Cánovas, F. M. (2003). Genetic modification of amino acid metabolism in woody plants. *Plant Physiology and Biochemistry*, *41*(6–7), 587–594. https://doi.org/10.1016/S0981-9428(03)00081-0.

Gálvez, S., Gallardo, F., & Cánovas, F. (1990). The occurrence of glutamine synthetase isoenzymes in tomato plants is correlated to plastid differentiation. *Journal of Plant Physiology*, *137*(1), 1–4. https://doi.org/10.1016/S0176-1617(11)80002-7.

García-Gutiérrez, Á., Cánovas, F. M., & Ávila, C. (2018). Glutamate synthases from conifers: Gene structure and phylogenetic studies. *BMC Genomics*, *19*(1), 1–12. https://doi.org/10.1186/s12864-018-4454-y.

Gent, L., & Forde, B. G. (2017). How do plants sense their nitrogen status? *Journal of Experimental Botany*, *68*(10), 2531–2540. https://doi.org/10.1093/jxb/erx013.

Glass, M., Barkwill, S., Unda, F., & Mansfield, S. D. (2015). Endo-β-1,4-glucanases impact plant cell wall development by influencing cellulose crystallization. *Journal of Integrative Plant Biology*, *57*(4), 396–410. https://doi.org/10.1111/jipb.12353.

Good, A. G., Johnson, S. J., De Pauw, M., Carroll, R. T., Savidov, N., Vidmar, J., et al. (2007). Engineering nitrogen use efficiency with alanine aminotransferase. *Canadian Journal of Botany*, *85*(3), 252–262. https://doi.org/10.1139/B07-019.

Good, A. G., Shrawat, A. K., & Muench, D. G. (2004). Can less yield more? Is reducing nutrient input into the environment compatible with maintaining crop production? *Trends in Plant Science*, *9*(12), 597–605. https://doi.org/10.1016/j.tplants.2004.10.008.

Häggman, H., Raybould, A., Borem, A., Fox, T., Handley, L., Hertzberg, M., et al. (2013). Genetically engineered trees for plantation forests: Key considerations for environmental risk assessment. *Plant Biotechnology Journal*, *11*(7), 785–798. https://doi.org/10.1111/pbi.12100.

Han, M., Okamoto, M., Beatty, P. H., Rothstein, S. J., & Good, A. G. (2015). The genetics of nitrogen use efficiency in crop plants. *Annual Review of Genetics*, *49*(1), 269–289. https://doi.org/10.1146/annurev-genet-112414-055037.

Harfouche, A., Meilan, R., & Altman, A. (2014). Molecular and physiological responses to abiotic stress in forest trees and their relevance to tree improvement. *Tree Physiology*, *34*(11), 1181–1198. https://doi.org/10.1093/treephys/tpu012.

Häusler, R. E., Ludewig, F., & Krueger, S. (2014). Plant science amino acids—A life between metabolism and signaling. *Plant Science*, *229*, 225–237. https://doi.org/10.1016/j.plantsci.2014.09.011.

Hirel, B., Le Gouis, J., Ney, B., & Gallais, A. (2007). The challenge of improving nitrogen use efficiency in crop plants: Towards a more central role for genetic variability and quantitative genetics within integrated approaches. *Journal of Experimental Botany*, *58*(9), 2369–2387. https://doi.org/10.1093/jxb/erm097.

IPCC. (2014). Climate Change 2014: Mitigation of climate change. Summary for policymakers and technical summary. In *Part of the working group III contribution to the fifth assessment report of the intergovernmental panel on climate change*. https://doi.org/10.1017/CBO9781107415416.005.

Ireland, R., & Lea, P. J. (1999). The enzymes of glutamine, glutamate, asparagine, and aspartate metabolism. In B. Singh (Ed.), *Plant amino acids: Biochemistry and bsiotechnology* (pp. 49–109). New York: Marcel Dekker, Inc.

Kalluri, U. C., Payyavula, R. S., Labbé, J. L., Engle, N., Bali, G., Jawdy, S. S., et al. (2016). Down-regulation of KORRIGAN-like endo-β-1,4-glucanase genes impacts carbon partitioning, mycorrhizal colonization and biomass production in Populus. *Frontiers in Plant Science*, 7(October), 1–12. https://doi.org/10.3389/fpls.2016.01455.

Kant, S., Bi, Y.-M., & Rothstein, S. J. (2011). Understanding plant response to nitrogen limitation for the improvement of crop nitrogen use efficiency. *Journal of Experimental Botany*, *62*(4), 1499–1509. https://doi.org/10.1093/jxb/erq297.

Kathiresan, A., Tung, P., Chinnappa, C. C., & Reid, D. M. (1997). Υ-Aminobutyric acid stimulates ethylene biosynthesis in sunflower. *Plant Physiology*, *115*, 129–135.

Katoch, R., & Thakur, N. (2013). Advances in RNA interference technology and its impact on nutritional improvement, disease and insect control in plants. *Applied Biochemistry and Biotechnology*, *169*(5), 1579–1605. https://doi.org/10.1007/s12010-012-0046-5.

Kinnersley, A. M., & Turano, F. J. (2000). Gamma aminobutyric acid (GABA) and plant responses to stress. *Critical Reviews in Plant Sciences*, *19*(6), 479–509. https://doi.org/10.1080/07352680091139277.

Kirby, E. G., Gallardo, F., Man, H., & El-Khatib, R. (2006). The overexpression of glutamine synthetase in transgenic poplar: A review. *Silvae Genetica*, *55*(6), 278–284. https://doi.org/10.1515/sg-2006-0036.

Lam, H.-M., Coschigano, K. T., Oliveira, I. C., Melo-Oliveira, R., & Coruzzi, G. M. (1996). The molecular-genetics of nitrogen assimilation into amino acids in higher plants. *Annual Review of Plant Physiology and Plant Molecular Biology*, *47*(1), 569–593. https://doi.org/10.1146/annurev.arplant.47.1.569.

Lancien, M., & Roberts, M. R. (2006). Regulation of Arabidopsis thaliana 14-3-3 gene expression by gamma-aminobutyric acid. *Plant, Cell and Environment*, *29*(7), 1430–1436. https://doi.org/10.1111/j.1365-3040.2006.01526.x.

Leple, J. C., Brasileiro, A. C. M., Michel, M. F., et al. (1992). Transgenic poplars: Expression of chimeric genes using four different constructs. *Plant Cell Reports*, *11*(3), 137–141. https://doi.org/10.1007/BF00232166.

Li, J., Brunner, A. M., Meilan, R., & Strauss, S. H. (2009). Stability of transgenes in trees: Expression of two reporter genes in poplar over three field seasons. *Tree Physiology*, *29*(2), 299–312. https://doi.org/10.1093/treephys/tpn028.

Li, H., Hu, B., & Chu, C. (2017). Nitrogen use efficiency in crops: Lessons from Arabidopsis and rice. *Journal of Experimental Botany*, *68*(10), 2477–2488. https://doi.org/10.1093/jxb/erx101.

Maheshwari, P., & Kovalchuk, I. (2016). Agrobacterium-mediated stable genetic transformation of Populus angustifolia and Populus balsamifera. *Frontiers in Plant Science*, 7(March), 1–12. https://doi.org/10.3389/fpls.2016.00296.

Man, H.-M., Boriel, R., El-Khatib, R., & Kirby, E. G. (2005). Characterization of transgenic poplar with ectopic expression of pine cytosolic glutamine synthetase under conditions of varying nitrogen availability. *The New Phytologist*, *167*(1), 31–39. https://doi.org/10.1111/j.1469-8137.2005.01461.x.

Man, H., Pollmann, S., Weiler, E. W., & Kirby, E. G. (2011). Increased glutamine in leaves of poplar transgenic with pine GS1a caused greater anthranilate synthetase α-subunit (ASA1) transcript and protein abundances: An auxin-related mechanism for enhanced growth in GS transgenics? *Journal of Experimental Botany*, *62*(13), 4423–4431. https://doi.org/10.1093/jxb/err026.

McAllister, C. H., Beatty, P. H., & Good, A. G. (2012). Engineering nitrogen use efficient crop plants: The current status. *Plant Biotechnology Journal*, *10*(9), 1011–1025. https://doi.org/10.1111/j.1467-7652.2012.00700.x.

McCune, A. R., & Schimenti, J. C. (2012). Using genetic networks and homology to understand the evolution of phenotypic traits. *Current Genomics*, *13*(1), 74–84. https://doi.org/10.2174/138920212799034785.

Meyer, S., Nowak, K., Sharma, V. K., Schulze, J., Mendel, R. R., & Hänsch, R. (2004). Vectors for RNAi technology in poplar. *Plant Biology*, *6*(1), 100–103. https://doi.org/10.1055/s-2004-815729.

Molina-Rueda, J. J.; L. Du; H.-Q. Wang; C. J. Tsai; E. G. Kirby. (in preparation). Transcriptomic analysis reveals insights into mechanisms for drought tolerance in transgenic poplars expressing pine glutamine synthetase (GS1a).

Molina-Rueda, J. J., & Kirby, E. G. (2015). Transgenic poplar expressing the pine GS1a show alterations in nitrogen homeostasis during drought. *Plant Physiology and Biochemistry*, *94*, 181–190. https://doi.org/10.1016/j.plaphy.2015.06.009.

Molina-Rueda, J. J., Tsai, C. J., & Kirby, E. G. (2013). The Populus superoxide dismutase gene family and its responses to drought stress in transgenic poplar overexpressing a pine cytosolic glutamine synthetase (GS1a). *PLoS One*, *8*(2), e56421. https://doi.org/10.1371/journal.pone.0056421.

Movahedi, A., Zhang, J., Amirian, R., & Zhuge, Q. (2014). An efficient agrobacterium-mediated transformation system for poplar. *International Journal of Molecular Sciences*, *15*(6), 10780–10793. https://doi.org/10.3390/ijms150610780.

Pagliai, M., Vignozzi, N., & Pellegrini, S. (2004). Soil structure and the effect of management practices. *Soil and Tillage Research*, *79*(2 SPEC.ISS), 131–143. https://doi.org/10.1016/j.still.2004.07.002.

Parsons, T. J., Sinkar, V. P., Stettler, R. F., Nester, E. W., & Gordon, M. P. (1986). Transformation of poplar by Agrobacterium tumefaciens. *Bio/Technology*, *4*(6), 533–536. https://doi.org/10.1038/nbt0686-533.

Pascual, M., Jing, Z., Kirby, E., Cánovas, F., & Gallardo, F. (2008). Response of transgenic poplar overexpressing cytosolic glutamine synthetase to phosphinothricin. *Phytochemistry*, *69*(2), 382–389. https://doi.org/10.1016/j.phytochem.2007.07.031.

Perez-Garcia, A., Canovas, F. M., Gallardo, F., Hirel, B., & Devicente, A. (1995). Differential expression of glutamine-synthetase isoforms in tomato detached leaflets infected with Pseudomonas-syringae Pv tomato. *Molecular Plant-Microbe Interactions*, *8*(1), 96–103. https://doi.org/10.1094/Mpmi-8-0096.

Polle, A., & Chen, S. (2015). On the salty side of life: Molecular, physiological and anatomical adaptation and acclimation of trees to extreme habitats. *Plant, Cell and Environment*, *38*(9), 1794–1816. https://doi.org/10.1111/pce.12440.

Polle, A., Janz, D., Teichmann, T., & Lipka, V. (2013). Poplar genetic engineering: Promoting desirable wood characteristics and pest resistance. *Applied Microbiology and Biotechnology*, *97*(13), 5669–5679. https://doi.org/10.1007/s00253-013-4940-8.

Porth, I., Klápště, J., Skyba, O., Friedmann, M. C., Hannemann, J., Ehlting, J., et al. (2013). Network analysis reveals the relationship among wood properties, gene expression levels and genotypes of natural Populus trichocarpa accessions. *The New Phytologist*, *200*(3), 727–742. https://doi.org/10.1111/nph.12419.

Ramoni, M. F., Sebastiani, P., & Kohane, I. S. (2002). Cluster analysis of gene expression dynamics. *Proceedings of the National Academy of Sciences of the United States of America*, *99*(14), 9121–9126. https://doi.org/10.1073/pnas.132656399.

Salvi, P., Kamble, N. U., & Majee, M. (2017). Stress-inducible galactinol synthase of chickpea (CaGolS) is implicated in heat and oxidative stress tolerance through reducing stress-induced excessive reactive oxygen species accumulation. *Plant and Cell Physiology*, *59*(November 2017), 155–166. https://doi.org/10.1093/pcp/pcx170.

Samanta, M. K., Dey, A., & Gayen, S. (2016). CRISPR/Cas9: An advanced tool for editing plant genomes. *Transgenic Research*, *25*(5), 561–573. https://doi.org/10.1007/s11248-016-9953-5.

Shelp, B. J., Bozzo, G. G., Trobacher, C. P., Zarei, A., Deyman, K. L., & Brikis, C. J. (2012). Hypothesis/review: Contribution of putrescine to 4-aminobutyrate (GABA) production in response to abiotic stress. *Plant Science: An International Journal of Experimental Plant Biology, 193–194*, 130–135. https://doi.org/10.1016/j.plantsci.2012.06.001.

Shelp, B. J., Mullen, R. T., & Waller, J. C. (2012). Compartmentation of GABA metabolism raises intriguing questions. *Trends in Plant Science, 17*(2), 57–59. https://doi.org/10.1016/j.tplants.2011.12.006.

Shi, S.-Q., Shi, Z., Jiang, Z.-P., Qi, L.-W., Sun, X.-M., Li, C.-X., et al. (2010). Effects of exogenous GABA on gene expression of Caragana intermedia roots under NaCl stress: Regulatory roles for H2O2 and ethylene production. *Plant, Cell & Environment, 33*(2), 149–162. https://doi.org/10.1111/j.1365-3040.2009.02065.x.

Shrawat, A. K., Carroll, R. T., DePauw, M., Taylor, G. J., & Good, A. G. (2008). Genetic engineering of improved nitrogen use efficiency in rice by the tissue-specific expression of alanine aminotransferase. *Plant Biotechnology Journal, 6*(7), 722–732. https://doi.org/10.1111/j.1467-7652.2008.00351.x.

Strauss, S. H., Jones, K. N., Lu, H., Petit, J. D., Klocko, A. L., Betts, M. G., et al. (2017). Reproductive modification in forest plantations: Impacts on biodiversity and society. *New Phytologist, 213*(3), 1000–1021. https://doi.org/10.1111/nph.14374.

Szechyńska-Hebda, M., Czarnocka, W., Hebda, M., & Karpiński, S. (2016). PAD4, LSD1 and EDS1 regulate drought tolerance, plant biomass production, and cell wall properties. *Plant Cell Reports, 35*(3), 527–539. https://doi.org/10.1007/s00299-015-1901-y.

Tark-Dame, M., Weber, B., De Sain, M., Anggoro, D. T., Bader, R., Walmsley, A., et al. (2018). Generating transgenic plants with single-copy insertions using BIBAC-GW binary vector. *Journal of Visualized Experiments, 2018*(133), 1–13. https://doi.org/10.3791/57295.

Taylor, G. (2002). Populus: Arabidopsis for forestry. Do we need a model tree? *Annals of Botany, 90*(6), 681–689. https://doi.org/10.1093/aob/mcf255.

Thimm, O., Bläsing, O., Gibon, Y., Nagel, A., Meyer, S., Krüger, P., et al. (2004). Mapman: A user-driven tool to display genomics data sets onto diagrams of metabolic pathways and other biological processes. *The Plant Journal, 37*(6), 914–939. https://doi.org/10.1111/j.1365-313X.2004.02016.x.

Thomsen, H. C., Eriksson, D., Møller, I. S., & Schjoerring, J. K. (2014). Cytosolic glutamine synthetase: A target for improvement of crop nitrogen use efficiency? *Trends in Plant Science, 19*(10), 656–663. https://doi.org/10.1016/j.tplants.2014.06.002.

Tsai, C. J., & Xue, L. J. (2015). CRISPRing into the woods. *GM Crops & Food, 6*(4), 206–215. https://doi.org/10.1080/21645698.2015.1091553.

Walsh, J., Wuebbles, D., Hayhoe, K., Kossin, J., Kunkel, K., Stephens, G., et al. (2014). In J. M. Melillo, T. (. T. C.). Richmond, & G. W. Yohe (Eds.), *Chapter 2: Our changing climate. Climate change impacts in the United States: The third national climate assessment*: U.S. Global Change Research Program.

Wang, X., Cai, X., Xu, C., Wang, Q., & Dai, S. (2016). Drought-responsive mechanisms in plant leaves revealed by proteomics. *International Journal of Molecular Sciences, 17*(10), 1706. https://doi.org/10.3390/ijms17101706.

Wang, C., Liu, S., Dong, Y., Zhao, Y., Geng, A., Xia, X., et al. (2016). PdEPF1 regulates water-use efficiency and drought tolerance by modulating stomatal density in poplar. *Plant Biotechnology Journal, 14*(3), 849–860. https://doi.org/10.1111/pbi.12434.

Wang, M., Shen, Q., Xu, G., & Guo, S. (2014). New insight into the strategy for nitrogen metabolism in plant cells. *International Review of Cell and Molecular Biology, 310*, 1–37.

Wang, L., Su, H., Han, L., Wang, C., Sun, Y., & Liu, F. (2014). Differential expression profiles of poplar MAP kinase kinases in response to abiotic stresses and plant hormones, and overexpression of PtMKK4 improves the drought tolerance of poplar. *Gene, 545*(1), 141–148. https://doi.org/10.1016/j.gene.2014.04.058.

Wang, H.-Q., Tuominen, L. K., & Tsai, C.-J. (2011). SLIM: A sliding linear model for estimating the proportion of true null hypotheses in datasets with dependence structures.

Bioinformatics (Oxford, England), *27*(2), 225–231. https://doi.org/10.1093/bioinformatics/btq650.

Wolt, J. D., & Wolt, J. (1994). *Soil solution chemistry: Applications to environmental science and agriculture.* New York: John Wiley & Sons.

Xu, G., Fan, X., & Miller, A. J. (2012). Plant nitrogen assimilation and use efficiency. *Annual Review of Plant Biology*, *63*(1), 153–182. https://doi.org/10.1146/annurev-arplant-042811-105532.

Xue, L. J., Frost, C. J., Tsai, C. J., & Harding, S. A. (2016). Drought response transcriptomes are altered in poplar with reduced tonoplast sucrose transporter expression. *Scientific Reports*, *6*(August), 1–14. https://doi.org/10.1038/srep33655.

Yadav, U. P., Ayre, B. G., & Bush, D. R. (2015). Transgenic approaches to altering carbon and nitrogen partitioning in whole plants: Assessing the potential to improve crop yields and nutritional quality. *Frontiers in Plant Science*, *6*(April), 1–13. https://doi.org/10.3389/fpls.2015.00275.

Ye, C.-Y., Li, T., Yin, H., Weston, D. J., Tuskan, G. A., Tschaplinski, T. J., et al. (2013). Evolutionary analyses of non-family genes in plants. *The Plant Journal: For Cell and Molecular Biology*, *73*(5), 788–797. https://doi.org/10.1111/tpj.12073.

Yevtushenko, D. P., & Misra, S. (2010). Efficient Agrobacterium-mediated transformation of commercial hybrid poplar Populus nigra L. × P. maximowiczii A. Henry. *Plant Cell Reports*, *29*(3), 211–221. https://doi.org/10.1007/s00299-009-0806-z.

Yu, J., Ge, H., Wang, X., Tang, R., Wang, Y., Zhao, F., et al. (2017). Overexpression of pyrabactin resistance-like abscisic acid receptors enhances drought, osmotic, and cold tolerance in transgenic poplars. *Frontiers in Plant Science*, *8*(October), 1–13. https://doi.org/10.3389/fpls.2017.01752.

Yu, J., Yang, L., Liu, X., Tang, R., Wang, Y., Ge, H., et al. (2016). Overexpression of poplar pyrabactin resistance-like abscisic acid receptors promotes abscisic acid sensitivity and drought resistance in transgenic Arabidopsis. *PLoS One*, *11*(12), 1–17. https://doi.org/10.1371/journal.pone.0168040.

Zhang, B., & Horvath, S. (2005). A general framework for weighted gene co-expression network analysis. *Statistical Applications in Genetics and Molecular Biology*, *4*(1), Article 17.

Zhang, J., Movahedi, A., Sang, M., Wei, Z., Xu, J., Wang, X., et al. (2017). Functional analyses of NDPK2 in Populus trichocarpa and overexpression of PtNDPK2 enhances growth and tolerance to abiotic stresses in transgenic poplar. *Plant Physiology and Biochemistry*, *117*, 61–74. https://doi.org/10.1016/j.plaphy.2017.05.019.

Zhang, J., Zhang, Y., Du, Y., Chen, S., & Tang, H. (2011). Dynamic metabonomic responses of tobacco (Nicotiana tabacum) plants to salt stress. *Journal of Proteome Research*, *10*(4), 1904–1914. https://doi.org/10.1021/pr101140n.

Zhong, P. J., Gallardo, F., Pascual, M. B., Sampalo, R., Romero, J., De Navarra, A. T., et al. (2004). Improved growth in a field trial of transgenic hybrid poplar overexpressing glutamine synthetase. *New Phytologist*, *164*(1), 137–145. https://doi.org/10.1111/j.1469-8137.2004.01173.x.

Zhou, X., Jacobs, T. B., Xue, L. J., Harding, S. A., & Tsai, C. J. (2015). Exploiting SNPs for biallelic CRISPR mutations in the outcrossing woody perennial Populus reveals 4-coumarate: CoA ligase specificity and redundancy. *New Phytologist*, *208*(2), 298–301. https://doi.org/10.1111/nph.13470.

Further reading

Song, J., Lu, S., Chen, Z. Z., Lourenco, R., & Chiang, V. L. (2006). Genetic transformation of Populus trichocarpa genotype Nisqually-1: A functional genomic tool for woody plants. *Plant and Cell Physiology*, *47*(11), 1582–1589. https://doi.org/10.1093/pcp/pcl018.

CHAPTER TEN

Phytoremediation with trees

Luis Gómez[a,b,*], Angela Contreras[a], David Bolonio[a], Julia Quintana[a], Luis Oñate-Sánchez[a], Irene Merino[a]

[a]Centro de Biotecnología y Genómica de Plantas (CBGP UPM-INIA), Universidad Politécnica de Madrid, Madrid, Spain
[b]Forestry School, Universidad Politécnica de Madrid, Madrid, Spain
*Corresponding author: e-mail address: luis.gomez@upm.es

Contents

Abstract

Phytoremediation is a proven technology for many organic and inorganic pollutants. Practical reasons have led trees to be neglected in laboratory studies in favour of herbaceous model species. However, trees are gaining momentum over herbaceous plants in field trials because of their greater effectiveness. Recent progress supports that fast-growing trees commonly used in short-rotation coppicing are naturally endowed with key advantages to fight pollution. If properly managed, these trees can also provide important ecological services that will contribute positively to cost-effectiveness and social perception. To fully exploit this potential, we need to better understand the biochemical networks underlying pollutant metabolism and the complex interactions between trees and their endophytic and rhizospheric communities. We examine here recent advances regarding the uptake and fate of relevant organic and inorganic pollutants in plants, with particular emphasis on results obtained with trees. The prospects offered by microbial inoculation and genetic transformation, two approaches successfully field tested, are also discussed. Tree planting for land restoration fits nicely in the new paradigm of phytomanagement.

Advances in Botanical Research, Volume 89
ISSN 0065-2296
https://doi.org/10.1016/bs.abr.2018.11.010

1. Introduction

A downside of industrialization, agriculture, urbanization, and other human activities is the dispersal of vast amount of pollutants into the environment, many of which accumulate at alarming rates. Despite increased public awareness, population growth and globalization are expected to worsen the problem (Landrigan et al., 2018). According to the latest estimates of the World Health Organization, environmental health risks are causing already 12.6 million deaths annually, almost one-fourth of all deaths, as well as countless cases of weakening diseases (Prüss-Ustün, Wolf, Corvalán, Bos, & Neira, 2016). The associated economic impact has been estimated in the range of trillions of dollars, posing a serious threat to social development (The World Bank & Institute for Health Metrics and Evaluation, 2016). No less significant is the ecological toll of pollution, for it impinges on aspects as fundamental for the biosphere as water and air quality, soil fertility, pest and disease proliferation, forest health, biodiversity, carbon sequestration, and climate change mitigation. In spite of these negative effects, the build-up of pollutants is still overlooked by many governments, especially in low- and moderate-income countries (Landrigan et al., 2018). Effective global actions are needed to improve this situation, including advances in dedicated research, education, and regulation. It is equally imperative to allocate more resources to remediate existing polluted sites.

Environmental cleanup is dominated nowadays by conventional engineering. Its methods rely mainly on soil removal and water pumping, followed by ad hoc combinations of physicochemical treatments such as incineration, filtration, adsorption, washing, multi-phase extraction, and mixing with surfactants, solvents, strong oxidants and other reactants (Suthersan, Horst, Schnobrich, Welty, & McDonough, 2016). Such methods are typically energy-intensive, environmentally disruptive and, above all, very expensive. The result is that remediation of many polluted sites is neglected, even in wealthy nations. The US Environmental Protection Agency estimates that there are approximately 450,000 brownfields in the USA (https://www.epa.gov/brownfields). About 250,000 is the appraisal of the European Environment Agency for its member countries (EEA, 2017). Most catalogued sites are polluted with mixtures of heavy metals and organic compounds, the latter mainly consisting of petrol hydrocarbons and persistent organochlorines.

The limitations of engineering-based methods have sparkled a search for alternative or complementary biological technologies. In this review we will focus on the exceptional potential of long-lived woody perennial plants, trees hereafter, to remediate organic and elemental pollutants. The utilization of vegetable organisms for environmental cleanup has been referred to as phytoremediation since Ilya Raskin coined the term nearly three decades ago. Since then, comprehensive reviews have been regularly published in the scientific literature (e.g. Doty, 2008; Kramer, 2010; Lee, 2013; Pilon-Smits, 2005; Robinson, Bañuelos, Conesa, Evangelou, & Schulin, 2009; Van Aken & Doty, 2010; Vangronsveld et al., 2009; etc.). For a recent compilation of field studies and a general overview of phytoremediation, the reader is kindly referred to Gerhardt, Gerwing, and Greenberg (2017) and references therein. Recent reviews on specific aspects treated here are referenced in the appropriate sections.

2. Key advantages and disadvantages of phytoremediation

Phytoremediation exploits the extraordinary metabolic and extractive capabilities of plants and their associated microorganisms to stabilize, remove, store, and even metabolize pollutants present in soil, water and air. Degradation is only feasible for organic pollutants, which can ideally be mineralized by appropriate biochemical pathways, or at least be broken down into less harmful molecules. This fate is not feasible for heavy metals and other elemental pollutants, which are instead stabilized by the root-rhizosphere system or stored within plant tissues, ideally in harvestable parts. Toxic organic compounds and a few elements can be volatilized as well (Limmer & Burken, 2016; Robinson et al., 2009). Root exudates have a particular influence in phytoremediation, as they alter the bioavailability of pollutants and affect the composition of the plant-associated microbiota. Another interesting phenomenon is hyper-accumulation, whereby some plants have the ability to store elemental pollutants at unusually high levels (Kramer, 2010). All these natural processes have given rise to a set of phytotechnologies aimed at improving environmental health. Their application essentially offers two possibilities: either pollutants are removed from the substrate (decontamination) or their bioavailability is reduced, resulting in diminished exposure (stabilization) (Gerhardt et al., 2017; Lee, 2013; Vangronsveld et al., 2009). Five major phytotechnologies are commonly applied: (i) phytoextraction refers to the accumulation of pollutants (primarily heavy metals) into harvestable plant tissues; (ii) rhizofiltration is the term specifically used when pollutants are removed from water sources; (iii) phytostabilization implies a reduction in the bioavailability of pollutants upon interaction with the root-rhizosphere system; (iv) phytodegradation refers to the metabolic transformation of organic pollutants, with varying degrees of deactivation depending on the chemical structures involved; and (v) phytovolatilization implies the translocation of ground pollutants into the atmosphere, ideally in a harmless form. According to Gerhardt et al. (2017), the phytoremediation of organic contaminants that can be degraded (either *in planta* or in the rhizosphere) has the highest potential for wide-scale deployment, with petrol hydrocarbons providing the best return on investment.

Commercial phytoremediation is often carried out with plant species or varieties that thrive on sites polluted above regulatory limits (Gerhardt et al., 2017; Vangronsveld et al., 2009). The use of dedicated tree plantations is gaining momentum, however, for reasons that will be discussed in the next section. The environmental benefits of such plantations will soon be expanded by novel developments in tree genetics, biochemistry and biotechnology as well as in silviculture (Neale, Martínez-García, De La Torre, Montanari, & Wei, 2017; Whitaker et al., 2018). Significant gains can also be expected from progress in the study of plant-microbial associations that result beneficial for environmental purposes (Levy et al., 2017; Mercado-Blanco et al., 2018; Thijs, Sillen, Rineau, Weyens, & Vangronsveld, 2016). Compared with alternative cleanup methods, phytoremediation offers several key advantages:

(1) The remediation process is essentially powered by sunlight, a major backer for public acceptance and also for cost-effectiveness. Whereas expenditure information is not usually provided in scientific papers, extensive data recently compiled by Gerhardt et al. (2017) suggest that phytoremediation costs are between 1.5 and 15 times cheaper than those of conventional methods.

(2) As photosynthetic organisms, plants are more resilient than microbes and require considerably less management in terms of infrastructure and nutritional supplements. In addition, the use of tree plantations can generate valuable biomass unconnected with food markets. These features improve cost-effectiveness and add to sustainability and public perception.

(3) Most activities take place in situ and are environmentally friendly. The growth of vegetable covers improves soil structure, prevents erosion by water and wind, and increases soil fertility. It also boosts local biodiversity below and aboveground by creating new habitats. A key difference with conventional engineering is that phytoremediation generates very few residues. The overall landscape usually improves, especially when the planned actions include ecological restoration (e.g. Hua et al., 2016; Isebrands et al., 2014).

(4) Many plant species or taxa have unique biochemical capabilities not present in microbes. These are essentially associated with secondary metabolism, with wood synthesis providing an outstanding example (Celedon & Bohlmann, 2018; Wang et al., 2018). The complex enzyme networks associated with such metabolism can contribute to counter pollution by acting on toxic compounds structurally-similar to natural substrates.

Like any other biotechnology, phytoremediation also has restrictions and pitfalls. Its biological foundation imposes limits to the range of environmental conditions under which it can be applied (Pilon-Smits, 2005; Vangronsveld et al., 2009). The potential of hyper-accumulator plants, for example, is usually hampered by their insufficient growth in heavily polluted sites. Recalcitrant xenobiotics pose similar problems when present at high concentrations, due to their deleterious effects on plant growth and development. Regarding soil features other than pollution, plants are especially sensitive to pH, chemical composition, redox status, and microbiota. Other environmental cues that significantly affect plant performance are water availability, temperature, photoperiod, and air humidity. Combinations of these factors can make phytoremediation time-consuming, only seasonally applicable or impractical altogether. Phytotechnologies are hardly perceived as a fast solution, especially during early stages, and regulation tends to require substantial advances in relatively short periods of time. From the scientific viewpoint, one of the key limitations of phytoremediation is that a large proportion of published data is fragmented and lacks biochemical detail. More dedicated research and the application of high-throughput molecular technologies should contribute to overcome these problems as well as to improve our understanding of how plants deal with pollution.

3. Why using trees?

An adequate selection of species is a major factor for the success of any phytoremediation project (Vangronsveld et al., 2009). Practical considerations and availability of experimental data explain why this scenario has been dominated by herbaceous species. They are also object of the vast majority of studies reported in the scientific literature. Trees have been largely ignored, on the other hand, with the only exception of a few species and hybrids (Campos, Merino, Casado, Pacios, & Gomez, 2008; Doty, 2008; Marmiroli et al., 2011; Peuke & Rennenberg, 2015; Yadav, Arora, Kumar, & Chaudhury, 2010). In this review we contend that tree plantations offer unique and important benefits for phytoremediation programmes, and an extraordinary potential for rational improvements that should no longer be neglected. Planting trees has advantages that extend beyond environmental cleanup.

3.1 Global benefits

It has been recently estimated that some 3 trillion trees populate our planet (Crowther et al., 2015). They all share the distinctive features of being perennial, often very long lived, and capable of producing wood. Since the dawn of human societies, trees have been key providers of renewable energy, food, and raw materials for tools, housing and other uses. Even today trees contribute to the food security of hundreds of millions of people. According to the latest global analyses of the UN Food and Agriculture Organization, some 3 billion square metres of wood are removed each year worldwide, contributing about USD 600 billion to the global economy (FAO, 2016a, 2016b; 2011 datasets). Partly to satisfy such demand and partly yielding to agricultural and urbanization pressures, 15 billion trees are cut down annually (Crowther et al., 2015). Most wood comes from production ($\sim$1.2 billion ha) and multiple-use forest areas ($\sim$1 billion ha), providing direct labour for >13 million people (full-time equivalent; see FAO, 2016a). The indirect welfare associated to forestry is both huge and complex to estimate. Forestry-related jobs are especially important in rural areas, where the sector has a relevant presence and alternatives are scarce (FAO, 2016b).

In addition to their immeasurable socioeconomic impact, trees play essential roles in ecologically sound processes, such as biogeochemical cycles, environmental health, climate regulation, soil conservation, and habitat protection (Bonan, 2016; Trumbore, Brando, & Hartmann, 2015). After all, tree growth accounts annually for half of the terrestrial net primary production and plays a major role in sequestering carbon emissions from fossil fuels (Bonan, 2016). This role has often been underrated due to the tendency of scientific reports to focus on net forest balances, overlooking that deforestation-driven emissions are much smaller than anthropogenic emissions. In any case, carbon sequestration can be boosted by optimizing cultural practices and genotype selection. There is a general consensus among experts that planting trees should be a priority under the Paris Climate Agreement (see Bonan, 2016; FAO, 2016a, 2016b). Plantations should help compensate the effects of deforestation and contribute to sustainable development, especially in developing countries.

3.2 Some trees are exceptionally endowed for phytoremediation

Trees exhibit many features that make them paradigmatic organisms for phytoremediation, with some taxa being particularly well endowed. They also constitute excellent systems to investigate plant responses to pollution. Relevant features include high biomass production, rapid growth, powerful root systems, high transpiration rates, great genetic variability, and an innate ability to tolerate and remediate a wide range of pollutants from soil, water and air sources (Campos et al., 2008; Marmiroli et al., 2011; Peuke & Rennenberg, 2015; Yadav et al., 2010).

3.2.1 Biomass

Trees are capable of producing huge amounts of biomass in relatively short periods of time, especially in dedicated plantations. Regardless of the variability due to genotypic effects, environmental stress and management (Dimitriou & Mola-Yudego, 2017; Lutter, Tullus, Kanal, Tullus, & Tullus, 2016), such biomass is in general orders of magnitude higher than that of any other organism used for remediation. This fact endows trees with a unique bulk capacity to remediate pollutants: wood offers a vast reservoir to safely allocate them; foliage, roots and living stem cells offer myriads of cells for degradation and/or sequestration in cell walls and vacuoles; and leaves present considerable surface for volatilization. Numerous assays conducted with *Populus*, *Salix*, *Picea*, *Pinus*, and other widely-planted trees (reviewed by Luo, He, Polle, & Rennenberg, 2016) confirm remarkable pollutant removals relative to background levels. The same holds true in comparisons with herbaceous plants growing under the same pollution regimes. Although detailed cellular measurements are lacking, these findings raise the possibility that sequestration is more efficient in trees than in small-sized plants, which would result in lower cytosolic concentrations of pollutants (i.e. reduced toxicity). Field experiments with herbaceous hyper-accumulators, such as *Thlaspi caerulescens* or *Arabidopsis halleri*, have shown that insufficient biomass production counterbalances their removal capacity (Chaney, Baker, & Morel, 2018; Kramer, 2010). Regarding tree plantations, it seems likely that biomass yields will be substantially improved in the foreseeable future (Harfouche, Meilan, & Altman, 2014; Neale et al., 2017), with positive consequences for phytoremediation.

3.2.2 Fast growth, big roots, high transpiration

Fast-growing trees usually develop extensive root systems and show high rates of transpiration, two important features to tackle pollution. After all, tree roots conform one of the largest abiotic–biotic boundaries in the biosphere, mediating a colossal exchange of pollutants. The fast-growing species belonging to the Salicaceae family, which comprises willows (*Salix*) and poplars (*Populus*), have attracted particular attention in the context of phytoremediation (Campos et al., 2008; Isebrands & Richardson, 2014; Marmiroli et al., 2011; Yadav et al., 2010; see also the web server of the FAO's International Poplar Commission: www.fao.org/forestry/ipc/32608/en). These trees are extensively used in short-rotation coppicing (SRC) in many regions. *Populus* has been hailed as the choice genus for Europe's temperate areas (Dillen, Rood, & Ceulemans, 2010), where Salicaceae trees reach biomass yields of up to 24t per ha and year and an average value close to 10t (e.g. Broeckx, Verlinden, & Ceulemans, 2012). Interestingly, these trees tolerate very high planting densities (Marmiroli et al., 2011), which has been shown to favour metal remediation (Pajević et al., 2016). Fig. 1 shows a high-density poplar plantation used for

Fig. 1 Poplars planted at high-density for phytoremediation studies performed by our research group. *Pictures generously provided by C. Salarnier.*

phytoremediation studies. Due to the increased competition for light, these densities promote primary over secondary growth, with metabolic consequences for phytoremediation that remain to be studied. Generally speaking, the large root systems of trees favour both the physical contact with pollutants and the development of complex rhizospheric communities, which play essential roles in phytoremediation (Thijs et al., 2016). Root depth is critical for reaching sources and leachates too distant for herbaceous plants.

The relationship between transpiration and phytoremediation has long been known (Pilon-Smits, 2005; Yadav et al., 2010). The uptake of soil pollutants occurs mainly through the root water stream, and many experimental studies have substantiated that the species and hybrids commonly used in SRC are capable to mobilize significant amounts of water (see Fischer et al., 2018). The impact of such ability on local water sources has been controversial. Although some studies suggested excessive consumption (e.g. Petzold, Schwarzel, & Feger, 2011), accurate field data have confirmed more recently that annual evapotranspiration is substantially lower in coppices than in reference grasslands across vast regions (Bloemen et al., 2017; Fischer et al., 2018). Water savings will even increase when innovative irrigation technologies are adopted by traditionally-managed plantations (Paris et al., 2018), still irrigated through gravity-fed and sprinkler systems. In sum, the impact on local water sources does not furnish a reliable criterion to exclude fast-growing trees from phytoremediation programmes.

3.2.3 Wood and tree secondary metabolism

The ability to synthesize wood is widespread within the plant kingdom, being present in about 60% of all botanical orders (Neale, Langley, Salzberg, & Wegrzyn, 2013). In a simplified view, a tree trunk can be considered a pillar of wood (xylem) covered by a thin layer of vascular cambium and an outer shell of bark (see Pallardy, 2008 for more detail). The relative abundance of living cells significantly changes over time, affecting phytoremediation capacities. Fig. 2 presents transversal sections of a young and a mature tree. Heartwood (mature xylem, dead cells) is darker than sapwood (young xylem), which contains a variable proportion of metabolically-active cells. The cambium and inner bark are formed by living tissues, but not the outer bark (rhytidome). Wood is made up essentially by the cell walls of xylem fibres, vessels and ray cells that deploy intricate metabolic networks during secondary differentiation. As a result, a set of monolignols (mainly *p*-coumaryl alcohol, coniferyl alcohol and sinapyl alcohol) are combinatorically linked in complex spatial scaffolds that pervade the polysaccharide matrix

Fig. 2 Sections of tree stems at different ages. Heartwood (mature xylem) is darker than sapwood (young xylem). The xylem is surrounded by concentric cylinders of cambium and inner bark. The outer bark has been mostly removed in the right panel. *Pictures generously provided by C. Salarnier (left) and Dr. M. Esteban (Forestry School, Universidad Politecnica de Madrid).*

(Mottiar, Vanholme, Boerjan, Ralph, & Mansfield, 2016; Ye & Zhong, 2015). Specific laccases and peroxidases are required for monolignols to react. In addition, wood also contains ill-defined arrays of secondary metabolites, which are responsible for important properties and sometimes have industrial application (reviewed for heartwood by Celedon & Bohlmann, 2018). A common drawback when obtaining extracts from wood is that the abundance of phenolic compounds causes bulk macromolecular aggregation and progressive darkening, interfering with downstream analyses. A second problem is that the underlying machinery shows remarkable variations across taxa, developmental stages, cell types, and even cell wall layers (Mottiar et al., 2016). And a third problem arises from the lack of detailed genomic studies in trees in general, compared with herbaceous model and agricultural systems (Neale et al., 2017). These pitfalls largely explain why fundamental details are still missing regarding wood biosynthesis and its regulation, despite the increasing application of state-of-the-art molecular methodology. For example, the apparently simple conversion of phenylalanine into monolignols has been recently analysed in *P. trichocarpa* through an integrative multi-omics approach (Wang et al., 2018). These authors identified a 'core' set of 21 dedicated enzymes mediating a network of 37 metabolic-fluxes, with a regulation so complex that it was only preliminarily studied. The production of other secondary metabolites is much less known.

Regardless the above limitations, there are two key issues connecting the rich and complex secondary metabolism of trees with phytoremediation applications. First, there are striking structural similarities between relevant xenobiotics and plant secondary metabolites or compounds produced by plant-associated bacteria. Many of them have been identified in tree species, where they are mostly connected with wood formation and/or stress responses (Fig. 3) (e.g. Constabel & Lindroth, 2010; Woolbright et al., 2018). The fortuitous degradation of xenobiotics by endogenous enzymes involved in natural biochemical processes is a phenomenon well known in microorganisms (Chakraborty & Das, 2016). A similar phenomenon

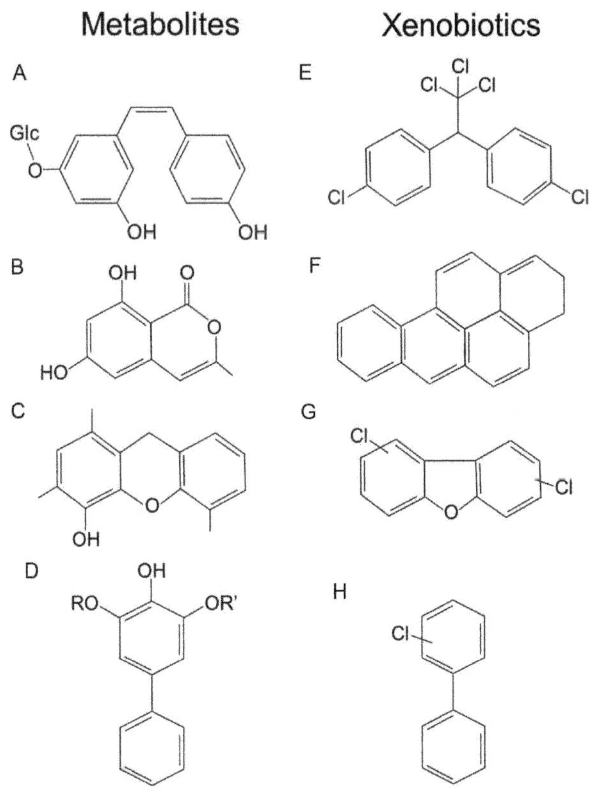

Fig. 3 Structural similarity between relevant xenobiotics and natural metabolites found in trees. Metabolites: stilbene (A), isocoumarin (B), xanthone (C), and aucuparin-like phytoalexin (D). Xenobiotics: DDT (E), benzo(*a*)pyrene (F), polychlorodibenzodioxins (G), and polychlorobiphenyls (H). Benzo(*a*)pyrene is converted into isocoumarin derivatives by endophytic microorganisms such as *Crinipellis* or *Cephalosporium*. Common sites for ring substitution are indicated by empty bonds. Glc, glucose.

seems likely to occur in trees, given their vastly complex secondary metabolism. This possibility represents an excellent target for future research, including genetic and enzyme engineering approaches. The xenobiotics shown in Fig. 3 include aromatic compounds of utmost concern (see below: Stockholm's Convention on Persistent Organic Pollutants). Besides wood formation, the morphological and physiological features associated with the tree phenotype impose particular requirements for which specific secondary compounds must have evolved—the chemical defence of trunks from insect or fungal attack is an obvious example. In this regard, we must keep in mind that the genetic variability associated with the biosynthesis of these metabolites is so large that finding genus- or species-specific compounds is not rare (Celedon & Bohlmann, 2018; Kroymann, 2011). And on the top of this, we must add the enormous genetic complexity characteristic of trees (Neale et al., 2017). As metabolite profiling improves and more tree genomes become available, novel compounds will certainly be identified that may be of interest for xenobiotic degradation.

Second, insofar as there is a specific tree biochemistry with no paragon in other organisms, not even in prokaryotes, we can envision trees as potential sources of novel genes for remediation applications. The work by Gordon et al. (1998) showing that poplars are unusually apt to extract and detoxify trichloroethene constitutes a pioneering example in this regard. Numerous field trials and greenhouse studies support that poplar trees can be established in sites heavily polluted with organochlorines and petroleum hydrocarbons and contribute to their remediation (Gerhardt et al., 2017; Marmiroli et al., 2011; Peuke & Rennenberg, 2015). Trees can therefore be envisioned not only as remarkable organisms to perform remediation tasks, but also as valuable biological systems to study molecular responses to pollution, including the interaction with suitable soil microorganisms and endophytes. This potential remains poorly explored for organic pollutants, despite the prospect of finding natural mechanisms with applied interest. Albeit more studied, our current understanding of how trees deal with heavy metals is rather incomplete as well (Luo et al., 2016). Recently, Yanitch et al. (2017) have demonstrated a major role for tannins in relieving the cellular toxicity caused by arsenic in willow trees, an observation that might be extensible to other elemental pollutants and/or secondary compounds.

3.2.4 Longevity and stress tolerance

Most tree species are capable of surviving for decades, even centuries. During its lifetime, a single tree can mobilize millions of litres of groundwater (Kunert et al., 2017), along with significant amounts of chemical elements and compounds. It will also incorporate a daunting amount of carbon dioxide into biomass, compared to alternative plants used for remediation purposes (Gerhardt et al., 2017; Vangronsveld et al., 2009). The benefits of using dedicated tree plantations for environmental cleanup will increase over time. There are three key aspects in this regard: (i) the remediation capacity per surface unit will increment every season for a fair number of years (genotypes and silviculture are key factors here); (ii) after initial establishment, the resilience and self-sufficiency of plantations will keep costs low; and (iii) if appropriate trees are chosen, an ecologic and aesthetic improvement of the polluted site can be expected (Isebrands & Richardson, 2014; Lutter et al., 2016; Tullus, Rytter, Tullus, Weih, & Tullus, 2012). Even the 'limited' longevity in practice of fast-growing SRC plantations can be sufficient for relevant cleanup (e.g. Doty et al., 2017). Another important aspect related with their extended longevity is the remarkable adaptability of trees to environmental stress (Harfouche et al., 2014; Neale et al., 2017). While this seems to be a general characteristic of the tree phenotype, probably associated to its perennial nature, the underlying genetic variability is huge and poorly understood (Harfouche et al., 2014; Neale et al., 2017). The application of novel molecular technologies is uncovering major mechanisms responsible for stress tolerance in a number of economically relevant tree species (e.g. Fox et al., 2018; Hamanishi, Barchet, Dauwe, Mansfield, & Campbell, 2015). In due time, this knowledge should help identify or even obtain genotypes better tailored for phytoremediation and other specific needs. For the moment, a few attempts have succeeded in improving stress tolerance in model trees (Harfouche et al., 2014). Tree breeders and biotechnologists should focus in particular on stresses that are difficult to manage under field conditions, especially increased carbon dioxide and warmer temperatures (Fares, Mugnozza, Corona, & Palahí, 2015). Merino et al. (2014) have recently demonstrated the feasibility of improving thermotolerance in hybrid poplars through genetic manipulation. Interestingly, the transgenic poplar lines obtained in this work exhibited more tolerance than controls to the herbicide phosphinotricine, an effect attributed to increased glutamate synthase activity.

3.3 Additional environmental services

Planting trees for phytoremediation can often provide additional services. Archaeological studies have revealed millennia-old uses of planted trees for environmental purposes, like soil and streamside stabilization, crop protection or landscape improvement (Isebrands & Richardson, 2014).

This aspect of tree planting is becoming today more important than ever, especially for species and varieties that can deliver results in relatively short timespans (Whitaker et al., 2018). Riparian corridors (also called riparian buffers) provide an excellent example. The main problem tackled by these corridors is agricultural runoff, a vastly important diffuse source of pollution. It carries biocides, fertilizers and organic farm wastes to water streams and reservoirs, causing a gradual deterioration of aquatic ecosystems (Pinay et al., 2018). One of the measures most widely recommended to mitigate this problem is to maintain riparian corridors in the riverbanks next to croplands (e.g. de Sosa, Glanville, Marshall, Prysor-Williams, & Jones, 2018; see also the documentation on riparian buffers at the USDA's web server, https://www.nrcs.usda.gov). In the Northern hemisphere, the natural vegetation of these corridors typically includes phreatophyte trees such as poplars, willows, alders (*Alnus*), and ashes (*Fraxinus*). They are fast-growing pioneers with potent roots and good tolerance to environmental stresses (Catford & Jansson, 2014). These features are well suited for phytoremediation purposes. Among riparian trees, poplars and willows show the best aptitudes for large-scale plantation (Dimitriou & Mola-Yudego, 2017; Isebrands & Richardson, 2014). Trees from both genera are increasingly planted to filter agricultural runoff and/or to prevent losses of fertile soil (Da Ros, Soolanayakanahally, Guy, & Mansfield, 2018; Fortier, Truax, Gagnon, & Lambert, 2016). Several interesting case-studies are extensively described in Isebrands et al. (2014), including the role played by riparian corridors in the Lower Mississippi Alluvial Valley, an 11 million ha site where both native and planted Salicaceae trees thrive abundantly. In addition to improving water quality, riparian trees provide ecological niches and shelter for local wildlife; control erosion by wind and rain; protect bank material from floods and strong seasonal currents; stabilize stream banks through their massive root systems; and their litter constitutes a primary food source for aquatic ecosystems.

Ecological restoration of polluted or degraded sites is another application of tree planting that has become increasingly important in recent years, especially for abandoned croplands. As in the previous examples, fast-growing trees have attracted most attention (Isebrands et al., 2014). The relevant issue is that, beyond pollutant cleanup, afforestation tends to improve the physicochemical and biological properties of degraded soils. Important properties affected by planting trees include soil porosity, bulk density, moisture, bioavailability of micro and macronutrients, and biodiversity (Berhongaray, Verlinden, Broeckx, Janssens, & Ceulemans, 2017; Wang et al., 2017;

Wu et al., 2018). The use of poplars and other trees for restoration purposes has not been without controversy, however, as discussed above for water resources. Concerns on local biodiversity have been raised, especially for tree monocultures, although there is now ample consensus that well-managed plantations can provide valuable habitat, even for endangered species (Hua et al., 2016). Restoring lost forest habitat often improves connectivity and buffers edge effects, two aspects with substantial positive effects on biodiversity (Aerts & Honnay, 2011; Pfeifer et al., 2017). The heterogeneity and ecological ambivalence of riparian trees is an important advantage in this context, as it facilitates the use of indigenous or naturalized genotypes rather than exotic ones.

4. Phytoremediation with trees

We will focus here on recent developments regarding the use of trees to remediate elemental and organic pollutants. A number of recent reviews, cited in the appropriate subsections, have addressed how plants respond in general to either type of pollution and how the success of current phytotechnologies can be improved through genetic engineering, the use of suitable microorganisms and other innovative strategies. The vast majority of studies to date have been conducted with herbaceous species and refer to the remediation of toxic metals and metalloids (metals hereafter).

4.1 Elemental pollution

4.1.1 Relevant targets

The build-up of toxic metals in croplands elicits particular concern, for it directly impinges on food security (Singh et al., 2011). It has been recently estimated that, only in Europe, metals have rendered 137,000 km^2 of cropland (>6%) unsafe for food production (Toth, Hermann, Da Silva, & Montanarella, 2016). Table 1 summarizes the main anthropogenic sources and health effects of the top toxic metals in the latest Substance Priority List of the US Agency for Toxic Substances and Disease Registry (ATSDR, https://www.atsdr.cdc.gov). Besides human action, toxic elements spread globally from natural sources, including volcanism, forest fires, rock erosion, microbial activity, and volatilization from soil and water.

A wealth of experimental data support that phytoremediation can significantly contribute to manage elemental pollution and restore croplands. For recent reviews focused on this type of pollution, the reader is kindly referred to Luo et al. (2016) and Sarwar et al. (2017). Because elemental pollutants

Table 1 Health effects and major anthropogenic sources of priority elemental pollutants for the US ATSDR.

Element	Anthropogenic sources[a]	Health effects[b]	Top producers[c]
As	Thermal power plants, fuel burning, ore smelting, biocides, dyes, pyrotechnics	Haematologic and vascular disorders, diabetes, hearing loss, neurologic disorders, teratogenic effects, carcinogenesis	(1) China (2) Chile (3) Kazakhstan
Pb	Pigments, electronic waste, ammunition, metallurgy, ore smelting, ceramics, fossil fuel burning	Multiorgan damage, pain, hypertension, systemic effects (haematopoiesis, endocrine regulation, reproduction), suspected carcinogen	(1) China (2) Australia (3) USA
Hg	Electronic waste, chemical industry, metallurgy, fungicides, wood preservation	Multiorgan damage (mainly kidney and liver), diarrhoea, headache, inflammation, teratogenic effects, suspected carcinogen	(1) China (2) Kyrgyzstan (3) Chile
Cr	Pigments, metal parts industry, refractory industry, chromate production, tannery, wood preservation	Multiorgan damage (mainly lungs), ulcers, skin irritation, asthma, carcinogenesis	(1) South Africa (2) Kazakhstan (3) India
Cd	Alloy metallurgy, batteries, pigments, waste incineration, fossil fuel burning	Lung disorders, teratogenic effects, carcinogenesis	(1) China (2) Republic of Korea (3) Japan

[a]Based on Tchounwou, Yedjou, Patlolla, and Sutton (2012) and Yewale et al. (2017).
[b]All relevant chemical species are considered here, regardless differences in their effects. See the ATSDR Toxic Substances Portal for detailed information on each specific form (https://www.atsdr.cdc.gov/substances/index.asp).
[c]Data from the UK Royal Society of Chemistry, available at www.rsc.org/periodic-table.

cannot be degraded, their remediation with plants relies mainly on either immobilization (phytostabilization) or extraction from soil (phytoextraction) or water sources (rhizofiltration), followed ideally by translocation to harvestable parts. Also amenable to successful phytoremediation are radionuclides (Gupta, Chatterjee, Datta, Voronina, & Walther, 2017; Lee, 2013; Vangronsveld et al., 2009) and airborne inorganic pollutants (e.g. NO_x, SO_2, CO_2, O_3) (Weyens et al., 2015). Another interesting possibility is the phytovolatilization of As, Hg, and Se, for which gaseous species naturally exist, via endogenous plant mechanisms (e.g. Abbas et al., 2018) or genetic modification (Bizily, Rugh, & Meagher, 2000; Che et al., 2003; Hussein, Ruiz, Terry, & Daniell, 2007; LeBlanc, McKinney, Meagher, & Smith, 2013). The above technologies are sometimes combined with conventional engineering methods, such as electrokinetic remediation, soil washing and soil amendment.

4.1.2 Plant uptake and fate of elemental pollutants

Like in herbaceous plants, trees take up most elemental pollutants from the soil water phase, which moves freely in the capillary spaces between the outer layers of root cells (cortex). Besides factors such as soil composition and pH, uptake is substantially influenced by root exudates and the plant-associated microbiota, two factors tightly connected (Doty, 2008; Thijs et al., 2016). Regarding pH, metals tend to form insoluble carbonates and phosphates in alkaline soils. This affects their bioavailability, as recently shown for *Salix* trees growing on adjacent soils with very different pH, shale overburden and acidic clay (Mosseler & Major, 2017). Tree rhizospheres are more acidic than neighbouring non-vegetated soils and, moreover, contain natural chelators and other compounds that promote metal uptake by root cells (Ullah, Heng, Munis, Fahad, & Yang, 2015; Wenzel, 2009). In fact, inoculation with appropriate ectomycorrhizal fungal strains has been shown to increase metal absorption by tree roots (Huang, Han, & Li, 2018; Ma et al., 2014). Also interesting for phytoremediation are the effects of plant growth-promoting bacteria (e.g. Durand et al., 2018). The application of soil amendments is another common strategy to improve metal bioavailability and even root-to-shoot translocation (Shahid et al., 2014). Caution must be exerted, however, as some compounds have proven toxic for local plants and/or mycorrhizae; they can also induce pollutant leaching into the soil (Agnello, Huguenot, Van Hullebusch, & Esposito, 2014; Lu et al., 2017).

The cell wall polysaccharide matrix is known to bind the ionic forms of various metals (Chen, Liu, Wang, Zhang, & Owens, 2013), acting as a first barrier for root cell uptake. A more selective barrier is the plasma membrane that contains specific transporters for nutritional ions (Fe^{2+}, Cu^{2+}, Mg^{2+}, Ca^{2+}, Mn^{2+}, and others) but not for non-nutritional ones (Kramer, 2010; Pilon-Smits, 2005). The same is true for the tonoplast, the third barrier (Peng & Gong, 2014). Membrane-mediated transport is unavoidable at some point if the pollutants are to reach aboveground parts, because the Casparian band blocks xylem loading through the apoplast. This band confers therefore a pivotal role to the endoderm cell membranes for the success of phytoremediation. Once inside root cells, toxic metals can accumulate intracellularly or be translocated into adjacent cells via membrane transporters or plasmodesmata. The latter make up the cellular continuum known as symplast. The study of Cd movement in roots of hybrid poplars has provided a well-characterized example of symplastic translocation (He et al., 2015). Ricachenevsky, Araújo, Fett, and Sperotto (2018) have recently reviewed the role of the tonoplast as a checkpoint for this type of transport. Regardless of the specific mechanism, intercellular transport allows metals to reach the root xylem vessels and, from there, aboveground tissues. Whether for root cell uptake, vacuolar sequestration or xylem loading, toxic metals highjack membrane transporters for nutritional ions. In recent years, genome-wide analyses and other molecular tools have allowed the identification of numerous metal transporters in trees (e.g. Fan et al., 2018; Migeon et al., 2010; Yanitch et al., 2017; Yildirim, 2017). Functional equivalences have been established with well-characterized transporters from herbaceous species (e.g. Tan et al., 2017). Albeit sparse, these studies have shed light on the molecular mechanisms involved in metal tolerance and detoxification in trees. Major proteins involved in toxic metal transport include ATP-binding cassette (ABC) transporters, heavy metal(loid) ATPase (HMA) transporters, metal transporters of the NRAMP family, transporters of the zinc/iron regulated protein (ZIP) family, and cation diffusion facilitators (CDFs), also called metal tolerance proteins (MTPs). The unusual heterogeneity detected for these transporters in trees (e.g. Fan et al., 2018; Migeon et al., 2010) may be indicative that metal translocation is more complex than in herbaceous plants. In any case, the extraordinary capacity of some fast-growing trees to remediate metals from polluted substrates deserves further study. New species and varieties should be analysed, given the vast genetic variability found in trees for most traits analysed.

When the intracellular accumulation of metals reaches toxic levels, plants develop symptoms like chlorosis, leaf deformities, defoliation, reduced growth and general weakness towards biotic and abiotic stress. Major protective mechanisms include pollutant sequestration/deactivation and ROS-scavenging machinery (Jalmi et al., 2018; Kramer, 2010; Peng & Gong, 2014; Pilon-Smits, 2005). It is noteworthy that the antioxidant systems are typically upregulated in hyperaccumulator plants, even in the absence of metal pollution (Kramer, 2010). The rationale of metal sequestration is to protect metalloenzymes and other vulnerable cell machinery. Cell walls represent a major sequestering site in trees, which accumulate metals in wood and outer bark (Chen et al., 2013; Evangelou, Robinson, Günthardt-Goerg, & Schulin, 2013; Zárubová et al., 2015). This capacity is probably beneficial against the attack of insects as well as pathogenic and wood-degrading microorganisms, i.e., constitutes an adaptive advantage. The notion that metal accumulation has a defensive role is supported by several lines of evidence (Hörger, Fones, & Preston, 2013), including rhizosphere studies (Lakshmanan, Cottone, & Bais, 2016). Besides cell walls, sequestration in vacuoles is an important strategy to cope with metal toxicity (Peng & Gong, 2014; Sharma, Dietz, & Mimura, 2016). Putative tonoplast isoforms of ZIP, CDF (MTP) and HMA transporters have been identified in trees, mostly through genomic comparisons. Studies with herbaceous plants have shown that, prior to vacuolar storage, metallic ions are chelated by glutathione (GSH), phytochelatins, metallothioneins, and other metabolites (amino acids, other organic acids, phenolics) (Pilon-Smits, 2005; Sharma et al., 2016; Song et al., 2014). Whereas experimental evidence is still scarce for tree species, genomic comparisons and reverse genetics experiments suggest that both trees and herbaceous plants have similar mechanisms of vacuolar sequestration (Ariani et al., 2015; Di Baccio et al., 2011; He et al., 2013, 2015). The narrow correlation found in hybrid poplars between vacuolar sequestration and Cd tolerance is of particular interest (He et al., 2013). This work also reported that sulphur-containing chelators are probably involved. Evidence for metal sequestration in vacuoles has been obtained in field studies with different poplar varieties (e.g. Luo et al., 2016; Marmiroli, Imperiale, Maestri, & Marmiroli, 2013). When a yeast ABC transporter (YCF1) that sequesters cytosolic GSH-conjugated cadmium into the vacuole was overexpressed in poplar (Shim et al., 2013), the resulting plants performed better than wild type controls in mine tailing soil contaminated with Pb, Cd, and As. The transgenic lines also showed enhanced extraction capacities. Improved vacuolar sequestration is the most likely reason why transgenic hybrid poplars overexpressing a *P. trichocarpa* ABC metal transporter show improved Hg tolerance and accumulation (Sun et al., 2018).

4.2 Organic pollution

4.2.1 Relevant targets

The variety of relevant elemental pollutants is paltry compared with the array of organic compounds for which adverse effects on human health and ecosystems have been reported. The vast majority of organic pollutants are of anthropogenic origin, being also referred to as xenobiotics. The threats posed globally by persistent xenobiotics are so important that the

Governing Council of the United Nations Environmental Programme (UNEP) promoted more than two decades ago different measures to reduce and eventually eliminate their release into the environment. This initiative crystallized in the Stockholm's Convention on Persistent Organic Pollutants (POPs), which entered into force in 2004 (see www.pops.int) and represents one of the most ambitious international treaties ever adopted to combat chemical pollution. The Convention maintains updated an inventory of particularly hazardous POPs that reunite three key characteristics: (i) they must remain unaltered in the environment for very long periods of time, a reflection of their outstanding chemical stability; (ii) they must be widely distributed over large areas, sometimes around the globe, by natural processes that encompass soil, water and air; and (iii) they must be magnified along trophic chains as a consequence of their stability and low solubility in water. The latter two features lie behind their long-term accumulation in fatty tissues, where they reach concentrations that surpass background levels up to four orders of magnitude. As one would expect, the highest levels are recorded in humans, along with top predatory fish, birds and mammals—migrant species help disperse these compounds. POPs can be found indeed in people and wildlife located thousands of kilometres away from production centres, with the presence of high levels of DDT and PCBs in Eskimo mother milk being the paradigmatic example. Table 2 summarizes the main sources and health effects of top xenobiotics included in both the latest Substance Priority List of the US ATSDR and the Stockholm's Convention. They fall essentially into three groups: chlorinated pesticides, industrial chemicals and by-products. There are additional classes of relevant organic pollutants, such as polycyclic aromatic hydrocarbons (PAHs), solvents (e.g. benzene, trichloroethylene), short-chained chlorinated paraffins (SCCPs), flame retardants (e.g. polybromo-biphenyls and polybromo-cyclododecanes), perfluorooctane derivatives, industrial and combustion by-products (e.g. benzopyrenes, polychlorinated benzenes and naphthalenes), explosives, and many other reagents. Exposure occurs mainly via industrial processes, waste disposal, food products, drinking water, air or physical contact with contaminated materials.

Phytoremediation of different classes of organic pollutants has been demonstrated in the greenhouse and in field trials. The latter is especially important when using trees, as greenhouse results are not always duplicated under field settings (e.g. Voelker, Lachenbruch, Meinzer, & Strauss, 2011). Current field data strongly support that phytoremediation of organic pollutants is not only feasible, but a cost-effective alternative to current

Table 2 Main sources, uses and health effects of top xenobiotics for the US ATSDR and the Stockholm's Convention on POPs.

Compound/family	Main sources and uses[a]	Long-term health effects[b]	SC annexes[c]
Polychlorinated biphenyls (PCBs)/arochlor	Electric transformers and capacitors, paints, plastics, old electric equipment and fluorescent lighting fixtures	Fatigue, nausea, immunological depression, behavioural and neurologic disorders, teratogenic effects, suspected carcinogens	A, C
DDT (and its sub-products DDD and DDE)	Contaminated food products. Still in use to control malaria	Liver damage, endocrine effects (via adrenal gland), reproductive disorders, suspected carcinogens	B
Aldrin/Dieldrin	Contaminated food products and materials (used to control termites and textile pests)	Neurological and kidney disorders, headache, dizziness, irritability, nausea, suspected carcinogens	A
Hexachlorobutadiene	Industrial by-product. Used to make rubber compounds and lubricants, also as a hydraulic fluid	Fatty liver degeneration, renal damage, central nervous system depression and cyanosis, suspected carcinogen	A, C
Chlordane (complex mixture)	Contaminated food products and materials (used to control termites)	Neurological and digestive disorders	A
Heptachlor	Contaminated food products	Neurological and reproductive disorders, suspected carcinogen	A
Toxaphene	Contaminated food products (used to control insects in crops and livestock)	Liver and kidney damage, suspected carcinogen	A

Continued

Table 2 Main sources, uses and health effects of top xenobiotics for the US ATSDR and the Stockholm's Convention on POPs.—cont'd

Compound/family	Main sources and uses[a]	Long-term health effects[b]	SC annexes[c]
Hexachloro cyclohexane	Contaminated food products and water, production of lindane (by-product), still used to treat lice and mites	Neurological, endocrine and immunological disorders, liver and kidney damage, suspected carcinogen	A
Endrin	Contaminated food products	Neurological problems, suspected teratogen, lethal upon exposure to high doses	A
Pentachlorophenol	General biocide, still used industrially as a wood preservative (utility poles, railroad ties, wharf pilings)	Liver, kidney and blood damage, multiple disorders (nervous, endocrine, immune, and reproductive system), suspected carcinogen	A[d]
Dioxins (PCDDs and PCDFs)	By-products of combustion (waste incineration, fires, transportation, etc.) and industrial processes (chlorinated substances)	Chloracne and other skin problems, liver damage, diabetes, endocrine disruption, suspected carcinogens	C

[a]Based on data from the ATSDR Toxic Substances Portal (https://www.atsdr.cdc.gov/substances/index.asp) and the Stockholm's Convention web server (www.pops.int).
[b]Consult the PUBCHEM Open Database (https://pubchem.ncbi.nlm.nih.gov) and the ATSDR Toxic Substances Portal for more detailed description and the effects of acute short-term exposure.
[c]Stockholm's Convention Annexes: A, elimination. B, restriction. C, unintentional production.
[d]While classified in Annex A, there are specific exemptions for its use as a wood preservative (Decision SC-7/13).

cleanup technologies (Gerhardt et al., 2017). Rhizodegradation and phytodegradation are the primary mechanisms of remediation of organic pollutants, which offer the prospect, unlike metals or radionuclides, of being mineralized or at least metabolized to some extent by plants and/or their associated microbiota. Phytovolatilization of organics has proven successful in some instances (Limmer & Burken, 2016). Due to the enormous heterogeneity of organic pollutants, many recent reviews focus on specific compounds and/or technical aspects, like plant-microbe partnerships or genetic engineering applications (Arslan, Imran, Khan, & Afzal, 2017; Eevers, White, Vangronsveld, & Weyens, 2017; Feng et al., 2017; Fester, Giebler, Wick, Schlosser, & Kästner, 2014; Gkorezis et al., 2016; Hussain et al., 2018).

4.2.2 Plant uptake and fate of organic pollutants

The uptake of organic pollutants by plants can occur through the roots or aboveground parts, depending on properties such as hydrophobicity, vapour pressure and acidity constant. Aerial uptake is important for volatile compounds, especially in trees (Weyens et al., 2015). Regarding soil compounds, adsorption to the soil matrix is the major limiting factor for bioavailability (Ren et al., 2018). The only pollutants amenable to efficient phytoremediation are those present in the 'rapid desorption' soil fraction, where adsorbed molecules are interchangeable with the water phase that penetrates into the roots. The desorption status is not unyielding, however, and might be altered by the biochemical events associated with the continuous and fractal-like growth of roots and rhizospheres. The large biomass and depth of tree roots represent again an obvious advantage with respect to herbaceous plants. In any case, the detailed micro-processes that occur in the rhizosphere, that is, the specific roles played by microbial and plant biochemical machineries in the metabolism of organic pollutants, remain in general poorly understood (Arslan et al., 2017; Gerhardt et al., 2017; Hussain et al., 2018).

Although the octanol-water partition coefficient (K_{ow}) has often been used to predict the absorption and translocation of specific organic pollutants *in planta*, some authors have questioned its validity (e.g. Limmer & Burken, 2014). Biosurfactants secreted by the roots and/or the plant-associated microbiota are known to greatly affect both pollutant desorption from soil aggregates and plant uptake (Fester et al., 2014; Khan, Afzal, Iqbal, & Khan, 2013)—rationally-designed soil amendments might play similar roles in the future (Agnello et al., 2014). Moreover, specific plant characteristics encompassing genotype, morphology and physiology have relevant effects on pollutant uptake (Arslan et al., 2017; Hussain et al., 2018; Isebrands & Richardson, 2014). Before entering root cells, organic molecules can suffer different chemical transformations which may improve their bioavailability, for example, by reducing steric hindrance. Laccases, peroxidases, transferases, hydrolases, and many other cell wall-bound enzymes are presumably involved in reactions of this sort (Bártíková et al., 2015; Lee, 2013), not to mention the complex plant-associated microbial arsenal (Levy et al., 2017; Mercado-Blanco et al., 2018; Thijs et al., 2016). The recent finding that peroxidases have a potential to increase lignin degradation and saccharification in transgenic tobacco (Ligaba-Osena et al., 2017) raises the interesting possibility of improving both phytoremediation features and biomass value in suitable trees.

As mentioned earlier, the free movement of water and solutes in the capillary spaces between root cortical cells is not sufficient for xylem upload. Translocation to aboveground parts requires that endodermal cell membranes are crossed at some point. Small hydrophobic pollutants appear to readily diffuse across lipid bilayers. Schneider and Nizzetto (2012) found for example that the relatively small hexachlorobenzene (HCB) molecule, which is markedly hydrophobic, readily reached the cytoplasm of internodal *Chara rudis* cells after diffusing through the protective layer of cortical cells (this macroalga has a complex body structure and cell walls that share many features with higher plants). Structurally-similar pollutants have been shown to translocate from polluted soil to aerial parts of adult trees (e.g. Gkorezis et al., 2016; Lei, Ran, Lu, Du, & Liu, 2015). Sufficiently large or hydrophilic molecules cannot diffuse across membranes and must highjack membrane machinery to enter root cells, another aspect that deserves further study. More than one-fourth of the proteome of higher plants consists of membrane-bound proteins, some of which participate in biochemical pathways as complex as phenylpropanoid biosynthesis (Wang & Zhao, 2018). The membrane machinery responsible for the uptake (or efflux) of xenobiotics has been much less characterized in plants than in other organisms. Once inside the cell, active toxic molecules can be inactivated by endogenous enzyme machinery or deactivated via vacuolar sequestration or efflux to the apoplast. These processes are usually preceded by the formation of conjugates with small hydrophilic molecules, like GSH, glucose or malonate, by the action of appropriate cytosolic transferases. The most relevant carriers in plants belong to the ABC family, the Major Facilitator Superfamily (MFS) and the Multidrug And Toxic compound Extrusion (MATE) family (Remy & Duque, 2014). They are all part of the ubiquitous Multiple Drug Resistance (MDR) response that confers resistance to numerous structurally-dissimilar cytotoxic compounds in all kinds of organisms.

Several xenobiotics have been reported to enter vacuoles as GSH-conjugates, a process involving MDR transporters (Lee, 2013; Rylott et al., 2015). The induction of glutathione-*S*-transferases (GSTs) and other detoxifying enzymes in response to a toxic explosive was first reported in poplar by Tanaka et al. (2007). Interestingly, MDR transporters form one of the largest protein superfamilies in trees and other plant species, showing a complexity unparalleled in other eukaryotes (Lane et al., 2015; Li et al., 2017; Yildirim, 2017; Zhiyi, Guijuan, Yu, Longjun, & Rizhong, 2015). Besides proven roles in cellular detoxification of metals and xenobiotics, individual MDR members participate in many other physiological processes, including wax transport to form cuticles, transport of lignin precursors, stomatal movement, regulation of ion homeostasis, root exudation of phytochemicals, biotic and abiotic stress tolerance, and hormone transport (Remy & Duque, 2014). Their biotechnological interest to improve phytoremediation of organic pollutants has been substantiated through reverse genetics approaches. For example, two yeast plasma-membrane transporters (ABC and MFS families) have been recently shown to separately improve tolerance to different herbicides in transgenic Arabidopsis plants (Remy et al., 2017). The MFS transporter also enhanced tolerance to Co^{2+}, Cu^{2+}, Ni^{2+}, Al^{3+} and Cd^{2+} cations, most likely by promoting their plant repartition—a backup for the biodilution effect posited in Section 3.2.1. Although the mechanisms for herbicide tolerance were unclear (several hormones appear to be involved), these findings prove the feasibility of improving phytoremediation features through the rational manipulation of MDR transporters. Ectopic overexpression of an ABC transporter from *P. trichocarpa* has been recently shown to increase tolerance to Hg in transgenic hybrid poplars and Arabidopsis plants (Sun et al., 2018). The exogenous addition of GSH to hybrid poplars enhances Cd tolerance (Ding et al., 2017), a finding correlated with increased transcription of genes for several membrane transporters (ABC, ZIP, NRAMP, and MTP families) and for γ-glutamylcysteine synthetase (ECS). Induced GSH biosynthesis is also a natural response to herbicide treatments, and overexpression of bacterial ECS was previously reported to enhance tolerance to chloroacetanilide herbicides in transgenic poplars (Gullner, Kömives, & Rennenberg, 2001). Although the underlying mechanisms were not analysed, enhanced GSH-conjugation followed by MDR-mediated sequestration seems a reasonable possibility. Similarly, overexpression of an Arabidopsis MDR transporter has been reported to improve tolerance to aminoglycoside antibiotics in transgenic poplar (Kang, Ye, Osburn, Stewart, & Cheng, 2010).

Besides being transported and sequestered, *in planta* chemical transformations have been described for a number of xenobiotics. TNT provides one of the best studied cases (Johnston et al., 2015; Panz & Miksch, 2012;

Rylott et al., 2015). Its detoxification in plants has common elements with the reductive pathway previously described in enteric bacteria, whereby TNT is reduced to hydroxylaminodinitrotoluene (HADNT) and amino-dinitrotoluene (ADNT) intermediates. This step is catalysed in plants by oxophytodienoate reductases (OPRs). UDP-glucosyl transferases and GSTs then promote the formation of various HADNT and ADNT conjugates, a process not described in bacteria. Among the identified conjugates, the denitrated 2-glutathionyl-4,6-dinitrotoluene has elicited particular interest because structurally-related dinitrotoluene compounds are known to be mineralized by bacteria. While the fate of these conjugates remains to be determined, the upregulation of MDR transporters by TNT treatments suggests that cell wall or vacuole sequestration might take place (Rylott et al., 2015). It is worth mentioning here that GSTs appear to directly bind and transport small organic ligands, in addition to their catalytic activity (Ahmad, Rylott, Bruce, Edwards, & Grogan, 2016). GSH conjugation reactions have been suggested to deplete plant GSH pools in TNT-contaminated sites (Rylott et al., 2015), thus limiting natural phytoremediation. This extreme has been bolstered by overexpressing a Drosophila GST in transgenic Arabidopsis plants (Tzafestas et al., 2017). While the transgenic plants were more resistant to TNT and showed increased capacity to remove it relative to control plants, excess GST activity was detrimental in the absence of pollutant. The above findings are in agreement with the observation that TNT forms a nitro radical in mitochondria that then spontaneously reacts with oxygen to generate superoxide (Johnston et al., 2015). This work also found that Arabidopsis plants deficient in the key enzyme, a monodehydroascorbate reductase, displayed improved TNT tolerance. Besides unveiling the toxicity mechanism of TNT, these results highlighted the complexity of plant antioxidant networks, where GSH plays a pivotal role. Another relevant pollutant is hexahydro-1,3,5-trinitro-1,3,5-triazine (RDX). Unlike TNT, which is mostly found in roots, ^{14}C-radiolabeling studies confirmed that RDX is readily translocated to aboveground tissues in poplar and switchgrass (Brentner, Mukherji, Walsh, & Schnoor, 2010). Its uptake is highly variable among plants, however, and its fate *in planta* is unclear (Panz & Miksch, 2012). Studies in poplars and herbaceous species have confirmed that RDX treatments induce enzymes typically involved in xenobiotic detoxification, such as GSTs, peroxidases, reductases, and cytochrome P450s (Rylott et al., 2015; Tanaka et al., 2007). Moreover, laboratory studies with *Populus deltoides* × *nigra* DN-34 support that RDX can be degraded *in planta* via reduction to hexahydrotriazines and further light-dependent conversion to formaldehyde, methanol and CO_2 (Van Aken, Yoon, Just, & Schnoor, 2004). By overexpressing two bacterial reductase genes for RDX degradation (*xplA* and *xplB*) and one for NT transformation (*nfsI*), Zhang et al. (2017) have recently obtained transgenic lines of *Panicum virgatum* and *Agrostis stolonifera* that extract and degrade RDX more efficiently than control plants. TNT, RDX, HMX and similar munition explosives are being replaced with 2,4-dinitroanisole (DNAN) and other insensitive compounds for which cleanup strategies have not yet been developed. Schroer, Li, Lehmler, and Just (2017) have demonstrated that Arabidopsis plants reduce separately the *p*- and *o*-nitro groups and efficiently accumulate glycosylated conjugates within leaves. More interestingly, evidence of leaf photolysis was obtained for a glycosylated metabolite and ($^{15}N_2$)-DNAN itself, with production of NO_2. Although artificial light was used, further research of this novel phytoremediation mechanism might open interesting avenues for trees.

The above studies and comparable reports for herbicides, aromatic hydrocarbons and other xenobiotics have underscored a general mechanism by which plants deal with organic pollutants (Bártíková et al., 2015; Lee, 2013). Regardless of compound-specific reactions, this general mechanism consists of three sequential phases: (i) hydrolysis, oxidation and other reactions that create or expose reactive sites in the offender (phase I, activation); (ii) conjugation reactions on the exposed sites that decrease toxicity, increase solubility and open the door for molecular recognition (phase II, conjugation); and (iii) binding of conjugates to membrane-anchored carriers (mainly of the MDR superfamily) that export them from the cytosol to vacuoles and the apoplast (phase III, sequestration). Once exported to either compartment, further chemical transformations can take place. Ideally, the organic pollutant will be mineralized.

4.3 Enhancing endogenous capacities

In field settings, phytoremediation is never an 'only plant' issue whereby plant genes are confronted with a pollution problem, not even in the greenhouse. Plants associate naturally with numerous bacteria and fungi and both play crucial roles in plant growth, development and adaptation to environmental changes (Coninx, Martinova, & Rineau, 2017; Levy et al., 2017; Mercado-Blanco et al., 2018). This microbiota also plays important roles when plants deal with pollutants (Doty, 2008; Feng et al., 2017; Fester et al., 2014; Thijs et al., 2016; Ullah et al., 2015; Van Aken & Doty, 2010;

Weyens et al., 2015). Salicaceae trees (mostly *Populus*) have been actively studied in this context, due to their ability to establish many endophytic and rhizospheric associations, including ectomycorrhizal and arbuscular mycorrhizal interactions (Beckers, Op De Beeck, Weyens, Boerjan, & Vangronsveld, 2017; Shakya et al., 2013). These studies have shown that endophytic communities are substantially different from those found in tree rhizospheres. The former communities hold more promise for phytoremediation, since competition with autochthonous microbiota makes rhizospheric habitats difficult to manipulate. Several instances have been documented where inoculation with fitting endophytes improves remediation, in poplars and other plant species (e.g. Doty, 2008; Feng et al., 2017; Khan et al., 2013). An outstanding example was recently provided by Doty et al. (2017), who found that a metabolically-suitable strain isolated from poplars was capable of improving the tolerance and degradative capacity of hybrid poplars (*P. deltoides* × *P. nigra*) towards trichloroethylene at a superfund site. Relative to controls, the inoculated trees exhibited both increased growth and improved degradation of the solvent, which was efficiently removed from groundwater. Although their recalcitrance to manipulation is higher, rhizosphere communities have also been successfully managed to enhance the phytoremediation of elemental and organic pollutants in trees (Hussain et al., 2018; Mishra, Singh, & Arora, 2017). Despite promising results, our understanding of plant colonization must greatly improve if we want to fully benefit of the possibilities offered by microbiota-assisted phytoremediation. Processes as important as the reciprocal plant-bacterial signalling or the molecular networks governing the dynamics and spatial distribution of endophytic and rhizospheric communities are still poorly understood (Levy et al., 2017).

Biotechnology provides another powerful way to improve the capabilities of trees for pollutant cleanup, and also of linking additional services or values to the phytoremediation process. The recent development of transgenic poplar plants able to recycle excess soil nitrate for biomass production constitutes a good example (Castro-Rodríguez et al., 2016). For almost three decades, overexpression of heterologous genes has proven its validity to enhance the natural abilities of plants to remediate organic and/or elemental pollutants. Whereas most experiments have been performed with herbaceous model species, the transfer of results to *Populus* is usually straightforward. Hg provides a well-known example and the first case of transgenic phytoremediation. The rationale was to generate *in planta* volatile elemental Hg(0) from its ionic species, to preclude the formation of highly toxic methylmercury. After promising results with transgenic model plants overexpressing bacterial Hg reductases (see Bizily et al., 2000), transgenic *P. deltoides* lines were finally obtained which efficiently reduced Hg(II) to Hg(0) (Che et al., 2003). Relative to controls, the transgenic poplar lines withstood higher Hg concentrations and evolved more Hg(0). Similar results have been reported in other plants, including chloroplastic expression of Hg-detoxifying genes. Biotechnological approaches are currently under way to improve phytoremediation of additional toxic metals. The three main avenues under consideration are (i) shifting the oxidation state of the offender, to generate less toxic and volatile species; (ii) altering cell uptake via membrane-anchored transporters; and (iii) enhancing conjugation (and thus sequestration) through appropriate enzymes (Koźmińska, Wiszniewska, Hanus-Fajerska, & Muszyńska, 2018; Luo et al., 2016). The results reported for transgenic trees have been mostly obtained with poplars kept in growth chambers or greenhouses, and refer to improved tolerance and aboveground accumulation of heavy metals (Zn, Cd, Hg, and Cu). Nonetheless, until systematic field trials are not conducted, the applicability of the above strategies will be a matter of speculation.

The viability of using transgenic approaches to phytoremediate organic pollutants has mostly focused so far on explosives, petrol hydrocarbons, volatile organic compounds, herbicides, and PCBs (recently reviewed by Feng et al., 2017; Hussain et al., 2018). The first genetic transformations were carried out again in herbaceous model plants. Promising results were later transferred to poplar genotypes suitable for genetic transformation, often *P. tremula* × *P. alba*. Among other reasons already discussed, such shift was motivated by the remarkable endogenous ability of poplars to tolerate and even metabolize relevant xenobiotics (Doty, 2008; Isebrands & Richardson, 2014; Marmiroli et al., 2011; Peuke & Rennenberg, 2015). By using poplars transformed with the mammalian gene for cytochrome P450 2E1 (CYP2E1), Legault et al. (2017) have recently obtained one of the soundest proofs in this research area that laboratory results hold under field conditions. Their conclusions also validate that transgenic approaches can work for phytoremediation. In order to determine mass balances, these authors added TCE to three hydraulically contained test beds, two of them planted with either transgenic or control poplars and a third unplanted. Whereas chlorinated ethene removal was rather similar in both planted beds (about 2.5-times higher than in the blank bed), the loss of TCE was considerably reduced in transgenic leaves (80%, evapotranspiration) and stems (90%, diffusion) relative to wild type trees. Microbial activity aside, experimental evidence was obtained that such differences were due to increased pollutant degradation in the transgenic trees. While new field tests are needed to further validate the viability of transgenic approaches, mounting evidence supports that this technology should be part of our toolkit for fighting pollution (Gerhardt et al., 2017; Hussain et al., 2018; Koźmińska et al., 2018; Van Aken & Doty, 2010; Yadav et al., 2010). In a 5-year field study recently conducted in China, the only country that has commercialized genetically engineered trees, no measurable effects were observed for Bt transgenic poplars on local bacterial diversity and the arthropod community (Zuo et al., 2018). Two decades have passed between the pioneering field study of Legault and co-workers and the first successful application of transgenesis to improve the ability of a plant to degrade xenobiotics. The number of available genes today is humongous, and will continue increasing as a result of advances in molecular technologies and genome annotation (Bell, Joly, Pitre, & Yergeau, 2014). In this respect, the anticipated benefits of using transgenic trees are too important to be sidestepped by considerations not aligned with sound scientific evidence (see Strauss & Sax, 2016). Promising candidate genes can and should be safely assessed under controlled conditions, so tree breeding programmes can integrate this knowledge to improve the phytoremediation capacity of valuable genotypes.

5. Conclusions and paths forward

Field trials and many greenhouse studies attest to the unmatched potential of fast-growing tree plantations for environmental cleanup, a technology that adds key ecological and economic values to classic remediation methods. Yet, our current knowledge of tree biology must significantly improve to fully develop such potential. Different molecular and imaging tools are being actively deployed to disentangle the mechanisms by which trees (and plants in general) respond to pollution. Similar strategies are being applied to dissect the complex relationships of trees with rhizospheric and endophytic communities. Members of both communities have important effects on the way trees deal with pollutants, and therefore on the success of phytotechnologies. Most studies to date have focused on the genus *Populus*, for it bears interesting features both as a model system (complete genomic sequence, omics tools, genome editing, high level of domestication, easy propagation, genetic transformation) and as a virtually ideal tree for phytoremediation (pioneer, fast growth, high transpiration, big roots, natural tolerance to pollution and stress, economic and ecological values). *Salix* is also an outstanding genus for environmental applications, as are probably other tree species little or not yet studied in this context. As bearers of a unique secondary metabolism that involves molecules very similar to relevant pollutants, it seems likely that trees possess interesting genes for phytodegradation, an aspect that deserves further exploration. In this respect, the genetic variability of trees is huge and poorly understood, providing a subject where experimental studies and bioinformatic tools will probably identify interesting candidates (genes and genotypes) in the near future. The potential of tree biotechnology for improving the efficiency and cost-effectiveness of phytoremediation has already been proven in the field. The same holds true for the inoculation with suitable microbial strains. A number of recent studies conducted in different settings have confirmed that short-rotation plantations save water, protect and improve the soil, and have in general positive ecological impacts, being an excellent choice to restore polluted and degraded lands. Novel management practices should help improve the sustainability of these plantations and the services they provide. Moreover, a culture of tree planting and restoration is in line with the Paris Climate Agreement and with general objectives posited by the international community for the present century.

There is a general consensus that phytoremediation is a proven technology with immense practical potential. Its many advantages notwithstanding, a remarkable breach stands today between laboratory results and field application. This phenomenon has attracted much attention, and a handful of dedicated studies and reflections have been recently published (Beans, 2017; Gerhardt et al., 2017; Stephenson & Black, 2014; Weir & Doty, 2016). Surprisingly, a generalized lack of understanding of this technology has been detected among environmental experts, including people directly involved in the selection of remediation technologies. To address this issue, Gerhardt et al. (2017) advocate for a strong track record of successes (additional to the scientific literature) as well as aggressive education campaigns in the public and private sectors, among other measures. Regarding the use of trees, the valorization of phytoremediation biomass and the ecological services provided by plantations should be strong assets and obvious targets for future research.

Acknowledgements

Our research on phytoremediation has been funded by Comunidad de Madrid's R&D Programme (grant S2009AMB-1478), the EU's LIFE+ Programme (grant LIFE/ENV/ES/505) and Spains' Ministry of Economy and Competitiveness (grant BIO2016-77840-R). We thank C. Salarnier (IAA) and M. Esteban (UPM), respectively, for the kind gift of pictures appearing in Figs 1 and 2.

References

Abbas, G., Murtaza, B., Bibi, I., Shahid, M., Niazi, N. K., Khan, M. I., et al. (2018). Arsenic uptake, toxicity, detoxification, and speciation in plants: Physiological, biochemical, and molecular aspects. *International Journal of Environmental Research and Public Health*, *15*, 59.

Aerts, R., & Honnay, O. (2011). Forest restoration, biodiversity and ecosystem functioning. *BMC Ecology*, *11*, 29.

Agnello, A. C., Huguenot, D., Van Hullebusch, E. D., & Esposito, G. (2014). Enhanced phytoremediation: A review of low molecular weight organic acids and surfactants used as amendments. *Critical Reviews in Environmental Science and Technology*, *44*, 2531–2576.

Ahmad, L., Rylott, E. L., Bruce, N. C., Edwards, R., & Grogan, G. (2016). Structural evidence for Arabidopsis glutathione transferase AtGSTF2 functioning as a transporter of small organic ligands. *FEBS Open Bio*, *7*, 122–132.

Ariani, A., Di Baccio, D., Romeo, S., Lombardi, L., Andreucci, A., Lux, A., et al. (2015). RNA sequencing of Populus x canadensis roots identifies key molecular mechanisms underlying physiological adaption to excess zinc. *PLoS One*, *10*, e0117571.

Arslan, M., Imran, A., Khan, Q. M., & Afzal, M. (2017). Plant-bacteria partnerships for the remediation of persistent organic pollutants. *Environmental Science and Pollution Research*, *24*, 4322–4336.

Bártíková, H., Skálová, L., Stuchlíková, L., Vokřál, I., Vaněk, T., & Podlipná, R. (2015). Xenobiotic-metabolizing enzymes in plants and their role in uptake and biotransformation of veterinary drugs in the environment. *Drug Metabolism Reviews*, *47*, 374–387.

Beans, C. (2017). Core concept: Phytoremediation advances in the lab but lags in the field. *Proceedings of the National Academy of Sciences USA*, *114*, 7475–7477.

Beckers, B., Op De Beeck, M., Weyens, N., Boerjan, W., & Vangronsveld, J. (2017). Structural variability and niche differentiation in the rhizosphere and endosphere bacterial microbiome of field-grown poplar trees. *Microbiome*, *5*, 25.

Bell, T. H., Joly, S., Pitre, F. E., & Yergeau, E. (2014). Increasing phytoremediation efficiency and reliability using novel omics approaches. *Trends in Biotechnology*, *32*, 271–280.

Berhongaray, G., Verlinden, M. S., Broeckx, L. S., Janssens, I. A., & Ceulemans, R. (2017). Soil carbon and belowground carbon balance of a short-rotation coppice: Assessments from three different approaches. *GCB Bioenergy*, *9*, 299–313.

Bizily, S. P., Rugh, C. L., & Meagher, R. B. (2000). Phytodetoxification of hazardous organomercurials by genetically engineered plants. *Nature Biotechnology*, *18*, 213–217.

Bloemen, J., Fichot, R., Horemans, J. A., Broeckx, L. S., Verlinden, M. S., Zenone, T., et al. (2017). Water use of a multigenotype poplar short-rotation coppice from tree to stand scale. *GCB Bioenergy*, *9*, 370–384.

Bonan, G. B. (2016). Forests, climate, and public policy: A 500-year interdisciplinary odyssey. *Annual Review of Ecology, Evolution, and Systematics*, *47*, 97–121.

Brentner, L. B., Mukherji, S. T., Walsh, S. A., & Schnoor, J. L. (2010). Localization of hexahydro-1,3,5-trinitro-1,3,5-triazine (RDX) and 2,4,6-trinitrotoluene (TNT) in poplar and switchgrass plants using phosphor imager autoradiography. *Environmental Pollution*, *158*, 470–475.

Broeckx, L. S., Verlinden, M. S., & Ceulemans, R. (2012). Establishment and two-year growth of a bio-energy plantation with fast-growing Populus trees in Flanders (Belgium): Effects of genotype and former land use. *Biomass and Bioenergy*, *42*, 151–163.

Campos, V. M., Merino, I., Casado, R., Pacios, L. F., & Gomez, L. (2008). Phytoremediation of organic pollutants: A review. *Spanish Journal of Agricultural Science*, *6*, 38–47.

Castro-Rodríguez, V., García-Gutiérrez, A., Canales, J., Cañas, R. A., Kirby, E. G., Avila, C., et al. (2016). Poplar trees for phytoremediation of high levels of nitrate and applications in bioenergy. *Plant Biotechnology Journal*, *14*, 299–312.

Catford, J. A., & Jansson, R. (2014). Drowned, buried and carried away: Effects of plant traits on the distribution of native and alien species in riparian ecosystems. *New Phytologist, 204*, 19–36.

Celedon, J. M., & Bohlmann, J. (2018). An extended model of heartwood secondary metabolism informed by functional genomics. *Tree Physiology, 38*, 311–319.

Chakraborty, J., & Das, S. (2016). Molecular perspectives and recent advances in microbial remediation of persistent organic pollutants. *Environmental Science and Pollution Research, 23*, 16883–16903.

Chaney, R. L., Baker, A. J. M., & Morel, J. L. (2018). The long road to developing agromining/phytomining. In A. Van der Ent, G. Echevarria, A. Baker, & J. Morel (Eds.), *Agromining: Farming for metals. mineral resource reviews* (pp. 1–17). Cham, Switzerland: Springer.

Che, D., Meagher, R. B., Heaton, A. C., Lima, A., Rugh, C. L., & Merkle, S. A. (2003). Expression of mercuric ion reductase in Eastern cottonwood (*Populus deltoides*) confers mercuric ion reduction and resistance. *Plant Biotechnology Journal, 1*, 311–319.

Chen, G. C., Liu, Y. Q., Wang, R. M., Zhang, J. F., & Owens, G. (2013). Cadmium adsorption by willow root: The role of cell walls and their subfractions. *Environmental Science and Pollution Research, 20*, 5665–5672.

Coninx, L., Martinova, V., & Rineau, F. (2017). Mycorrhiza-assisted phytoremediation. *Advances in Botanical Research, 83*, 127–188.

Constabel, C. P., & Lindroth, L. R. (2010). The impact of genomics on advances in herbivore defense and secondary metabolism in Populus. In S. Jansson, R. Bhalerao, & A. Groover (Eds.), *Plant genetics and genomics—Crops and models: Vol. 8. Genetics and genomics of Populus* (pp. 279–305). New York: Springer.

Crowther, T. W., Glick, H. B., Covey, K. R., Bettigole, C., Maynard, D. S., Thomas, S. M., et al. (2015). Mapping tree density at a global scale. *Nature, 525*, 201–205.

Da Ros, L. M., Soolanayakanahally, R. Y., Guy, R. D., & Mansfield, S. D. (2018). Phosphorus storage and resorption in riparian tree species: Environmental applications of poplar and willow. *Environmental and Experimental Botany, 149*, 1–8.

de Sosa, L. L., Glanville, H. C., Marshall, M. R., Prysor-Williams, A., & Jones, D. L. (2018). Quantifying the contribution of riparian soils to the provision of ecosystem services. *Science of the Total Environment, 624*, 807–819.

Di Baccio, D., Galla, G., Bracci, T., Andreucci, A., Barcaccia, G., Tognetti, R., et al. (2011) Transcriptome analyses of *Populus* × *euramericana* clone I-214 leaves exposed to excess zinc. *Tree Physiology, 31*, 1293–1308.

Dillen, S. Y., Rood, S. B., & Ceulemans, R. (2010). Growth and physiology. In S. Jansson, S. Bhalerao, & A. Groover (Eds.), *Plant genetics and genomics: Vol. 8. Genetics and genomics of* Populus (pp. 39–63). New York: Springer Science & Business Media.

Dimitriou, I., & Mola-Yudego, B. (2017). Poplar and willow plantations on agricultural land in Sweden: Area, yield, groundwater quality and soil organic carbon. *Forest Ecology and Management, 383*, 99–107.

Ding, S., Ma, C., Shi, W., Liu, W., Lu, Y., Liu, Q., et al. (2017). Exogenous glutathione enhances cadmium accumulation and alleviates its toxicity in *Populus* × *canescens*. *Tree Physiology, 37*, 1697–1712.

Doty, S. (2008). Enhancing phytoremediation through the use of transgenics and endophytes. *New Phytologist, 179*, 318–333.

Doty, S. L., Freeman, J. L., Cohu, C. M., Burken, J. G., Firrincieli, A., Simon, A., et al. (2017). Enhanced degradation of TCE on a superfund site using endophyte-assisted poplar tree phytoremediation. *Environmental Science & Technology, 51*, 10050–10058.

Durand, A., Maillard, F., Alvarez-Lopez, V., Guinchard, S., Bertheau, C., Valot, B., et al. (2018). Bacterial diversity associated with poplar trees grown on a Hg-contaminated site: Community characterization and isolation of Hg-resistant plant growth-promoting bacteria. *Science of the Total Environment, 622*, 1165–1177.

EEA. (2017). *Progress in management of contaminated sites*. Copenhagen, Denmark: European Environment Agency. https://www.eea.europa.eu/data-and-maps/indicators/progress-in-management-of-contaminated-sites/progress-in-management-of-contaminated-1.

Eevers, N., White, J. C., Vangronsveld, J., & Weyens, N. (2017). Bio- and phytoremediation of pesticide-contaminated environments: A review. *Advances in Botanical Research*, *83*, 277–318.

Evangelou, M. W., Robinson, B. H., Günthardt-Goerg, M. S., & Schulin, R. (2013). Metal uptake and allocation in trees grown on contaminated land: Implications for biomass production. *International Journal of Phytoremediation*, *15*, 77–90.

Fan, W., Liu, C., Cao, B., Qin, M., Long, D., Xiang, Z., et al. (2018). Genome-wide identification and characterization of four gene families putatively involved in cadmium uptake, translocation and sequestration in mulberry. *Frontiers in Plant Science*, *9*, 879.

FAO. (2016a). *Global forest resources assessment 2015: How are the world's forests changing?* (2nd ed.). Rome: The Food and Agriculture Organization of the United Nations.

FAO. (2016b). *State of the World's Forests 2016. Forests and agriculture: Land-use challenges and opportunities*. Rome: The Food and Agriculture Organization of the United Nations.

Fares, S., Mugnozza, G. S., Corona, P., & Palahí, M. (2015). Sustainability: Five steps for managing Europe's forests. *Nature*, *519*, 407–409.

Feng, N. X., Yu, J., Zhao, H. M., Cheng, Y. T., Mo, C. H., Cai, Q. Y., et al. (2017). Efficient phytoremediation of organic contaminants in soils using plant-endophyte partnerships. *Science of the Total Environment*, *583*, 352–368.

Fester, T., Giebler, J., Wick, L. Y., Schlosser, D., & Kästner, M. (2014). Plant-microbe interactions as drivers of ecosystem functions relevant for the biodegradation of organic contaminants. *Current Opinion in Biotechnology*, *27*, 168–175.

Fischer, M., Zenone, T., Trnka, M., Orság, M., Montagnani, L., Ward, E. J., et al. (2018). Water requirements of short rotation poplar coppice: Experimental and modelling analyses across Europe. *Agricultural and Forest Meteorology*, *250–251*, 343–360.

Fortier, J., Truax, B., Gagnon, D., & Lambert, F. (2016). Potential for hybrid poplar riparian buffers to provide ecosystem services in three watersheds with contrasting agricultural land use. *Forests*, *7*, 37.

Fox, H., Doron-Faigenboim, A., Kelly, G., Bourstein, R., Attia, Z., Zhou, J., et al. (2018). Transcriptome analysis of *Pinus halepensis* under drought stress and during recovery. *Tree Physiology*, *38*, 423–441.

Gerhardt, K. E., Gerwing, P. B., & Greenberg, B. M. (2017). Opinion: Taking phytoremediation from proven technology to accepted practice. *Plant Science*, *256*, 170–185.

Gkorezis, P., Daghio, M., Franzetti, A., Van Hamme, J. D., Sillen, W., & Vangronsveld, J. (2016). The interaction between plants and bacteria in the remediation of petroleum hydrocarbons: An environmental perspective. *Frontiers in Microbiology*, *7*, 1836.

Gordon, M., Choe, N., Duffy, J., Ekuan, G., Heilman, P., Muiznieks, I., et al. (1998). Phytoremediation of trichloroethylene with hybrid poplars. *Environmental Health Perspectives*, *106*, 1001–1004.

Gullner, G., Kömives, T., & Rennenberg, H. (2001). Enhanced tolerance of transgenic poplar plants overexpressing gamma-glutamylcysteine synthetase towards chloroacetanilide herbicides. *Journal of Experimental Botany*, *52*, 971–979.

Gupta, D. K., Chatterjee, S., Datta, S., Voronina, A. V., & Walther, C. (2017). Radionuclides: Accumulation and transport in plants. *Reviews of Environmental Contamination and Toxicology*, *241*, 139–160.

Hamanishi, E. T., Barchet, G. L., Dauwe, R., Mansfield, S. D., & Campbell, M. M. (2015). Poplar trees reconfigure the transcriptome and metabolome in response to drought in a genotype- and time-of-day-dependent manner. *BMC Genomics*, *21*, 329.

Harfouche, A., Meilan, R., & Altman, A. (2014). Molecular and physiological responses to abiotic stress in forest trees and their relevance to tree improvement. *Tree Physiology*, *34*, 1181–1198.

He, J. L., Li, H., Luo, J., Ma, C. F., Li, S. J., Qu, L., et al. (2013). A transcriptomic network underlies microstructural and physiological responses to cadmium in *Populus × canescens*. *Plant Physiology*, *162*, 424–439.

He, J. L., Li, H., Ma, C. F., Zhang, Y. L., Polle, A., Rennenberg, H., et al. (2015). Overexpression of bacterial γ-glutamyl cysteine synthetase mediates changes in cadmium influx, allocation, and detoxification in poplar. *New Phytologist*, *205*, 240–254.

Hörger, A. C., Fones, H. N., & Preston, G. M. (2013). The current status of the elemental defense hypothesis in relation to pathogens. *Frontiers in Plant Science*, *4*, 395.

Hua, F., Wang, X., Zheng, X., Fisher, B., Wang, L., Zhu, J., et al. (2016). Opportunities for biodiversity gains under the world's largest reforestation programme. *Nature Communications*, 7, 12717.

Huang, J., Han, Q., & Li, J. (2018). Soil propagule bank of ectomycorrhizal fungi associated with Masson pine (*Pinus massoniana*) grown in a manganese mine wasteland. *PLoS One*, *13*, e0198628.

Hussain, I., Aleti, G., Naidu, R., Puschenreiter, M., Mahmood, Q., Rahman, M. M., et al. (2018). Microbe and plant assisted-remediation of organic xenobiotics and its enhancement by genetically modified organisms and recombinant technology: A review. *Science of the Total Environment*, *628*, 1582–1599.

Hussein, H. S., Ruiz, O. N., Terry, N., & Daniell, H. (2007). Phytoremediation of mercury and organomercurials in chloroplast transgenic plants: Enhanced root uptake, translocation to shoots, and volatilization. *Environmental Science & Technology*, *41*, 8439–8446.

Isebrands, J. G., Aronsson, P., Carlson, M., Ceulemans, R., Coleman, M., Dickinson, N., et al. (2014). Environmental applications of poplars and willows. In J. G. Isebrands & J. Richardson (Eds.), *Poplars and willows: Trees for society and the environment* (pp. 258–335). Oxfordshire, UK: CABI.

Isebrands, J. G., & Richardson, J. (2014). *Poplars and willows: Trees for society and the environment*. Oxfordshire, UK: CABI.

Jalmi, S. K., Bhagat, P. K., Verma, D., Noryang, S., Tayyeba, S., Singh, K., et al. (2018). Traversing the links between heavy metal stress and plant signaling. *Frontiers in Plant Science*, *9*, 12.

Johnston, E. J., Rylott, E. L., Beynon, E., Lorenz, A., Chechik, V., & Bruce, N. C. (2015). Monodehydroascorbate reductase mediates TNT toxicity in plants. *Science*, *349*, 1072–1075.

Kang, B. G., Ye, X., Osburn, L. D., Stewart, C. N., Jr., & Cheng, Z. M. (2010). Transgenic hybrid aspen overexpressing the Atwbc19 gene encoding an ATP-binding cassette transporter confers resistance to four aminoglycoside antibiotics. *Plant Cell Reports*, *9*, 643–650.

Khan, S., Afzal, M., Iqbal, S., & Khan, Q. M. (2013). Plant-bacteria partnerships for the remediation of hydrocarbon contaminated soils. *Chemosphere*, *90*, 1317–1332.

Koźmińska, A., Wiszniewska, A., Hanus-Fajerska, E., & Muszyńska, E. (2018). Recent strategies of increasing metal tolerance and phytoremediation potential using genetic transformation of plants. *Plant Biotechnology Reports*, *12*, 1–14.

Kramer, U. (2010). Metal hyperaccumulation in plants. *Annual Review of Plant Biology*, *61*, 517–534.

Kroymann, J. (2011). Natural diversity and adaptation in plant secondary metabolism. *Current Opinion in Plant Biology*, *14*, 246–251.

Kunert, N., Aparecido, L. M. T., Wolff, S., Higuchi, N., dos Santos, J., Carioca de Araujo, A., et al. (2017). A revised hydrological model for the Central Amazon: The

importance of emergent canopy trees in the forest water budget. *Agricultural and Forest Meteorology*, *239*, 47–57.
Lakshmanan, V., Cottone, J., & Bais, H. P. (2016). Killing two birds with one stone: Natural rice rhizospheric microbes reduce arsenic uptake and blast infections in rice. *Frontiers in Plant Science*, *7*, 1514.
Landrigan, P. J., Fuller, R., Acosta, N. J. R., Adeyi, O., Arnold, R., Basu, N. N., et al. (2018). The Lancet Commission on pollution and health. *Lancet*, *391*, 462–512.
Lane, T. S., Rempe, C. S., Davitt, J., Staton, M. E., Peng, Y., Soltis, D. E., et al. (2015). Diversity of ABC transporter genes across the plant kingdom and their potential utility in biotechnology. *BMC Biotechnology*, *16*, 47.
LeBlanc, M. S., McKinney, E. C., Meagher, R. B., & Smith, A. P. (2013). Hijacking membrane transporters for arsenic phytoextraction. *Journal of Biotechnology*, *163*, 1–9.
Lee, J. H. (2013). An overview of phytoremediation as a potentially promising technology for environmental pollution control. *Biotechnology and Bioprocess Engineering*, *18*, 431–439.
Legault, E. K., James, C. A., Stewart, K., Muiznieks, I., Doty, S. L., & Strand, S. E. (2017). A field trial of TCE phytoremediation by genetically modified poplars expressing cytochrome P450 2E1. *Environmental Science & Technology*, *51*, 6090–6099.
Lei, X., Ran, D., Lu, J., Du, Z., & Liu, Z. (2015). Concentrations and distribution of organochlorine pesticides in pine needles of typical regions in Northern Xinjiang. *Environmental Science and Pollution Research*, *22*, 1705–1712.
Levy, A., Gonzalez, I. S., Mittelviefhaus, M., Clingenpeel, S., Paredes, S. H., Miao, J., et al. (2017). Genomic features of bacterial adaptation to plants. *Nature Genetics*, *50*, 138–150.
Li, N., Meng, H., Xing, H., Liang, L., Zhao, X., & Luo, K. (2017). Genome-wide analysis of MATE transporters and molecular characterization of aluminum resistance in *Populus*. *Journal of Experimental Botany*, *68*, 5669–5683.
Ligaba-Osena, A., Hankoua, B., Dimarco, K., Pace, R., Crocker, M., McAtee, J., et al. (2017). Reducing biomass recalcitrance by heterologous expression of a bacterial peroxidase in tobacco (*Nicotiana benthamiana*). *Scientific Reports*, *7*, 17104.
Limmer, M., & Burken, J. (2014). Plant translocation of organic compounds: Molecular and physicochemical predictors. *Environmental Science & Technology Letters*, *1*, 156–161.
Limmer, M., & Burken, J. (2016). Phytovolatilization of organic contaminants. *Environmental Science & Technology*, *50*, 6632–6643.
Lu, Y., Luo, D., Lai, A., Liu, G., Liu, L., Long, J., et al. (2017). Leaching characteristics of EDTA-enhanced phytoextraction of Cd and Pb by Zea mays L. in different particle-size fractions of soil aggregates exposed to artificial rain. *Environmental Science and Pollution Research*, *24*, 1845–1853.
Luo, Z.-B., He, J., Polle, A., & Rennenberg, H. (2016). Heavy metal accumulation and signal transduction in herbaceous and woody plants: Paving the way for enhancing phytoremediation efficiency. *Biotechnology Advances*, *34*, 1131–1148.
Lutter, R., Tullus, A., Kanal, A., Tullus, T., & Tullus, H. (2016). The impact of short-rotation hybrid aspen (*Populus tremula* L. × *P. tremuloides* Michx.) plantations on nutritional status of former arable soils. *Forest Ecology and Management*, *362*, 184–193.
Ma, Y. L., He, J. L., Ma, C. F., Luo, J., Li, H., Liu, T. X., et al. (2014). Ectomycorrhizas with *Paxillus involutus* enhance cadmium uptake and tolerance in *Populus* × *canescens*. *Plant, Cell & Environment*, *37*, 627–642.
Marmiroli, M., Imperiale, D., Maestri, E., & Marmiroli, N. (2013). The response of *Populus* spp. to cadmium stress: Chemical, morphological and proteomics study. *Chemosphere*, *93*, 1333–1344.
Marmiroli, M., Pietrini, F., Maestri, E., Zacchini, M., Marmiroli, N., & Massacci, A. (2011). Growth, physiological and molecular traits in Salicaceae trees investigated for phytoremediation of heavy metals and organics. *Tree Physiology*, *31*, 1319–1334.

Mercado-Blanco, J., Abrantes, I., Barra Caracciolo, A., Bevivino, A., Ciancio, A., Grenni, P., et al. (2018). Belowground microbiota and the health of tree crops. *Frontiers in Microbiology, 9*, 1006.

Merino, I., Contreras, A., Jing, Z. P., Gallardo, F., Cánovas, F. M., & Gomez, L. (2014). Plantation forestry under global warming: Hybrid poplars with improved thermotolerance provide new insights on the in vivo function of small heat shock protein chaperones. *Plant Physiology, 164*, 978–991.

Migeon, A., Blaudez, D., Wilkins, O., Montanini, B., Campbell, M. M., Richaud, P., et al. (2010). Genome-wide analysis of plant metal transporters, with an emphasis on poplar. *Cellular and Molecular Life Sciences, 67*, 3763–3784.

Mishra, J., Singh, R., & Arora, N. K. (2017). Alleviation of heavy metal stress in plants and remediation of soil by rhizosphere microorganisms. *Frontiers in Microbiology, 8*, 1706.

Mosseler, A., & Major, J. E. (2017). Phytoremediation efficacy of *Salix discolor* and *S. eriocephela* on adjacent acidic clay and shale overburden on a former mine site: Growth, soil, and foliage traits. *Forests, 8*, 475.

Mottiar, Y., Vanholme, R., Boerjan, W., Ralph, J., & Mansfield, S. D. (2016). Designer lignins: Harnessing the plasticity of lignification. *Current Opinion in Biotechnology, 37*, 190–200.

Neale, D. B., Langley, C. H., Salzberg, S. L., & Wegrzyn, J. L. (2013). Open access to tree genomes: The path to a better forest. *Genome Biology, 14*, 120.

Neale, D. B., Martínez-García, P. J., De La Torre, A. R., Montanari, S., & Wei, X. X. (2017). Novel insights into tree biology and genome evolution as revealed through genomics. *Annual Review of Plant Biology, 68*, 457–483.

Pajević, S., Borišev, M., Nikolić, N., Arsenov, D. D., Orlović, S., & Župunski, M. (2016). Phytoextraction of heavy metals by fast- growing trees: A review. In A. Ansari, S. Singh, G. Ritu, G. Guy, R. Lanza, & L. Newman (Eds.), *Phytoremediation* (pp. 29–64). Switzerland: Springer.

Pallardy, S. G. (2008). *Physiology of woody plants* (3rd ed.). New York: Academic Press.

Panz, K., & Miksch, K. (2012). Phytoremediation of explosives (TNT, RDX, HMX) by wild-type and transgenic plants. *Journal of Environmental Management, 113*, 85–92.

Paris, P., Di Matteo, G., Tarchi, M., Tosi, L., Spaccino, L., & Lauteri, M. (2018). Precision subsurface drip irrigation increases yield while sustaining water-use efficiency in Mediterranean poplar bioenergy plantations. *Forest Ecology and Management, 409*, 749–756.

Peng, J. S., & Gong, J. M. (2014). Vacuolar sequestration capacity and long-distance metal transport in plants. *Frontiers in Plant Science, 5*, 19.

Petzold, R., Schwarzel, K., & Feger, K. H. (2011). Transpiration of a hybrid poplar plantation in Saxony (Germany) in response to climate and soil conditions. *European Journal of Forest Research, 130*, 695–706.

Peuke, A. D., & Rennenberg, H. (2015). Phytoremediation with transgenic trees. *Zeitschrift für Naturforschung C—Journal of Biosciences, 60*, 199–207.

Pfeifer, M., Lefebvre, V., Peres, C. A., Banks-Leite, C., Wearn, O. R., & Marsh, C. J. (2017). Creation of forest edges has a global impact on forest vertebrates. *Nature, 551*, 187–191.

Pilon-Smits, E. (2005). Phytoremediation. *Annual Review of Plant Biology, 56*, 15–39.

Pinay, G., Bernal, S., Abbott, B. W., Lupon, A., Marti, E., Sabater, F., et al. (2018). Riparian corridors: A new conceptual framework for assessing nitrogen buffering across biomes. *Frontiers in Environmental Science, 6*, 47.

Prüss-Ustün, A., Wolf, J., Corvalán, C., Bos, R., & Neira, M. (2016). *Preventing disease through healthy environments: A global assessment of the burden of disease from environmental risks.* Geneva: World Health Organization. Available at http://apps.who.int/iris/bitstream/10665/204585/1/9789241565196_eng.pdf?ua=1.

Remy, E., & Duque, P. (2014). Beyond cellular detoxification: A plethora of physiological roles for MDR transporter homologs in plants. *Frontiers in Physiology*, *5*, 201.

Remy, E., Niño-González, M., Godinho, C. P., Cabrito, T. R., Teixeira, M. C., Sá-Correia, I., et al. (2017). Heterologous expression of the yeast Tpo1p or Pdr5p membrane transporters in Arabidopsis confers plant xenobiotic tolerance. *Scientific Reports*, *7*, 4529.

Ren, X., Zeng, G., Tang, L., Wang, J., Wan, J., Liu, Y., et al. (2018). Sorption, transport and biodegradation: An insight into bioavailability of persistent organic pollutants in soil. *Science of the Total Environment*, *610*, 1154–1163.

Ricachenevsky, F. K., Araújo, A. T., Fett, L. P., & Sperotto, R. A. (2018). You shall not pass: Root vacuoles as a symplastic checkpoint for metal translocation to shoots and possible application to grain nutritional quality. *Frontiers in Plant Science*, *9*, 412.

Robinson, B. H., Bañuelos, G., Conesa, H. M., Evangelou, M. W. H., & Schulin, R. (2009). The phytomanagement of trace elements in soil. *Critical Reviews in Plant Sciences*, *28*, 240–266.

Rylott, E. L., Gunning, V., Tzafestas, K., Sparrow, H., Johnston, E. J., Brentnall, A. S., et al. (2015). Phytodetoxification of the environmental pollutant and explosive 2,4,6-trinitrotoluene. *Plant Signaling & Behavior*, *10*, e977714.

Sarwar, N., Imran, M., Rashid-Shaheen, R. M., Ishaque, W., Asif-Kamran, M., et al. (2017). Phytoremediation strategies for soils contaminated with heavy metals: Modifications and future perspectives. *Chemosphere*, *171*, 710–721.

Schneider, S. C., & Nizzetto, L. (2012). Bioconcentration and intracellular storage of hexachlorobenzene in *Charophytes* and their potential role in monitoring and remediation actions. *Environmental Science & Technology*, *46*, 12427–12434.

Schroer, H. W., Li, X., Lehmler, H. J., & Just, C. L. (2017). Metabolism and photolysis of 2,4-dinitroanisole in Arabidopsis. *Environmental Science & Technology*, *51*, 13714–13722.

Shahid, M., Austruy, A., Echevarria, G., Arshad, M., Sanaullah, M., Aslam, M., et al. (2014). EDTA-enhanced phytoremediation of heavy metals: A review. *Soil and Sediment Contamination*, *23*, 389–416.

Shakya, M., Gottel, N., Castro, H., Yang, Z. K., Gunter, L., Labbé, J., et al. (2013). A multifactor analysis of fungal and bacterial community structure in the root microbiome of mature *Populus deltoides* trees. *PLoS One*, *8*, e76382.

Sharma, S. S., Dietz, K. J., & Mimura, T. (2016). Vacuolar compartmentalization as indispensable component of heavy metal detoxification in plants. *Plant, Cell & Environment*, *39*, 1112–1126.

Shim, D., Kim, S., Choi, Y.-I., Song, W.-Y., Park, J., Youk, E. S., et al. (2013). Transgenic poplar trees expressing yeast cadmium factor 1 exhibit the characteristics necessary for the phytoremediation of mine tailing soil. *Chemosphere*, *90*, 1478–1486.

Singh, B. R., Gupta, S. K., Azaizeh, H., Shilev, S., Sudre, D., Song, W. Y., et al. (2011). Safety of food crops on land contaminated with trace elements. *Journal of the Science of Food and Agriculture*, *91*, 1349–1366.

Song, W. Y., Mendoza-Cozatl, D. G., Lee, Y., Schroeder, J. I., Ahn, S. N., Lee, H. S., et al. (2014). Phytochelatin-metal(loid) transport into vacuoles shows different substrate preferences in barley and Arabidopsis. *Plant, Cell & Environment*, *37*, 1192–1201.

Stephenson, C., & Black, C. R. (2014). One step forward, two steps back: The evolution of phytoremediation into commercial technologies. *Bioscience Horizons*, *7*, 1–15.

Strauss, S. H., & Sax, J. K. (2016). Ending event-based regulation of GMO crops. *Nature Biotechnology*, *34*, 474–477.

Sun, L., Ma, Y., Wang, H., Huang, W., Wang, X., Han, L., et al. (2018). Overexpression of PtABCC1 contributes to mercury tolerance and accumulation in Arabidopsis and poplar. *Biochemical and Biophysical Research Communications*, *497*, 997–1002.

Suthersan, S. S., Horst, H., Schnobrich, M., Welty, N., & McDonough, J. (2016). *Remediation engineering: Design concepts* (2nd ed.). Boca Raton, FL: CRC Press.

Tan, M., Cheng, D., Yang, Y., Zhang, G., Qin, M., Chen, J., et al. (2017). Co-expression network analysis of the transcriptomes of rice roots exposed to various cadmium stresses reveals universal cadmium-responsive genes. *BMC Plant Biology, 17*, 194.

Tanaka, S., Brentner, L. B., Merchie, K. M., Schnoor, J. L., Yoon, J. M., & Van Aken, B. (2007). Analysis of gene expression in poplar trees (*Populus deltoides* × *nigra*, DN34) exposed to the toxic explosive hexahydro-1,3,5-trinitro-1,3,5-triazine (RDX). *International Journal of Phytoremediation, 9*, 15–30.

Tchounwou, P. B., Yedjou, C. G., Patlolla, A. K., & Sutton, D. J. (2012). Heavy metal toxicity and the environment. In A. Luch (Ed.), *Experientia supplementum: Vol. 101. Molecular, clinical and environmental toxicology* (pp. 133–164). Basel: Springer.

The World Bank & Institute for Health Metrics and Evaluation. (2016). *The cost of air pollution: Strengthening the economic case for action.* Washington, DC: The World Bank.

Thijs, S., Sillen, W., Rineau, F., Weyens, N., & Vangronsveld, J. (2016). Towards an enhanced understanding of plant-microbiome interactions to improve phytoremediation: Engineering the metaorganism. *Frontiers in Microbiology, 7*, 341.

Toth, G., Hermann, T., Da Silva, M. R., & Montanarella, L. (2016). Heavy metals in agricultural soils of the European Union with implications for food safety. *Environment International, 88*, 299–309.

Trumbore, S., Brando, P., & Hartmann, H. (2015). Forest health and global change. *Science, 349*, 814–818.

Tullus, A., Rytter, L., Tullus, T., Weih, M., & Tullus, H. (2012). Short-rotation forestry with hybrid aspen (*Populus tremula* L. × *P. tremuloides* Michx.) in Northern Europe. *Scandinavian Journal of Forest Research, 27*, 10–29.

Tzafestas, K., Razalan, M. M., Gyulev, I., Mazari, A. M., Mannervik, B., Rylott, E. L., et al. (2017). Expression of a Drosophila glutathione transferase in Arabidopsis confers the ability to detoxify the environmental pollutant, and explosive, 2,4,6-trinitrotoluene. *New Phytologist, 214*, 294–303.

Ullah, A., Heng, S., Munis, M. F. H., Fahad, S., & Yang, X. (2015). Phytoremediation of heavy metals assisted by plant growth promoting (PGP) bacteria: A review. *Environmental and Experimental Botany, 117*, 28–40.

Van Aken, B., & Doty, S. L. (2010). Transgenic plants and associated bacteria for phytoremediation of chlorinated compounds. *Biotechnology and Genetic Engineering Reviews, 26*, 43–64.

Van Aken, B., Yoon, J. M., Just, C. L., & Schnoor, J. L. (2004). Metabolism and mineralization of hexahydro-1,3,5-trinitro-1,3,5-triazine inside poplar tissues (*Populus deltoides* × nigra DN-34). *Environmental Science & Technology, 38*, 4572–4579.

Vangronsveld, J., Herzig, R., Weyens, N., Boulet, J., Adriaensen, K., Ruttens, A., et al. (2009). Phytoremediation of contaminated soils and groundwater: Lessons from the field. *Environmental Science and Pollution Research, 16*, 16765–16794.

Voelker, S. L., Lachenbruch, B., Meinzer, F. C., & Strauss, S. H. (2011). Reduced wood stiffness and strength, and altered stem form, in young antisense 4CL transgenic poplars with reduced lignin contents. *New Phytologist, 189*, 1096–1109.

Wang, J. P., Matthews, M. L., Williams, C. M., Shi, R., Yang, C., Tunlaya-Anukit, S., et al. (2018). Improving wood properties for wood utilization through multi-omics integration in lignin biosynthesis. *Nature Communications, 9*, 1579.

Wang, Q., Wang, W., He, X., Zheng, Q., Wang, H., Wu, Y., et al. (2017). Changes in soil properties, X-ray-mineral diffractions and infrared-functional groups in bulk soil and fractions following afforestation of farmland, Northeast China. *Scientific Reports, 7*, 12829.

Wang, B., & Zhao, Q. (2018). Membrane-bound metabolons. *Nature Plants, 4*, 245–246.

Weir, E., & Doty, S. L. (2016). Social acceptability of phytoremediation: The role of risk and values. *International Journal of Phytoremediation, 18*, 1029–1036.
Wenzel, W. (2009). Rhizosphere processes and management in plant-assisted bioremediation (phytoremediation) of soils. *Plant and Soil, 321*, 385–408.
Weyens, N., Thijs, S., Popek, R., Witters, N., Przybysz, A., Espenshade, J., et al. (2015). The role of plant–microbe interactions and their exploitation for phytoremediation of air pollutants. *International Journal of Molecular Sciences, 16*, 25576–25604.
Whitaker, J., Field, J. L., Bernacchi, C. J., Cerri, C. E. P., Ceulemans, R., Davies, C. A., et al. (2018). Consensus, uncertainties and challenges for perennial bioenergy crops and land use. *GCB Bioenergy, 10*, 150–164.
Woolbright, S. A., Rehill, B. J., Lindroth, R. L., DiFazio, S. P., Martinsen, G. D., Zinkgraf, M. S., et al. (2018). Large effect quantitative trait loci for salicinoid phenolic glycosides in *Populus*: Implications for gene discovery. *Ecology and Evolution, 8*, 3726–3737.
Wu, Y., Wang, W., Wang, Q., Zhong, Z., Pei, Z., Wang, H., et al. (2018). Impact of poplar shelterbelt plantations on surface soil properties in northeastern China. *Canadian Journal of Forest Research, 48*, 559–567.
Yadav, R., Arora, P., Kumar, S., & Chaudhury, A. (2010). Perspectives for genetic engineering of poplars for enhanced phytoremediation abilities. *Ecotoxicology, 19*, 1574–1588.
Yanitch, A., Brereton, N. J., Gonzalez, E., Labrecque, M., Joly, S., & Pitre, F. E. (2017). Transcriptomic response of purple willow (*Salix purpurea*) to arsenic stress. *Frontiers in Plant Science, 8*, 1115.
Ye, Z. H., & Zhong, R. (2015). Molecular control of wood formation in trees. *Journal of Experimental Botany, 66*, 4119–4131.
Yewale, P. P., Rahman, A., Nahar, N., Saha, A., Jass, J., Mandal, A., et al. (2017). Sources of metal pollution, global status, and conventional bioremediation practices. In S. Das & H. R. Dash (Eds.), *Handbook of metal-microbe interactions and bioremediation* (pp. 25–40). Boca Raton, FL: CRC Press.
Yildirim, K. (2017). Transcriptomic and hormonal control of boron uptake, accumulation and toxicity tolerance in poplar. *Environmental and Experimental Botany, 141*, 60–73.
Zárubová, P., Hejcman, M., Vondráčková, S., Mrnka, L., Száková, J., & Tlustoš, P. (2015). Distribution of P, K, Ca, Mg, Cd, Cu, Fe, Mn, Pb and Zn in wood and bark age classes of willows and poplars used for phytoextraction on soils contaminated by risk elements. *Environmental Science and Pollution Research, 22*, 18801–18813.
Zhang, L., Routsong, R., Nguyen, Q., Rylott, E. L., Bruce, N. C., & Strand, S. E. (2017). Expression in grasses of multiple transgenes for degradation of munitions compounds on live-fire training ranges. *Plant Biotechnology Journal, 15*, 624–633.
Zhiyi, N., Guijuan, K., Yu, L., Longjun, D., & Rizhong, Z. (2015). Whole-transcriptome survey of the putative ATP-binding cassette (ABC) transporter family genes in the latex-producing laticifers of *Hevea brasiliensis*. *PLoS One, 10*, e0116857.
Zuo, L., Yang, R., Zhen, Z., Liu, J., Huang, L., & Yang, M. (2018). A 5-year field study showed no apparent effect of the Bt transgenic 741 poplar on the arthropod community and soil bacterial diversity. *Scientific Reports, 8*, 1956.

Further reading

Cuypers, A., & Vangronsveld, J. (2017). Advances in botanical research. *Phytoremediation: Vol. 83*. Amsterdam: Elsevier.
Jia, J., Li, S., Cao, X., Li, H., Shi, W., Polle, A., et al. (2016). Physiological and transcriptional regulation in poplar roots and leaves during acclimation to high temperature and drought. *Physiologia Plantarum, 157*, 38–53.

Jia, J., Zhou, J., Shi, W., Cao, X., Luo, J., Polle, A., et al. (2017). Comparative transcriptomic analysis reveals the roles of overlapping heat-/drought-responsive genes in poplars exposed to high temperature and drought. *Scientific Reports*, 7, 43215.

Keeran, N. S., Ganesan, G., & Parida, A. K. (2017). A novel heavy metal ATPase peptide from *Prosopis juliflora* is involved in metal uptake in yeast and tobacco. *Transgenic Research*, *26*, 247–261.

World Health Organization. (2016). Public health, environmental and social determinants of health. PHE e-News, special issue, December 2016, Available at www.who.int/phe/PHE-e-News-special-issue_301116.pdf?ua=1.

Zhu, J. K. (2016). Abiotic stress signaling and responses in plants. *Cell*, *167*, 313–324.

CPI Antony Rowe
Chippenham, UK
2019-02-04 23:40

Decision-Making for Biomass-Based Production Chains